INTRODUCTION TO DERIVATIVE-FREE OPTIMIZATION

MPS-SIAM Series on Optimization

This series is published jointly by the Mathematical Programming Society and the Society for Industrial and Applied Mathematics. It includes research monographs, books on applications, textbooks at all levels, and tutorials. Besides being of high scientific quality, books in the series must advance the understanding and practice of optimization. They must also be written clearly and at an appropriate level.

Editor-in-Chief

Philippe Toint, *University of Namur (FUNDP)*

Editorial Board

Oktay Gunluk, *IBM T.J. Watson Research Center*
Matthias Heinkenschloss, *Rice University*
C.T. Kelley, *North Carolina State University*
Adrian S. Lewis, *Cornell University*
Pablo Parrilo, *Massachusetts Institute of Technology*
Daniel Ralph, *University of Cambridge*
Mike Todd, *Cornell University*
Laurence Wolsey, *Université Catholique de Louvain*
Yinyu Ye, *Stanford University*

Series Volumes

Conn, Andrew R., Scheinberg, Katya, and Vicente, Luis N., *Introduction to Derivative-Free Optimization*

Ferris, Michael C., Mangasarian, Olvi L., and Wright, Stephen J., *Linear Programming with MATLAB*

Attouch, Hedy, Buttazzo, Giuseppe, and Michaille, Gérard, *Variational Analysis in Sobolev and BV Spaces: Applications to PDEs and Optimization*

Wallace, Stein W. and Ziemba, William T., editors, *Applications of Stochastic Programming*

Grötschel, Martin, editor, *The Sharpest Cut: The Impact of Manfred Padberg and His Work*

Renegar, James, *A Mathematical View of Interior-Point Methods in Convex Optimization*

Ben-Tal, Aharon and Nemirovski, Arkadi, *Lectures on Modern Convex Optimization: Analysis, Algorithms, and Engineering Applications*

Conn, Andrew R., Gould, Nicholas I. M., and Toint, Phillippe L., *Trust-Region Methods*

INTRODUCTION TO DERIVATIVE-FREE OPTIMIZATION

Andrew R. Conn
Katya Scheinberg
IBM
Yorktown Heights, New York

Luis N. Vicente
Universidade de Coimbra
Coimbra, Portugal

Society for Industrial and Applied Mathematics
Philadelphia

Mathematical Programming Society
Philadelphia

Library of Congress Cataloging-in-Publication Data

Conn, Andrew R.
  Introduction to derivative-free optimization / Andrew R. Conn, Katya Scheinberg, Luis N. Vicente.
    p. cm. — (MPS-SIAM series on optimization ; 8)
  Includes bibliographical references and index.
  ISBN 978-0-898716-68-9
  1. Mathematical models. 2. Mathematical optimization—Industrial applications. 3. Engineering mathematics. 4. Industrial engineering—Mathematics. 5. Search theory. 6. Nonlinear theories. I. Scheinberg, Katya. II. Vicente, Luis N. III. Title.

TA342.C67 2008
519.6—dc22

                                                                                            2008038005

About the Cover

Cover art © Masha Ryskin, coffee, graphite, ink on board, 61 cm x 61 cm (24" x 24"). Title: *Inside*. Private collection. Photographer: Masha Ryskin.

Masha Ryskin is a Russian-born painter, printmaker, and installation artist who lives and works in the United States. She holds degrees from the Rhode Island School of Design and the University of Michigan. Her work has been exhibited nationally and internationally, and she has worked in the United States, Europe, Costa Rica, and Indonesia. Her art deals with a sense of place, history, and memory through the use of landscape and its elements. *Inside* addresses the concept of landscape as an amalgamation of fragments fitting together to make an integral whole.

Cover art reprinted with permission from the artist.

 is a registered trademark.

 is a registered trademark.

To Oscar
Who introduced me to much that matters.
(Andrew)

To my parents
Who support me in every endeavour.
(Katya)

To my parents
À memória do meu Pai e à minha Mãe que sempre me apoiou.
(Luis)

Contents

Preface

For many years all three of us have been interested in, and have tried to make contributions to, derivative-free optimization. Our motivation for writing this book resulted from various circumstances. We had the opportunity to work closely with the leading contributors to the field, including especially John Dennis, Michael Powell, Philippe Toint, and Virginia Torczon. We had some knowledge of various facets of the recent developments in the area, and yet felt there was a need for a unified view, and we hoped thereby to gain a better understanding of the field. In addition we were enticed by the growing number of applications. We also felt very strongly that there was a considerable need for a textbook on derivative-free optimization, especially since the foundations, algorithms, and applications have become significantly enriched in the past decade. Finally, although the subject depends upon much that is true for, and was developed for, optimization with derivatives, the issues that arise are new. The absence of computable derivatives naturally prohibits the use of Taylor models—so common in derivative-based optimization. The fact that typical applications involve expensive function evaluations shifts the emphasis from the cost of the linear algebra, or other contributors to the iteration complexity, to simply the number of function evaluations. Also, the noise in the function values affects the local convergence expectations. Thus, the area is both simpler, in the sense of diminished expectations, and harder, in the sense that one is trying to achieve something with considerably less information. It is definitely fun and challenging and, not incidentally, very useful.

Although we do make considerable effort to give a sense of the current state of the art, we do not attempt to present a comprehensive treatise on all the work in the area. This is in part because we think that the subject is not yet mature enough for such a treatise. For similar reasons our emphasis is on the unconstrained problem, although we include a review on the work done so far in constrained derivative-free optimization. The constrained problem is in fact very important for applications, but theoretical treatment of derivative-free methods for constrained problems is very limited in the literature published to date, and thus, for the present volume at least, we are content to concentrate on the unconstrained case.

The book is meant to be reasonably self-contained and is addressed to readers at the level of a graduate student or a senior undergraduate with a background in calculus, linear algebra, and numerical analysis. Some elementary notions of optimization would be helpful but are not necessary. It is certainly our intent that practitioners would find the material covered to be both accessible and reasonably complete for their needs, whether their emphasis is on the algorithms or the applications. We have also made an effort to include figures and exercises when appropriate. The major aims include giving any interested reader a good idea of the state of the art of derivative-free optimization, with a detailed description of the basic theory to the extent that the reader can well understand what is needed to ensure convergence, how this affects algorithm design, and what kind of success one can expect and where. Thus, it is certainly our goal that the material be of interest to those who want to do research in the area.

As we state in our introduction, due to the growing sophistication and efficiency of computer simulations as well as of other applications, there is an increasing number of instances where one wishes to perform optimization of a complex system and the derivative information of the resulting objective functions is not available. This book is intended to help the reader to study and select, if necessary, suitable approaches to do exactly that. It is also intended to extract and emphasize the common theoretical features used by modern derivative-free algorithms, as well as highlight the differences.

We would be remiss if we ended our preface without some indicator of what the future holds in this area. There is still much waiting to be discovered. Undoubtedly, researchers and practitioners, perhaps soon, will discover ways to tackle much larger problems, whether through the use of massively parallel architectures or through advances in hardware yet to be realized, or through breakthroughs in the theory, or likely all three. The theory of constrained derivative-free optimization is also likely to advance significantly in the near future. Certainly, and especially because of the broad availability of difficult and important applications, this promises to be an exciting, interesting, and challenging area for many years to come.

Aside from the colleagues we mentioned in the first paragraph of this preface, there are many others we would like to thank who have provided us their valuable feedback in specific parts of the book, including, in particular, Natalia Alexandrov, Charles Audet, and Paul Tseng. We are very grateful to Ana Luísa Custódio for her tireless proofreading of the manuscript. We would also like to thank the reviewers who have refereed the version originally submitted to SIAM, in particular Tim Kelley and Jorge Nocedal, for their many interesting comments and suggestions.

Chapter 1

Introduction

1.1 Why derivative-free optimization

It is well known that extensive useful information is contained in the derivatives of any function one wishes to optimize. After all, the "standard" mathematical characterization of a local minimum, given by the first-order necessary conditions, requires, for continuously differentiable functions, that the first-order derivatives are zero. However, for a variety of reasons there have always been many instances where (at least some) derivatives are unavailable or unreliable. Nevertheless, under such circumstances it may still be desirable to carry out optimization. Consequently, a class of nonlinear optimization techniques called derivative-free optimization methods has always been needed. In fact, we consider optimization without derivatives one of the most important, open, and challenging areas in computational science and engineering, and one with enormous practical potential. The reason that it is challenging is that, from the point of view of optimization, one gives up so much information by not having derivatives. The source of its current, practical importance is the ever growing need to solve optimization problems defined by functions for which derivatives are unavailable or available at a prohibitive cost. Increasing complexity in mathematical modeling, higher sophistication of scientific computing, and an abundance of legacy codes are some of the reasons why derivative-free optimization is currently an area of great demand.

In earlier days of nonlinear optimization perhaps one of the most common reasons for using derivative-free methods was the lack of sophistication or perseverance of the user. The users knew they wanted to improve on their current "solution," but they wanted to use something simple that they could understand, and so they used (and, unfortunately, sometimes continue to use) nonderivative methods, like the method by Nelder and Mead [177], even when more appropriate algorithms were available. In defense of the practitioner, we should remember that until relatively recently computing derivatives was the single most common source of user error in applying optimization software (see, for example, [104, Chapter 8, page 297]). As the scale and difficulty of the applications increased, more sophisticated derivative-based optimization methods became more essential. With the growth and development of derivative-based nonlinear optimization methods it became evident that large-scale problems can be solved efficiently, but only if there is accurate derivative information at hand. Users started either to provide such derivatives (by hand-coding them or applying automatic differentiation tools to their software) or to estimate the derivatives by finite differences. Some optimization software packages perform the finite-difference gradient evaluation internally, but it is usually better to leave this to the user since the ideal size of the finite-difference step may depend on the application.

However, there are situations where none of these approaches for obtaining the derivatives works. For example, in the case of legacy or proprietary codes, i.e., codes that have been written in the past and which have not been maintained by the original authors, rewriting such a code now, or adding to it what would be required to provide first-order derivatives, can be an extremely time-consuming task. The problem might also depend on a code owned by a company for which there is access only to the binary files. Automatic differentiation techniques (see, for example, [111, 112]) also cannot be applied in all cases. In particular, if the objective function is computed using a black-box simulation package, automatic differentiation is typically impossible, and even in the case where the computation is more accessible, legacy or proprietary issues may make such an approach unacceptable.

There are also two situations where applying finite-difference derivative approximation is inappropriate: when the function evaluations are costly and when they are noisy. In the first case, it may be prohibitive to perform the necessary number of function evaluations (normally no less than the number of variables plus one) to provide a single gradient estimation. In the second case, the gradient estimation may be completely useless. Ironically, with the growing sophistication of computer hardware and mathematical algorithms and software, situations like these are becoming more, rather than less, frequent. The reason is simply that while, before, simulation of complex systems was a difficult and costly task and did not provide a sufficiently good setting for optimization, now such simulations are becoming more routine and also more accurate; hence optimization of complex systems is becoming a reasonable possibility. The growing demand for sophisticated derivative-free optimization methods has triggered the development of a relatively wide range of approaches. In recent years, the field has undergone major changes with improvements in theory and practice and increased visibility in the nonlinear optimization community. By writing this book we hope to provide a unifying theoretical view of the derivative-free methods existing today. We discuss briefly the practical performance, and what can be expected, but the main purpose of the book is to give the general picture of why and how the algorithms work.

The methods we will consider do not rely on derivative information of the objective function or constraints, nor are the methods designed explicitly to approximate these derivatives. Rather, they build models of the functions based on sample function values or they directly exploit a sample set of function values without building an explicit model. Not surprisingly, as we already suggested, there are considerable disadvantages in not having derivative information, so one cannot expect the performance of derivative-free methods to be comparable to those of derivative-based methods. In particular, the scale of the problems that can currently be efficiently solved by derivative-free methods is still relatively small and does not exceed a few hundred variables even in easy cases. Stopping criteria are also a challenge in the absence of derivatives, when the function evaluations are noisy and/or expensive. Therefore, a near-optimal solution obtained by a derivative-free method is often less accurate than that obtained by a derivative-based method, assuming derivative information is available. That said, for many of the applications that exist today these limitations are acceptable.

1.2 Examples of problems where derivatives are unavailable

There is really an enormous number of problems where derivatives are unavailable but one has imperative practical reasons for wanting to do some optimization. It is almost a natural perversity that practical problems today are often complex, nonlinear, and not sufficiently explicitly defined to give reliable derivatives. Indeed, such problems were always numerous, but, 30 years ago, when nonlinear optimization techniques were relatively more naive than they are today, even the most optimistic practitioners would not try to optimize such complex problems. Not so in 2008! The diversity of applications includes problems in engineering, mathematics, physics, chemistry, economics, finance, medicine, transportation, computer science, business, and operations research.

As examples we include a subset of known applications and references. We start by some illustrations (like algorithmic parameter tuning and automatic error analysis) which may be atypical of the applications of derivative-free optimization but are easy to understand.

Tuning of algorithmic parameters and automatic error analysis

An interesting (and potentially useful) application of derivative-free optimization has been explored in [22] to tune parameters of nonlinear optimization methods, with promising results. Most numerical codes (for simulation, optimization, estimation, or whatever) depend on a number of parameters. Everybody implementing numerical algorithms knows how critical the choices of these parameters are and how much they influence the performance of solvers. Typically, these parameters are set to values that either have some mathematical justification or have been found by the code developers to perform well. One way to automate the choice of the parameters (in order to find possibly optimal values) is to consider an optimization problem whose variables are the parameters and whose objective function measures the performance of the solver for a given choice of parameters (measured by CPU time or by some other indicator such as the number of iterations taken by the solver). Such problems might have constraints like upper and lower bounds on the values of the solver parameters, and look like

$$\min_{p \in \mathbb{R}^{n_p}} f(p) = CPU(solver; p) \quad \text{s.t.} \quad p \in P,$$

where n_p is the number of parameters to be tuned and P is of the form $\{p \in \mathbb{R}^{n_p} : \ell \le p \le u\}$. Not only is it hard to calculate derivatives for such a function f, but numerical noise and some form of nondifferentiability are likely to take place.

Derivative-free optimization has also been used for automatic error analysis [126, 127], a process in which the computer is used to analyze the accuracy or stability of a numerical computation. One example of automatic error analysis is to analyze how large the growth factor for Gaussian elimination can be for a specific pivoting strategy. The relevance of such a study results from the influence of the growth factor in the stability of Gaussian elimination. Given a pivoting strategy and a fixed matrix dimension n, the optimization problem posed is to determine a matrix that maximizes the growth factor for

Gaussian elimination:

$$\max_{A \in \mathbb{R}^{n \times n}} f(A) = \frac{\max_{i,j,k} |a_{ij}^{(k)}|}{\max_{i,j} |a_{ij}|},$$

where the $a_{ij}^{(k)}$ are the intermediate elements generated during the elimination. A starting point could be the identity matrix of order n. When no pivoting is chosen, f is defined and continuous at all points where elimination does not break down. There could be a lack of differentiability when ties occur at the maxima that define the growth factor expression. For partial pivoting, the function f is defined everywhere (because the elimination cannot break down), but it can be discontinuous when a tie occurs at the choice of the pivot element. Other examples of automatic error analysis where derivative-free optimization has been used are the estimation of the matrix condition number and the analysis of numerical stability for fast matrix inversion and polynomial root finding [126, 127].

Engineering design

A case study in derivative-free optimization is the helicopter rotor blade design problem [38, 39, 206]. The goal is to find the structural design of the rotor blades to minimize the vibration transmitted to the hub. The variables are the mass, center of gravity, and stiffness of each segment of the rotor blade. The simulation code is multidisciplinary, including dynamic structures, aerodynamics, and wake modeling and control. The problem includes upper and lower bounds on the variables, and some linear constraints have been considered such as an upper bound on the sum of masses. Each function evaluation requires simulation and can take from minutes to days of CPU time. A surrogate management framework based on direct-search methods (see Section 12.2) led to encouraging results. Other multidisciplinary or complex design problems have been reported to be successfully solved by derivative-free optimization methods and/or surrogate management frameworks, like wing platform design [16], aeroacoustic shape design [164, 165], and hydrodynamic design [87].

Circuit design

Derivative-free methods have also been used for tuning parameters of relatively small circuits. In particular, they have been used in the tuning of clock distribution networks. Operations on a chip are done in cycles (e.g., eight cycles per operation), and every part of the chip has to be synchronized at the beginning of each cycle. The clock distribution network serves to synchronize the chip by sending a signal to every part, and the signal has to arrive everywhere with minimum distortion and delay in spite of the fact that there is always some of both. The resulting optimization problem includes as possible parameters wire length, wire width, wire shields, and buffer size. The possible constraints and objectives considered include delay, power, slew over/undershoot, the duty cycle, and the slew. The function values are computed by the `PowerSpice` package [2], a well-known accurate simulation package for circuits that, however, provides no derivative computation. Derivative-free optimization is applied to assist circuit designers in the optimization of their (small) circuits. In the particular case of clock distribution networks, it makes it possible for the circuit designers to try different objective functions and constraint combinations, relatively quickly, without manual tuning. This leads to a much better understanding of the problem and thereby enables discovery of alternative designs.

Molecular geometry

Another area where it is not unusual to use derivative-free methods of the type we are interested in is in the optimization of molecular geometries and related problems. An example of this would be considering the geometry of a cluster of N atoms (which amounts to $3N - 6$ variables). The aim is then the unconstrained minimization of the cluster's total energy computed by numerical simulation of the underlying dynamics. We point out that for these classes of problems there is the presence of several local minimizers (although the more realistic the computation of the total energy, the lower the number of local minima), and so there must be some caution when using the type of derivative-free methods covered in this book (see Section 13.3 for extensions to global optimization). The gradient might be available, but it could be either expensive or affected by noise or even undesirable to use given the presence of nonconvexity. Derivative-free optimization methods, in particular direct-search methods (see Chapter 7), have been appropriately adapted to handle classes of these problems (see [10, 170]).

Other applications

Many other recent applications of derivative-free optimization could be cited in diverse areas such as groundwater community problems [99, 174, 233], medical image registration [179, 180], and dynamic pricing [150].

1.3 Limitations of derivative-free optimization

Perhaps foremost among the limitations of derivative-free methods is that, on a serial machine, it is usually not reasonable to try and optimize problems with more than a few tens of variables, although some of the most recent techniques can handle unconstrained problems in hundreds of variables (see, for example, [192]). Also, even on relatively simple and well-conditioned problems it is usually not realistic to expect accurate solutions. Convergence is typically rather slow. For example, in order to eventually achieve something like a quadratic rate of local convergence, one needs either implicit or explicit local models that are reasonable approximations for a second-order Taylor series model in the current neighborhood. With 100 variables, if one was using interpolation, a local quadratic function would require 5151 function evaluations (see Table 1.1). Just as one can achieve good convergence using approximations to the Hessian matrix (such as quasi-Newton methods [76, 178]) in the derivative-based case, it is reasonable to assume that one can achieve similar fast convergence in the derivative-free case by using incomplete quadratic models or quadratic models based on only first-order approximations. These ideas are successfully used in [52, 59, 61, 191, 192]. For instance, the algorithm in NEWUOA [192] typically uses $2n + 1$ function evaluations to build its models. However, unless the function looks very similar to a quadratic in the neighborhood of the optimal solution, in order to progress, the models have to be recomputed frequently as the step size and the radius of sampling converge to zero. Even in the case of linear models this can be prohibitive when function evaluations are expensive. Thus typically one can expect a local convergence rate that is closer to linear than quadratic, and one may prefer early termination.

As for being able to tackle only problems in around 20 or moderately more variables (currently the largest unconstrained problems that can be tackled on a serial machine appear

Table 1.1. *Number of points needed to build a "fully quadratic" polynomial interpolant model.*

n	10	20	50	100	200
$(n+1)(n+2)/2$	66	231	1326	5151	20301

to be in several hundred variables), the usual remedy is to use statistical methods like analysis of variance (see, for example, [203]) to determine, say, the most critical 20 variables, and optimizing only over them. It may also be reasonable to take advantage of the relative simplicity of some of the derivative-free algorithms, like directional direct-search methods (explained in Chapter 7), and compute in a parallel, or even massively parallel, environment.

Another limitation of derivative-free optimization may occur when minimizing nonconvex functions. However, and although nothing has been proved to support this statement, it is generally accepted that derivative-free optimization methods have the ability to find "good" local optima in the following sense. If one has a function with an apparently large number of local optimizers, perhaps because of noise, then derivative-free approaches, given their relative crudeness, have a tendency to initially go to generally low regions in the early iterations (because of their myopia, or one might even say near-blindness) and in later iterations they still tend to smooth the function, whereas a more sophisticated method may well find the closest local optima to the starting point. The tendency to "smooth" functions is also why they are effective for moderately noisy functions. There are many situations where derivative-free methods are the only suitable approach, capable of doing better than heuristic or other "last-resort" algorithms and providing a supporting convergence theory.

There are, however, classes of problems for which the use of derivative-free methods that we address here are not suitable. Typically, rigorous methods for such problems would require an inordinate amount of work that grows exponentially with the size of the problem. This category of problems includes medium- and large-scale general global optimization problems with or without derivatives, problems which not only do not have available derivatives but which are not remotely like smooth problems, general large nonlinear problems with discrete variables including many combinatorial optimization ones (so-called NP hard problems), and stochastic optimization problems. Although relatively specialized algorithms can find good solutions to many combinatorial optimization problems perfectly adequately, this is not the case for general large nonlinear problems with integer variables. We do not address algorithmic approaches designed specifically for combinatorial or stochastic optimization problems.

For some of the extreme cases mentioned above, heuristics are frequently used, such as simulated annealing [143], genetic and other evolutionary algorithms [108, 129], artificial neural networks [122], tabu-search methods [107], and particle swarm or population-based methods [142], including (often very sophisticated variations of) enumeration techniques. The authors think of these as methods of a last resort (that is, applicable to problems where the search space is necessarily large, complex, or poorly understood and more sophisticated mathematical analysis is not applicable) and would use them only if nothing better is available. We do not address such methods in this book. However, sometimes these approaches are combined with methods that we do address; see, for example, [13, 135, 222] and Section 13.3 of this book.

It is sometimes perceived that derivative-free optimization methods should be simple and easy to implement. However, such methods are typically inferior in theory and in practice. The Nelder–Mead algorithm, however, can work very well and it is expected to survive a very long time. Nevertheless it is seriously defective: it is almost never the best method and indeed it has no general convergence results, because it can easily not converge (although modifications of it are provably convergent, as we will explain in our book). Since the authors' combined research experience in optimization is over 60 years, we believe that ultimately more sophisticated and successful methods will earn their rightful place in practical implementations once a comprehensive description of such methods is widely available. This, in part, is motivation for this book.

Finally, we want to make a strong statement that often councils against the use of derivative-free methods: if you can obtain clean derivatives (even if it requires considerable effort) and the functions defining your problem are smooth and free of noise you should not use derivative-free methods.

1.4 How derivative-free algorithms should work

This book is mostly devoted to the study of derivative-free algorithms for unconstrained optimization problems, which we will write in the form

$$\min_{x \in \mathbb{R}^n} f(x). \tag{1.1}$$

We are interested in algorithms that are globally convergent to stationary points (of first- or second-order type), in other words, algorithms that regardless of the starting point are able to generate sequences of iterates asymptotically approaching stationary points.

Some of the main ingredients

Perhaps this is oversimplifying, but one could say that there are three features present in all globally convergent derivative-free algorithms:

1. They incorporate some mechanism to impose descent away from stationarity. The same is done by derivative-based algorithms to enforce global convergence, so this imposition is not really new. It is the way in which this is imposed that makes the difference. Direct-search methods of directional type, for instance, achieve this goal by using *positive bases or spanning sets* (see Chapter 7) and moving in the direction of the points of the pattern with the best function value. Simplex-based methods (Chapter 8) ensure descent from *simplex operations* like *reflections*, by moving in the direction *away* from the point with the worst function value. Methods like the implicit-filtering method (Chapter 9) aim to get descent along negative *simplex gradients*, which are intimately related to polynomial models. Trust-region methods, in turn, minimize trust-region subproblems defined by *fully linear* or *fully quadratic models*, typically built from polynomial interpolation or regression—see Chapters 10 and 11.

   In every case, descent is guaranteed away from stationarity by combining such mechanisms with a possible reduction of the corresponding step size parameter. Such a parameter could be a mesh size parameter (directional direct search), a simplex diameter (simplicial direct search), a line-search parameter, or a trust-region radius.

2. They must guarantee some form of control of the geometry of the sample sets where the function is evaluated. Essentially, such operations ensure that any indication of stationarity (like model stationarity) is indeed a true one. Not enforcing good geometry explains the lack of convergence of the original Nelder–Mead method.

Examples of measures of geometry are (i) the cosine measure for positive spanning sets; (ii) the normalized volume of simplices (both to be kept away from zero); (iii) the Λ-poisedness constant, to be maintained moderately small and bounded from above when building interpolation models and simplex derivatives.

3. They must drive the step size parameter to zero. We know that most optimization codes stop execution when the step size parameter passes below a given small threshold. In derivative-based optimization such terminations may be premature and an indication of failure, perhaps because the derivatives are either not accurate enough or wrongly coded. The best stopping criteria when derivatives are available are based on some form of stationarity indicated by the first-order necessary conditions.

In derivative-free optimization the step size serves a double purpose: besides bounding the size of the minimization step it also controls the size of the local area where the function is sampled around the current iterate. For example, in direct-search methods the step size and the mesh size (defining the pattern) are the same or constant multiples of each other. In a model-based derivative-free method, the size of the trust region or line-search step is typically intimately connected with the radius of the sample set. Clearly, the radius of the sample set or mesh size has to converge to zero in order to ensure the accuracy of the objective function representation. It is possible to decouple the step size from the size of the sample set; however, so far most derivative-free methods (with the exception of the original Nelder–Mead method) connect the two quantities. In fact, the convergence theory of derivative-free methods that we will see in this book shows that the sequence (or a subsequence) of the step size parameters do converge to zero (see, e.g., Theorem 7.1 or Lemma 10.9). It is an implicit consequence of the mechanisms of effective algorithms and should not (or does not have to) be enforced explicitly. Thus, a stopping criterion based on the size of the step is a natural one.

The details of the ingredients listed above—it is hoped—will be made much clearer to the reader in the chapters to follow.

With the current state-of-the-art derivative-free optimization methods one can expect to successfully address problems (i) which do not have more than, say, a hundred variables; (ii) which are reasonably smooth; (iii) in which the evaluation of the function is expensive and/or computed with noise (and for which accurate finite-difference derivative estimation is prohibitive); (iv) in which rapid asymptotic convergence is not of primary importance.

An indication of typical behavior

We chose two simple problems to illustrate some of the validity of the above comments and some of the advantages and disadvantages of the different classes of methods for derivative-free optimization. We do not want to report a comprehensive comparison on how the different methods perform since such a numerical study is not an easy task to perform well, or even adequately, in derivative-free optimization given the great diversity of the application problems. A comparison among different methods is easier, however, if we focus on a particular application. Another possibility is to select a test set of problems with known derivatives and treat them as if the derivatives were unavailable. Moré and Wild [173] suggested the use of data profiles to compare a group of derivative-free algorithms on a test set. In any case we leave such comparisons to others.

The first problem consists of the minimization of the Rosenbrock function

$$\min_{(x_1,x_2)\in\mathbb{R}^2} 100(x_1^2 - x_2)^2 + (1 - x_1)^2,$$

which has a unique minimizer at $(1,1)$. The level curves of this function describe a strongly curved valley with steep sides (see Figure 1.1). Depending on the starting point picked, methods which do not explore the curvature of the function might be extremely slow. For instance, if one starts around $(-1,1)$, one has to follow a curved valley with relatively steep sides in order to attain the minimum.

The second problem involves the minimization of a deceptively simple, perturbed quadratic function. The perturbation involves cosine functions with periods of $2\pi/70$ and $2\pi/100$:

$$\min_{(x_1,x_2)\in\mathbb{R}^2} 10(x_1^2)(1 + 0.75\cos(70x_1)/12) + \cos(100x_1)^2/24$$
$$+ 2(x_2^2)(1 + 0.75\cos(70x_2)/12) + \cos(100x_2)^2/24 + 4x_1x_2.$$

The unique minimizer is at $(0,0)$ (see Figure 1.1). As opposed to the Rosenbrock function, the underlying smooth function here has a mild curvature. However, the perturbed function has been contaminated with noise which will then pose different difficulties to algorithms.

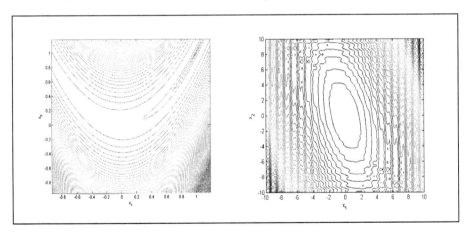

Figure 1.1. *Level curves of the Rosenbrock function (left) and a perturbed quadratic (right).*

We selected four methods to run on these two problems as representatives of the four main classes of methods addressed in this book (see Chapters 7–10). The first method is coordinate search, perhaps the simplest of all direct-search methods of directional type (see Chapter 7). In its simplest form it evaluates the function at $2n$ points around a current iterate defined by displacements along the coordinate directions and their negatives and a step size parameter (a process called *polling*). This set of directions forms a *positive basis*. The method is slow but robust and capable of handling noise in the objective function. The implementation used is reported in [70]. The second choice is the Nelder–Mead method [177], a simplicial direct-search method (see Chapter 8). Based on simplex operations such as reflections, expansions, and contractions (inside or outside), the Nelder–Mead method attempts to replace the simplex vertex that has the worst function value. The simplices generated by Nelder–Mead may adapt well to the curvature of the function. However, the original, unsafeguarded version of Nelder–Mead (as coded in the MATLAB® [1] routine we used in our testing) is not robust or reliable since the simplices' shapes might deteriorate arbitrarily. Coordinate search, on the contrary, is guaranteed to converge globally.

The other two methods follow a different approach. One of them is the implicit-filtering algorithm (see Chapter 9). This is a line-search algorithm that imposes *sufficient decrease* along a quasi-Newton direction. The main difference from derivative-based methods is that the true gradient is replaced by the simplex gradient. So, to some extent, the implicit-filtering method resembles a quasi-Newton approach based on finite-difference approximations to the gradient of the objective function. However, the method and its implementation [141] are particularly well equipped to handle noisy functions. The last method is a trust–region-based algorithm in which the quadratic models are built from polynomial interpolation or regression. We used the implementation from the DFO code of Scheinberg (see Chapter 11 and the appendix). Both methods can adapt well to the curvature of the function, the implicit-filtering one being less efficient than the trust-region one but more capable of filtering the noise (perhaps due to the use of the simplex gradient, which corresponds to the gradient of a linear interpolation or regression model, and to an inaccurate line search).

The results are reported in Figure 1.2 for the two functions of Figure 1.1 and follow the tendency known for these classes of methods. The type of plots of Figure 1.2 is widely used in derivative-free optimization. The horizontal axis marks the number of function evaluations as the optimization process evolves. The vertical axis corresponds to the value of the function, which typically decreases. We set the stopping tolerance for all four methods to 10^{-3} (related to the different parameters used for controlling the step sizes).

On the Rosenbrock function, the trust-region method performs the best because of its ability to incorporate curvature information into the models (from the start to the end of the iteration process). We chose the initial point $(1.2, 0)$ for all methods. As expected, coordinate search is reasonably fast at the beginning but is rather slow on convergence. Attempts to make it faster (like improving the poll process) are not very successful in this problem where curvature is the dominant factor (neither would other fancier directional direct-search methods of poll type). If we start coordinate search form $(1, -1)$ (curvature more favorable), coordinate search will do much better. But if we start it from $(-1, 1)$, it will perform much worse. The other two methods perform relatively well. It is actually remarkable how well a *direct-search* method like Nelder–Mead performed and seems to be able to exploit the curvature in this problem.

The results for the perturbed quadratic are less clear, as happens many times in derivative-free optimization. The Nelder–Mead method failed for the initial point chosen $(0.1, 0.1)$, although it would do better if we started not so close to the solution (but not always...). The method performs many inside contractions which are responsible for its lack of convergence (see Chapter 8 for more details). The implicit-filtering method does a

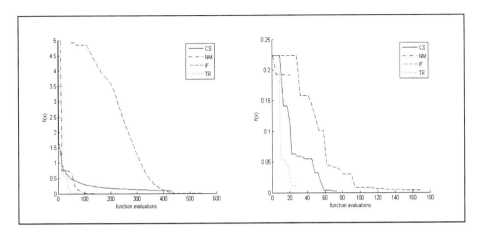

Figure 1.2. *Results of coordinate-search, Nelder–Mead, implicit-filtering, and trust-region interpolation-based methods for the Rosenbrock function (left) and a perturbed quadratic (right).*

good job in filtering the noise for this problem and progressing towards the solution (and its performance is not affected by the starting point). The interpolation-based trust-region approach is very efficient at the beginning but soon stagnates—and reducing the tolerance does not help in this case. Coordinate search performs relatively well (remember that the curvature is mild) and does not seem affected by the noise.

In summary, we would say that model-based methods (like interpolation with trust regions) are more efficient. Direct search loses in comparison when the function is smooth and the curvature is adverse, but it offers a valid alternative for noisy or nonsmooth problems and can be successfully combined with model-based techniques. The ease of parallelization is also one of its strongest points. There is certainly room for other methods like modified, safeguarded Nelder–Mead methods and for approaches particularly tailored for noisy problems and easy to parallelize like implicit filtering.

1.5 A short summary of the book

Thus, having convinced you, perhaps, that there is a need for good derivative-free methods we hope there is also some intellectual encouragement for pursuing research in the area, or at least continuing to read this book. The basic ideas, being reasonably simple, lead to new interesting results. Hence, besides being useful, it is fun.

After the introduction in this chapter, we begin the first part of the book dedicated to *Sampling and Modeling*. Thus Chapter 2 introduces the reader to positive spanning sets and bases, linear interpolation and regression models, simplex gradients, and the importance of geometry. It also includes error bounds in the linear case. Chapters 3, 4, and 5 then consider nonlinear polynomial interpolation models in a determined, regression and underdetermined form, respectively. They give due consideration to the concept of well poisedness, Lagrange polynomials, the conditioning of appropriate matrices, and Taylor-type error bounds. Chapter 6 is devoted to constructive ways to ensure that well poisedness holds and to prepare the material on derivative-free models for use in model-based algorithms such as the trust-region ones.

The second part of the book on *Frameworks and Algorithms* begins in Chapter 7, which addresses direct-search methods where sampling is guided by desirable sets of directions. Included is global convergence with integer lattices and sufficient decrease. The next chapter covers direct-search methods based on simplices and operations over simplices, of which a classical example is the Nelder–Mead method mentioned earlier, for which we include a globally convergent variant. Chapter 9 is devoted to line-search methods based on simplex derivatives, establishing a connection with the implicit-filtering method. Chapter 10 presents trust–region-based methods, including the relationship with derivative-based methods, the abstraction to fully linear and fully quadratic models, and a comprehensive global convergence analysis. The more practical aspects of derivative-free trust-region methods with particular examples of modified versions of existing methods are covered in Chapter 11, in connection with the material of Chapter 6.

Finally, Chapters 12 and 13 (the third part of the book) are concerned with some relevant topics not covered in full detail. In Chapter 12 we review surrogate models built by techniques different from polynomial interpolation or regression, and we describe rigorous optimization frameworks to handle surrogates. A survey of constrained derivative-free optimization is presented in Chapter 13, where we also discuss extension to other classes of problems, in particular global optimization. The book ends with an appendix reviewing the existent software for derivative-free optimization.

The reader is also referred to a number of survey papers on derivative-free optimization, namely the more recent ones by Conn, Scheinberg, and Toint [60], Kelley [141], Kolda, Lewis, and Torczon [145], Lewis, Torczon, and Trosset [156], Nocedal and Wright [178, Chapter 9], Powell [187], and Wright [231], as well as older ones by Brent [47], Fletcher [94], and Shawn [207].

Notation

In this book we have tried to use intuitive notation to simplify reading. This inevitably implies some abuse, but we hope the meaning will nevertheless be clear to the reader. For instance, vectors can be considered row vectors or column vectors according to context without always changing the notation or using transposes.

The uppercase letters typically denote sets or matrices. There are several constants denoted by κ with a subscript acronym; for instance, κ_{bhm} stands for the constant bound on the Hessian of the model. Each acronym subscript is intended to help the reader to remember the meaning of the constant.

The big-$\mathcal{O}$ notation is used in an intuitive way, and the meaning should be clear from the text. All balls in the book are considered closed. All norms are ℓ_2-norms unless stated otherwise.

Part I

Sampling and modeling

Chapter 2

Sampling and linear models

2.1   Positive spanning sets and positive bases

Positive spanning sets and positive bases are used in directional direct-search methods. As we will see later, the main motivation to look at a positive spanning set $D \subset \mathbb{R}^n$ is the guarantee that, given any nonzero vector v in $\mathbb{R}^n$, there is at least one vector d in D such that v and d form an acute angle. The implication in optimization is then obvious. Suppose that the nonzero vector v is the negative gradient, $-\nabla f(x)$, of a continuously differentiable function f at a given point x. Any vector d that forms an acute angle with $-\nabla f(x)$ is a descent direction.[1] In order to decrease $f(x)$, it might be required to evaluate the points $x + \alpha d$ (for all $d \in D$), where $\alpha > 0$, and to repeat this evaluation for smaller positive values of α. But, since the gradient $\nabla f(x)$ is nonzero, there exist a positive value for α and a vector d in D for which $f(x + \alpha d) < f(x)$, which shows that such a scheme should be terminated after a finite number of reductions of the parameter α.

In this section, we will review some of the basic properties of positive spanning sets and positive bases and show how to construct simple positive bases. Most of the basic properties about positive spanning sets are extracted from the theory of positive linear dependence developed by Davis [74] (see also the paper by Lewis and Torczon [153]).

Definitions and properties

The *positive span*[2] of a set of vectors $[v_1 \cdots v_r]$ in $\mathbb{R}^n$ is the convex cone

$$\left\{ v \in \mathbb{R}^n : v = \alpha_1 v_1 + \cdots + \alpha_r v_r, \quad \alpha_i \geq 0, \ i = 1, \ldots, r \right\}.$$

(Many times it will be convenient for us in this book to regard a set of vectors as a matrix whose columns are the vectors in the set.)

---

[1]By a descent direction for f at x we mean a direction d for which there exists an $\bar{\alpha} > 0$ such that $f(x + \alpha d) < f(x)$ for all $\alpha \in (0, \bar{\alpha}]$.

[2]Strictly speaking we should have written *nonnegative* instead of positive, but we decided to follow the notation in [74, 153]. We also note that by *span* we mean *linear span*.

Definition 2.1. *A positive spanning set in $\mathbb{R}^n$ is a set of vectors whose positive span is $\mathbb{R}^n$.*

The set $[v_1 \cdots v_r]$ is said to be positively dependent if one of the vectors is in the convex cone positively spanned by the remaining vectors, i.e., if one of the vectors is a positive combination of the others; otherwise, the set is positively independent.

A positive basis in $\mathbb{R}^n$ is a positively independent set whose positive span is $\mathbb{R}^n$.

Equivalently, a positive basis for $\mathbb{R}^n$ can be defined as a set of nonzero vectors of $\mathbb{R}^n$ whose positive combinations span $\mathbb{R}^n$ but for which no proper set exhibits the same property.

The following theorem due to [74] indicates that a positive spanning set contains at least $n + 1$ vectors in $\mathbb{R}^n$.

Theorem 2.2. *If $[v_1 \cdots v_r]$ spans $\mathbb{R}^n$ positively, then it contains a subset with $r - 1$ elements that spans $\mathbb{R}^n$.*

Proof. The set $[v_1 \cdots v_r]$ is necessarily linearly dependent (otherwise, it would be possible to construct a basis for $\mathbb{R}^n$ that would span $\mathbb{R}^n$ positively). As a result, there are scalars $\bar{a}_1, \ldots, \bar{a}_r$ (not all zero) such that $\bar{a}_1 v_1 + \cdots + \bar{a}_r v_r = 0$. Thus, there exists an $i \in \{1, \ldots, r\}$ for which $\bar{a}_i \neq 0$.

Now let v be an arbitrary vector in $\mathbb{R}^n$. Since $[v_1 \cdots v_r]$ spans $\mathbb{R}^n$ positively, there exist nonnegative scalars $a_1, \ldots, a_r$ such that $v = a_1 v_1 + \cdots + a_r v_r$.

As a result, we get

$$v = \sum_{j=1}^{r} a_j v_j = \sum_{\substack{j=1 \\ j \neq i}}^{r} \left(a_j - \frac{\bar{a}_j}{\bar{a}_i} a_i \right) v_j.$$

Since v is arbitrary, we have proved that $\{v_1, \ldots, v_r\} \backslash \{v_i\}$ spans $\mathbb{R}^n$. $\square$

It can also be shown that a positive basis cannot contain more than $2n$ elements (see [74]). Positive bases with $n + 1$ and $2n$ elements are referred to as *minimal* and *maximal* positive bases, respectively.

The positive basis formed by the vectors of the canonical basis and their negative counterparts is the most simple maximal positive basis one can think of. In $\mathbb{R}^2$, this positive basis is defined by the columns of the matrix

$$D_1 = \begin{bmatrix} 1 & 0 & -1 & 0 \\ 0 & 1 & 0 & -1 \end{bmatrix}.$$

Later in the book we will designate this basis by $D_\oplus$.

A simple minimal basis in $\mathbb{R}^2$ is formed by the vectors of the canonical basis and the negative of their sum:

$$\begin{bmatrix} 1 & 0 & -1 \\ 0 & 1 & -1 \end{bmatrix}.$$

For convenience, we normalize the third vector and write

$$D_2 = \begin{bmatrix} 1 & 0 & -\sqrt{2}/2 \\ 0 & 1 & -\sqrt{2}/2 \end{bmatrix}.$$

If we add one more vector to this positive basis, we get a positive spanning set that is not a positive basis:

$$D_3 = \begin{bmatrix} 1 & 0 & -\sqrt{2}/2 & 0 \\ 0 & 1 & -\sqrt{2}/2 & -1 \end{bmatrix}.$$

In Figure 2.1, we plot the positive bases D_1 and D_2 and the positive spanning set D_3.

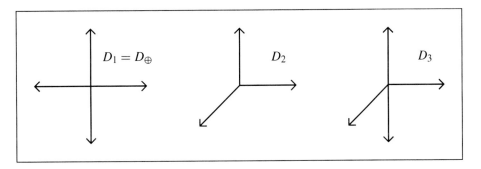

Figure 2.1. *A maximal positive basis (left), a minimal positive basis (center), and a positive spanning set that is not a positive basis (right).*

We now present three necessary and sufficient characterizations for a set that spans $\mathbb{R}^n$ to also span $\mathbb{R}^n$ positively (see also [74]).

Theorem 2.3. *Let $[v_1 \cdots v_r]$, with $v_i \neq 0$ for all $i \in \{1,\ldots,r\}$, span $\mathbb{R}^n$. Then the following are equivalent:*

(i) *$[v_1 \cdots v_r]$ spans $\mathbb{R}^n$ positively.*

(ii) *For every $i = 1,\ldots,r$, the vector $-v_i$ is in the convex cone positively spanned by the remaining $r-1$ vectors.*

(iii) *There exist real scalars $\alpha_1,\ldots,\alpha_r$ with $\alpha_i > 0$, $i \in \{1,\ldots,r\}$, such that $\sum_{i=1}^r \alpha_i v_i = 0$.*

(iv) *For every nonzero vector $w \in \mathbb{R}^n$, there exists an index i in $\{1,\ldots,r\}$ for which $w^\top v_i > 0$.*

Proof. The proof is made by showing the following implications: (i) $\Rightarrow$ (ii), (ii) $\Rightarrow$ (iii), (iii) $\Rightarrow$ (i), (i) $\Rightarrow$ (iv), and (iv) $\Rightarrow$ (i).

(i) $\Rightarrow$ (ii) Since $[v_1 \cdots v_r]$ spans $\mathbb{R}^n$ positively, the vector $-v_i$, with i in $\{1,\ldots,r\}$, can be written as

$$-v_i = \sum_{j=1}^r \lambda_{ij} v_j,$$

where the scalars $\lambda_{i1}, \ldots, \lambda_{ir}$ are nonnegative. As a consequence, we obtain

$$-v_i - \lambda_{ii} v_i = \sum_{\substack{j=1 \\ j \neq i}}^{r} \lambda_{ij} v_j$$

and

$$-v_i = \sum_{\substack{j=1 \\ j \neq i}}^{r} \frac{\lambda_{ij}}{1 + \lambda_{ii}} v_j = \sum_{\substack{j=1 \\ j \neq i}}^{r} \tilde{\lambda}_{ij} v_j,$$

where $\tilde{\lambda}_{ij} = \frac{\lambda_{ij}}{1+\lambda_{ii}} \geq 0$ for all $j \in \{1, \ldots, r\} \setminus \{i\}$. This shows that $-v_i$ is in the convex cone positively spanned by the remaining $r - 1$ vectors.

(ii) $\Rightarrow$ (iii) From the assumption (ii), there exist nonnegative scalars $\bar{\lambda}_{ij}$, $i, j = 1, \ldots, r$, such that

$$\begin{aligned}
v_1 + \bar{\lambda}_{12} v_2 + \cdots + \bar{\lambda}_{1r} v_r &= 0, \\
\bar{\lambda}_{21} v_1 + v_2 + \cdots + \bar{\lambda}_{2r} v_r &= 0, \\
&\vdots \\
\bar{\lambda}_{r1} v_1 + \bar{\lambda}_{r2} v_2 + \cdots + v_r &= 0.
\end{aligned}$$

By adding these r equalities, we get

$$\left(1 + \sum_{i=2}^{r} \bar{\lambda}_{i1} \right) v_1 + \cdots + \left(1 + \sum_{i=1}^{r-1} \bar{\lambda}_{ir} \right) v_r = 0,$$

which can be rewritten as

$$\alpha_1 v_1 + \cdots + \alpha_r v_r = 0,$$

with $\alpha_j = 1 + \sum_{\substack{i=1 \\ i \neq j}}^{r} \bar{\lambda}_{ij} > 0$, $j \in \{1, \ldots, r\}$.

(iii) $\Rightarrow$ (i) Let $\alpha_1, \ldots, \alpha_r$ be positive scalars such that $\alpha_1 v_1 + \cdots + \alpha_r v_r = 0$, and let v be an arbitrary vector in $\mathbb{R}^n$. Since $[v_1 \cdots v_r]$ spans $\mathbb{R}^n$, there exist scalars $\lambda_1, \ldots, \lambda_r$ such that $v = \lambda_1 v_1 + \cdots + \lambda_r v_r$. By adding to the right-hand side of this equality a sufficiently large multiple of $\alpha_1 v_1 + \cdots + \alpha_r v_r$, one can show that v can be expressed as a positive linear combination of $v_1, \ldots, v_r$. Thus, $[v_1 \cdots v_r]$ spans $\mathbb{R}^n$ positively.

(i) $\Rightarrow$ (iv) Let w be a nonzero vector in $\mathbb{R}^n$. From the assumption (i), there exist nonnegative scalars $\lambda_1, \ldots, \lambda_r$ such that

$$w = \lambda_1 v_1 + \cdots + \lambda_r v_r.$$

Since $w \neq 0$, we get that

$$\begin{aligned}
0 < w^\top w &= (\lambda_1 v_1 + \cdots + \lambda_r v_r)^\top w \\
&= \lambda_1 v_1^\top w + \cdots + \lambda_r v_r^\top w,
\end{aligned}$$

from which we conclude that at least one of the scalars $w^\top v_1, \ldots, w^\top v_r$ has to be positive.

(iv) $\Rightarrow$ (i) If the convex cone positively spanned by $v_1, \ldots, v_r$ is not $\mathbb{R}^n$, then there exists a hyperplane $H = \{v \in \mathbb{R}^n : v^\top h = 0\}$, with $h \neq 0$, such that this convex cone (and so all of its generators) is contained in either $\{v \in \mathbb{R}^n : v^\top h \geq 0\}$ or $\{v \in \mathbb{R}^n : v^\top h \leq 0\}$; see [200, Corollary 11.7.3]. The assumption would then be contradicted with either $w = h$ or $w = -h$. $\quad \square$

As mentioned before, the characterization (iv) of Theorem 2.3 is at the heart of directional direct-search methods. It implies that, given a continuously differentiable function f at some given point x where $\nabla f(x) \neq 0$, there must always exist a vector d in a given positive spanning set (or in a positive basis) such that

$$-\nabla f(x)^\top d > 0.$$

In other words, there must always exist a direction of descent in such a set. In Figure 2.2, we identify such a vector d for the three spanning sets D_1, D_2, and D_3 given before.

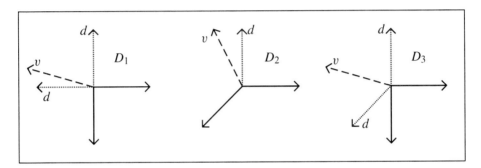

Figure 2.2. *Given a positive spanning set and a vector $v = -\nabla f(x)$ (dashed), there exists at least one element d (dotted) of the set such that $v^\top d > 0$.*

Simple positive bases

Now we turn our attention to the construction of positive bases. The following result (given in [153]) provides a simple mechanism for generating different positive bases.

Theorem 2.4. *Suppose $[v_1 \cdots v_r]$ is a positive basis for $\mathbb{R}^n$ and $W \in \mathbb{R}^{n \times n}$ is a nonsingular matrix. Then $[Wv_1 \cdots Wv_r]$ is also a positive basis for $\mathbb{R}^n$.*

Proof. It is obvious that $[v_1 \cdots v_r]$ spans $\mathbb{R}^n$ since it does it positively. Since W is nonsingular, $[Wv_1 \cdots Wv_r]$ also spans $\mathbb{R}^n$. Thus we can apply Theorem 2.3 for both $[v_1 \cdots v_r]$ and $[Wv_1 \cdots Wn_r]$.

Now let w be a nonzero vector in $\mathbb{R}^n$. Since $[v_1 \cdots v_r]$ spans $\mathbb{R}^n$ positively and W is nonsingular, we get from (iv) in Theorem 2.3 that

$$(W^\top w)^\top v_i > 0$$

for some i in $\{1, \ldots, r\}$. In other words,

$$w^\top (W v_i) > 0$$

for some i in $\{1, \ldots, r\}$, from which we conclude that $[W v_1 \cdots W v_r]$ also spans $\mathbb{R}^n$ positively.

It is a direct consequence of the definition of positive dependence that if $[W v_1 \cdots W v_r]$ was positively dependent, then $[v_1 \cdots v_r]$ would also be positively dependent, which concludes the proof of the theorem.   □

One can easily prove that $D_\oplus = [I \ -I]$ is a (maximal) positive basis. The result just stated in Theorem 2.4 allows us to say that $[W \ -W]$ is also a (maximal) positive basis for any choice of the nonsingular matrix $W \in \mathbb{R}^{n \times n}$.

From Theorems 2.3 and 2.4, we can easily deduce the following corollary. The proof is left as an exercise.

Corollary 2.5.

   (i) $[I \ -e]$ *is a (minimal) positive basis.*

   (ii) *Let* $W = [w_1 \cdots w_n] \in \mathbb{R}^{n \times n}$ *be a nonsingular matrix. Then* $[W \ -\sum_{i=1}^n w_i]$ *is a (minimal) positive basis for* $\mathbb{R}^n$.

Positive basis with uniform angles

Consider $n + 1$ vectors $v_1, \ldots, v_{n+1}$ in $\mathbb{R}^n$ for which all the angles between pairs v_i, v_j ($i \neq j$) have the same amplitude α. Assuming that the $n + 1$ vectors are normalized, this requirement is expressed as

$$a = \cos(\alpha) = v_i^\top v_j, \qquad i, j \in \{1, \ldots, n+1\}, i \neq j, \tag{2.1}$$

where $a \neq 1$. One can show that $a = -1/n$ (see the exercises).

Now we seek a set of $n + 1$ normalized vectors $[v_1 \cdots v_{n+1}]$ satisfying the property (2.1) with $a = -1/n$. Let us first compute $v_1, \ldots, v_n$; i.e., let us compute a matrix $V = [v_1 \cdots v_n]$ such that

$$V^\top V = A,$$

where A is the matrix given by

$$A = \begin{bmatrix} 1 & -1/n & -1/n & \cdots & -1/n \\ -1/n & 1 & -1/n & \cdots & -1/n \\ \vdots & & \ddots & & \\ \vdots & & & \ddots & \\ -1/n & -1/n & -1/n & \cdots & 1 \end{bmatrix}. \tag{2.2}$$

The matrix A is symmetric and diagonally dominant with positive diagonal entries, and, therefore, it is positive definite [109]. Thus, we can make use of its Cholesky decomposition

$$A = CC^\top,$$

where $C \in \mathbb{R}^{n \times n}$ is a lower triangular matrix of order n with positive diagonal entries. Given this decomposition, one can easily see that a choice for V is determined by

$$V = [v_1 \cdots v_n] = C^\top.$$

The vector v_{n+1} is then computed by

$$v_{n+1} = -\sum_{i=1}^{n} v_i. \tag{2.3}$$

One can easily show that $v_i^\top v_{n+1} = -1/n$, $i = 1, \ldots, n$, and $v_{n+1}^\top v_{n+1} = 1$.

Since V is nonsingular and v_{n+1} is determined by (2.3), we can apply Corollary 2.5(ii) to establish that $[v_1 \cdots v_{n+1}]$ is a (minimal) positive basis. The angles between any two vectors in this positive basis exhibit the same amplitude. We summarize this result below.

Corollary 2.6. *Let $V = C^\top = [v_1 \cdots v_n] \in \mathbb{R}^{n \times n}$, where $A = CC^\top$ and A is given by (2.2). Let $v_{n+1} = -\sum_{i=1}^{n} v_i$.*
  Then $[v_1 \cdots v_{n+1}]$ is a (minimal) positive basis for $\mathbb{R}^n$ satisfying $v_i^\top v_j = -1/n$, $i, j \in \{1, \ldots, n+1\}$, $i \neq j$, and $\|v_i\| = 1$, $i = 1, \ldots, n+1$.

A minimal positive basis with uniform angles is depicted in Figure 2.3, computed as in Corollary 2.6.

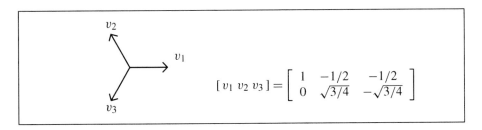

$$[v_1 \; v_2 \; v_3] = \begin{bmatrix} 1 & -1/2 & -1/2 \\ 0 & \sqrt{3/4} & -\sqrt{3/4} \end{bmatrix}$$

Figure 2.3. *A minimal positive basis with uniform angles.*

2.2 Gradient estimates used in direct search

What we now show is that if we sample $n+1$ points of the form $x + \alpha d$ defined by a positive basis D, and their function values are no better than the function value at x, then the size of the gradient (considered Lipschitz continuous) of the function at x is of the order of the distance between x and the sample points $x + \alpha d$ and, furthermore, the order constant depends only upon the nonlinearity of f and the geometry of the sample set.

To prove this result, used in the convergence theory of directional direct-search methods, we must first introduce the notion of cosine measure for positive spanning sets.

Definition 2.7. *The cosine measure of a positive spanning set (with nonzero vectors) or of a positive basis* D *is defined by*

$$\mathrm{cm}(D) \;=\; \min_{0 \neq v \in \mathbb{R}^n} \; \max_{d \in D} \; \frac{v^\top d}{\|v\|\|d\|}.$$

Given any positive spanning set, it necessarily happens that

$$\mathrm{cm}(D) \;>\; 0.$$

Values of the cosine measure close to zero indicate a deterioration of the positive spanning property. For example, the maximal positive basis $D_\oplus = [I \; -I]$ has cosine measure equal to $1/\sqrt{n}$. When $n = 2$ we have $\mathrm{cm}(D_\oplus) = \sqrt{2}/2$.

Now let us consider the following example. Let θ be an angle in $(0, \pi/4]$ and D_θ be a positive basis defined by

$$D_\theta \;=\; \begin{bmatrix} 1 & 0 & -\cos(\theta) \\ 0 & 1 & -\sin(\theta) \end{bmatrix}.$$

Observe that $D_{\frac{\pi}{4}}$ is just the positive basis D_2 considered before. The cosine measure of D_θ is given by $\cos((\pi - \theta)/2)$, and it converges to zero when θ tends to zero. Figure 2.4 depicts this situation.

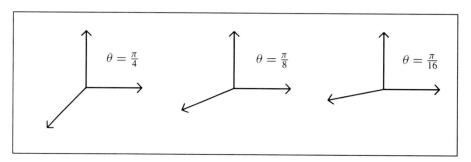

Figure 2.4. *Positive bases D_θ for three values of θ. The cosine measure is approaching zero.*

Another key point, related to the descent properties of positive spanning sets, is that, given any vector $v \neq 0$, we have

$$\mathrm{cm}(D) \;\leq\; \max_{d \in D} \; \frac{v^\top d}{\|v\|\|d\|}.$$

Thus, there must exist a $d \in D$ such that

$$\mathrm{cm}(D) \;\leq\; \frac{v^\top d}{\|v\|\|d\|}$$

or, equivalently,

$$\mathrm{cm}(D)\|v\|\|d\| \leq v^\top d.$$

Given a positive spanning set D, a point x, and a positive value for the parameter α, we are interested in looking at the points of the form $x + \alpha d$ for all $d \in D$. These points are in a ball centered at x, of radius Δ, defined by

$$\Delta = \alpha \max_{d \in D} \|d\|.$$

We point out that if only a finite number of positive spanning sets are used in an algorithm, then Δ tends to zero if and only if α tends to zero. The following result is taken from [80, 145].

Theorem 2.8. *Let D be a positive spanning set and $\alpha > 0$ be given. Assume that ∇f is Lipschitz continuous (with constant $v > 0$) in an open set containing the ball $B(x; \Delta)$. If $f(x) \leq f(x + \alpha d)$, for all $d \in D$, then*

$$\|\nabla f(x)\| \leq \frac{v}{2} \mathrm{cm}(D)^{-1} \max_{d \in D} \|d\| \alpha.$$

Proof. Let d be a vector in D for which

$$\mathrm{cm}(D)\|\nabla f(x)\|\|d\| \leq -\nabla f(x)^\top d.$$

Now, from the integral form of the mean value theorem and the fact that $f(x) \leq f(x + \alpha d)$, we get, for all $d \in D$, that

$$0 \leq f(x + \alpha d) - f(x) = \int_0^1 \nabla f(x + t\alpha d)^\top (\alpha d) dt.$$

By multiplying the first inequality by α and by adding it to the second one, we obtain

$$\mathrm{cm}(D)\|\nabla f(x)\|\|d\|\alpha \leq \int_0^1 (\nabla f(x + t\alpha d) - \nabla f(x))^\top (\alpha d) dt \leq \frac{v}{2}\|d\|^2 \alpha^2,$$

and the proof is completed. ☐

If a directional direct-search method is able to generate a sequence of points x satisfying the conditions of Theorem 2.8 for which α (and thus Δ) tends to zero, then clearly the gradient of the objective function converges to zero along this sequence.

The bound proved in Theorem 2.8 can be rewritten in the form

$$\|\nabla f(x)\| \leq \kappa_{eg} \Delta,$$

where $\kappa_{eg} = v \, \mathrm{cm}(D)^{-1}/2$. We point out that this bound has the same structure as other bounds used in different derivative-free methods. The bound is basically given by Δ times a constant that depends on the nonlinearity of the function (expressed by the Lipschitz constant v) and on the geometry of the sample set (measured by $\mathrm{cm}(D)^{-1}$).

2.3   Linear interpolation and regression models

Now we turn our attention to sample sets not necessarily formed by a predefined set of directions. Consider a sample set $Y = \{y^0, y^1, \ldots, y^p\}$ in $\mathbb{R}^n$. The simplest model based on $n+1$ sample points ($p = n$) that we can think of is an interpolation model.

Linear interpolation

Let $m(x)$ denote a polynomial of degree $d = 1$ interpolating f at the points in Y, i.e., satisfying the interpolation conditions

$$m(y^i) = f(y^i), \quad i = 0, \ldots, n. \tag{2.4}$$

We can express $m(x)$ in the form

$$m(x) = \alpha_0 + \alpha_1 x_1 + \cdots + \alpha_n x_n,$$

using, as a basis for the space $\mathcal{P}_n^1$ of linear polynomials of degree 1, the polynomial basis $\bar{\phi} = \{1, x_1, \ldots, x_n\}$. However, we point out that other bases could be used, e.g., $\{1, 1 + x_1, 1 + x_1 + x_2, \ldots, 1 + x_1 + x_2 + \cdots + x_n\}$. We can then rewrite (2.4) as

$$\begin{bmatrix} 1 & y_1^0 & \cdots & y_n^0 \\ 1 & y_1^1 & \cdots & y_n^1 \\ \vdots & \vdots & \vdots & \vdots \\ 1 & y_1^n & \cdots & y_n^n \end{bmatrix} \begin{bmatrix} \alpha_0 \\ \alpha_1 \\ \vdots \\ \alpha_n \end{bmatrix} = \begin{bmatrix} f(y^0) \\ f(y^1) \\ \vdots \\ f(y^n) \end{bmatrix}.$$

The matrix of this linear system is denoted by

$$M = M(\bar{\phi}, Y) = \begin{bmatrix} 1 & y_1^0 & \cdots & y_n^0 \\ 1 & y_1^1 & \cdots & y_n^1 \\ \vdots & \vdots & \vdots & \vdots \\ 1 & y_1^n & \cdots & y_n^n \end{bmatrix}. \tag{2.5}$$

In this book we write M as $M(\bar{\phi}, Y)$ to highlight the dependence of M on the basis $\bar{\phi}$ and on the sample set Y.

Definition 2.9. *The set $Y = \{y^0, y^1, \ldots, y^n\}$ is poised for linear interpolation in $\mathbb{R}^n$ if the corresponding matrix $M(\bar{\phi}, Y)$ is nonsingular.*

The definition of poisedness is independent of the basis chosen. In other words, if Y is poised for a basis ϕ, then it is poised for any other basis in $\mathcal{P}_n^1$. The definition of $m(x)$ is also independent of the basis chosen. These issues are covered in detail in Chapter 3. It is straightforward to see that the sample set Y is poised for linear interpolation if and only if the linear interpolating polynomial $m(x) = \alpha_0 + \alpha_1 x_1 + \cdots + \alpha_n x_n$ is uniquely defined.

Linear regression

When the number $p+1$ of points in the sample set exceeds by more than 1 the dimension n of the sampling space, it might not be possible to fit a linear polynomial. In this case,

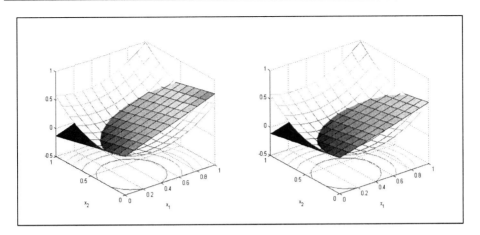

Figure 2.5. *At the left, the linear interpolation model of* $f(x_1, x_2) = (x_1 - 0.3)^2 +$
$(x_2 - 0.3)^2$ *at* $y^0 = (0,0)$, $y^1 = (0.2, 0.6)$, *and* $y^2 = (0.8, 0.7)$. *At the right, the linear
regression model of* $f(x_1, x_2) = (x_1 - 0.3)^2 + (x_2 - 0.3)^2$ *at* $y^0 = (0,0)$, $y^1 = (0.2, 0.6)$,
$y^2 = (0.8, 0.7)$, *and* $y^3 = (0.5, 0.5)$. *The inclusion of* y^3 *pushes the model down.*

one option is to use linear regression and to compute the coefficients of the linear (least-squares) regression polynomial $m(x) = \alpha_0 + \alpha_1 x_1 + \cdots + \alpha_n x_n$ as the least-squares solution of the system

$$
\begin{bmatrix}
1 & y_1^0 & \cdots & y_n^0 \\
1 & y_1^1 & \cdots & y_n^1 \\
\vdots & \vdots & \vdots & \vdots \\
\vdots & \vdots & \vdots & \vdots \\
1 & y_1^p & \cdots & y_n^p
\end{bmatrix}
\begin{bmatrix}
\alpha_0 \\
\alpha_1 \\
\vdots \\
\alpha_n
\end{bmatrix}
=
\begin{bmatrix}
f(y^0) \\
f(y^1) \\
\vdots \\
\vdots \\
f(y^p)
\end{bmatrix}.
\tag{2.6}
$$

Again, we denote the matrix of this (possibly overdetermined) system of linear equations by $M = M(\bar{\phi}, Y)$.

The definition of poisedness generalizes easily from linear interpolation to linear regression.

Definition 2.10. *The set* $Y = \{y^0, y^1, \ldots, y^p\}$, *with* $p > n$, *is poised for linear regression in* $\mathbb{R}^n$ *if the corresponding matrix* $M(\bar{\phi}, Y)$ *has full (column) rank.*

It is also possible to prove, in the regression case, that if a set Y is poised for a basis ϕ, then it is also poised for any other basis in $\mathcal{P}_n^1$ and that $m(x)$ is independent of the basis chosen. Finally, we point out that the sample set is poised for linear regression if and only if the linear regression polynomial $m(x)$ is uniquely defined. These issues are covered in detail in Chapter 4.

2.4   Error bounds for linear interpolation and regression

Error bounds for linear interpolation

Let us rewrite the linear interpolating polynomial in the form

$$m(y) = c + g^\top y$$

by setting $\alpha_0 = c$ and $\alpha_i = g_i$, $i = 1, \ldots, n$.

One of the natural questions that arises in interpolation is how to measure the quality of $m(y)$ as an approximation to $f(y)$. We start by looking at the quality of the gradient $g = \nabla m(y)$ of the model as an approximation to $\nabla f(y)$. We consider that the interpolation points $y^0, y^1, \ldots, y^n$ are in a ball of radius Δ centered at y^0. In practice, one might set

$$\Delta = \Delta(Y) = \max_{1 \le i \le n} \|y^i - y^0\|.$$

We are interested in the quality of $\nabla m(y)$ and $m(y)$ in the ball of radius Δ centered at y^0. The assumptions needed for this result are summarized below.

Assumption 2.1. *We assume that $Y = \{y^0, y^1, \ldots, y^n\} \subset \mathbb{R}^n$ is a poised set of sample points (in the linear interpolation sense) contained in the ball $B(y^0; \Delta(Y))$ of radius $\Delta = \Delta(Y)$.*

Further, we assume that the function f is continuously differentiable in an open domain Ω containing $B(y^0; \Delta)$ and ∇f is Lipschitz continuous in Ω with constant $\nu > 0$.

As we can observe from the proof of Theorem 2.11 below, the derivation of the error bounds is based on the application of one step of Gaussian elimination to the matrix $M = M(\bar{\phi}, Y)$ in (2.5). After performing such a step we arrive at the matrix

$$\begin{bmatrix} 1 & y_1^0 & \cdots & y_n^0 \\ 0 & y_1^1 - y_1^0 & \cdots & y_n^1 - y_n^0 \\ \vdots & \vdots & \vdots & \vdots \\ 0 & y_1^n - y_1^0 & \cdots & y_n^n - y_n^0 \end{bmatrix},$$

which can be expressed by blocks as

$$\begin{bmatrix} 1 & y_0^\top \\ 0 & L \end{bmatrix},$$

with

$$L = \begin{bmatrix} y^1 - y^0 \cdots y^n - y^0 \end{bmatrix}^\top = \begin{bmatrix} (y^1 - y^0)^\top \\ \vdots \\ (y^n - y^0)^\top \end{bmatrix} = \begin{bmatrix} y_1^1 - y_1^0 & \cdots & y_n^1 - y_n^0 \\ \vdots & \vdots & \vdots \\ y_1^n - y_1^0 & \cdots & y_n^n - y_n^0 \end{bmatrix}.$$

It is evident to see that L is nonsingular if and only if M is nonsingular, since $\det(L) = \det(M)$. Notice that the points appear listed in L by rows, which favors factorizations by rows.

It turns out that the error bounds for the approximation which we derive are in terms of the scaled matrix

$$\hat{L} = \frac{1}{\Delta}L = \frac{1}{\Delta}\left[y^1 - y^0 \cdots y^n - y^0\right]^{\top} = \begin{bmatrix} \frac{y_1^1 - y_1^0}{\Delta} & \cdots & \frac{y_n^1 - y_n^0}{\Delta} \\ \vdots & \vdots & \vdots \\ \frac{y_1^n - y_1^0}{\Delta} & \cdots & \frac{y_n^n - y_n^0}{\Delta} \end{bmatrix}.$$

This matrix $\hat{L}$ corresponds to a scaled sample set

$$\hat{Y} = \{y^0/\Delta, y^1/\Delta, \ldots, y^n/\Delta\} \subset B(y^0/\Delta; 1),$$

which is contained in a ball of radius 1 centered at y^0/Δ.

Theorem 2.11. *Let Assumption* 2.1 *hold. The gradient of the linear interpolation model satisfies, for all points y in* $B(y^0; \Delta)$, *an error bound of the form*

$$\|\nabla f(y) - \nabla m(y)\| \leq \kappa_{eg} \Delta, \tag{2.7}$$

where $\kappa_{eg} = \nu(1 + n^{\frac{1}{2}}\|\hat{L}^{-1}\|/2)$ *and* $\hat{L} = L/\Delta$.

Proof. If the set Y is poised, then the $(n+1) \times (n+1)$ matrix $M = M(\bar{\phi}, Y)$ is nonsingular and so is the $n \times n$ matrix L.

We look initially at the gradient of f at the point y^0. Subtracting the first interpolating condition from the remaining n, we obtain

$$(y^i - y^0)^{\top} g = f(y^i) - f(y^0), \quad i = 1, \ldots, n.$$

Then, if we use the integral form of the mean value theorem

$$f(y^i) - f(y^0) = \int_0^1 (y^i - y^0)^{\top} \nabla f(y^0 + t(y^i - y^0)) dt,$$

we obtain, from the Lipschitz continuity of ∇f, that

$$(y^i - y^0)^{\top}(\nabla f(y^0) - g) \leq \frac{\nu}{2}\|y^i - y^0\|^2, \quad i = 1, \ldots, n.$$

Then, from these last n inequalities, we derive

$$\|L(\nabla f(y^0) - g)\| \leq (n^{\frac{1}{2}}\nu/2)\Delta^2,$$

from which we conclude that

$$\|\nabla f(y^0) - g\| \leq (n^{\frac{1}{2}}\|\hat{L}^{-1}\|\nu/2)\Delta.$$

The error bound for any point y in the ball $B(y^0; \Delta)$ is easily derived from the Lipschitz continuity of the gradient of f:

$$\|\nabla f(y) - g\| \leq \|\nabla f(y) - \nabla f(y^0)\| + \|\nabla f(y^0) - g\| \leq (\nu + n^{\frac{1}{2}}\|\hat{L}^{-1}\|\nu/2)\Delta. \quad \square$$

Assuming a uniform bound on $\|\hat{L}^{-1}\|$ independent of Δ, the error in the gradient is linear in Δ. One can see also that the error in the interpolation model $m(x)$ itself is quadratic in Δ.

Theorem 2.12. *Let Assumption 2.1 hold. The interpolation model satisfies, for all points y in $B(y^0; \Delta)$, an error bound of the form*

$$|f(y) - m(y)| \leq \kappa_{ef} \Delta^2, \tag{2.8}$$

where $\kappa_{ef} = \kappa_{eg} + \nu/2$ and κ_{eg} is given in Theorem 2.11.

Proof. We use the same arguments as in the proof of Theorem 2.11 to obtain

$$f(y) - f(y^0) \leq \nabla f(y^0)^\top (y - y^0) + \frac{\nu}{2} \|y - y^0\|^2.$$

From this we get

$$f(y) - f(y^0) - g^\top(y - y^0) \leq (\nabla f(y^0) - g)^\top (y - y^0) + \frac{\nu}{2} \|y - y^0\|^2.$$

The error bound (2.8) comes from combining this inequality with (2.7) and from noting that the constant term in the model can be written as $c = f(y^0) - g^\top y^0$. $\square$

Error bounds for linear regression

In the regression case we are considering a sample set $Y = \{y^0, y^1, \ldots, y^p\}$ with more than $n + 1$ points contained in the ball $B(y^0; \Delta(Y))$ of radius

$$\Delta = \Delta(Y) = \max_{1 \leq i \leq p} \|y^i - y^0\|.$$

We start by also rewriting the linear regression polynomial in the form

$$m(y) = c + g^\top y,$$

where $c = \alpha_0$ and $g_i = \alpha_i$, $i = 1, \ldots, n$, are the components of the least-squares solution of the system (2.6).

Assumption 2.2. *We assume that $Y = \{y^0, y^1, \ldots, y^p\} \subset \mathbb{R}^n$, with $p > n$, is a poised set of sample points (in the linear regression sense) contained in the ball $B(y^0; \Delta(Y))$ of radius $\Delta = \Delta(Y)$.*

Further, we assume that the function f is continuously differentiable in an open domain Ω containing $B(y^0; \Delta)$ and ∇f is Lipschitz continuous in Ω with constant $\nu > 0$.

The error bounds for the approximation are also derived in terms of the scaled matrix

$$\hat{L} = \frac{1}{\Delta} L = \frac{1}{\Delta} \left[y^1 - y^0 \cdots y^p - y^0 \right]^\top,$$

which corresponds to a scaled sample set contained in a ball of radius 1 centered at y^0/Δ, i.e.,

$$\hat{Y} = \{y^0/\Delta, y^1/\Delta, \ldots, y^p/\Delta\} \subset B(y^0/\Delta; 1).$$

The proof of the bounds follows exactly the same steps as the proof for the linear interpolation case. For example, for the gradient approximation, once we are at the point in the proof where

$$\|L(\nabla f(y^0) - g)\| \leq (p^{\frac{1}{2}} \nu/2) \Delta^2,$$

or, equivalently,

$$\|\hat{L}(\nabla f(y^0) - g)\| \leq (p^{\frac{1}{2}} \nu/2) \Delta, \tag{2.9}$$

we "pass" $\hat{L}$ to the right-hand side by means of its left inverse $\hat{L}^\dagger$.[3] We can then state the bounds in the following format.

Theorem 2.13. *Let Assumption 2.2 hold. The gradient of the linear regression model satisfies, for all points y in $B(y^0; \Delta)$, an error bound of the form*

$$\|\nabla f(y) - \nabla m(y)\| \leq \kappa_{eg} \Delta,$$

where $\kappa_{eg} = \nu(1 + p^{\frac{1}{2}} \|\hat{L}^\dagger\|/2)$ and $\hat{L} = L/\Delta$.

The linear regression model satisfies, for all points y in $B(y^0; \Delta)$, an error bound of the form

$$|f(y) - m(y)| \leq \kappa_{ef} \Delta^2,$$

where $\kappa_{ef} = \kappa_{eg} + \nu/2$.

2.5 Other geometrical concepts

The notion of poisedness for linear interpolation is closely related to the concept of affine independence in convex analysis.

Affine independence

We will follow Rockafellar [200] to introduce affine independence as well as other basic concepts borrowed from convex analysis.

The affine hull of a given set $S \subset \mathbb{R}^n$ is the smallest affine set containing S (meaning that it is the intersection of all affine sets containing S). The affine hull of a set is always uniquely defined and consists of all linear combinations of elements of S whose scalars sum up to one (see [200]).

Definition 2.14. *A set of $m + 1$ points $Y = \{y^0, y^1, \ldots, y^m\}$ is said to be affinely independent if its affine hull $\mathrm{aff}(y^0, y^1, \ldots, y^m)$ has dimension m.*

The dimension of an affine set is the dimension of the linear subspace parallel to it. So, we cannot have an affinely independent set in $\mathbb{R}^n$ with more than $n + 1$ points.

---

[3] $A^\dagger$ denotes the Moore–Penrose generalized inverse of a matrix A, which can be expressed by the singular value decomposition of A for any real or complex matrix A. In the current context, where L has full column rank, we have $\hat{L}^\dagger = (\hat{L}^\top \hat{L})^{-1} \hat{L}^\top$.

Given an affinely independent set of points $\{y^0, y^1, \ldots, y^m\}$, we have that

$$\text{aff}(y^0, y^1, \ldots, y^m) = y^0 + \mathcal{L}(y^1 - y^0, \ldots, y^m - y^0),$$

where $\mathcal{L}(y^1 - y^0, \ldots, y^m - y^0)$ is the linear subspace of dimension m generated by the vectors $y^1 - y^0, \ldots, y^m - y^0$.

We associate with an affinely independent set of points $\{y^0, y^1, \ldots, y^m\}$ the matrix

$$\left[y^1 - y^0 \cdots y^m - y^0 \right] \in \mathbb{R}^{n \times m},$$

whose rank must be equal to m. This matrix is exactly the transpose of the matrix L that appeared when we linearly interpolated a function f on the set $Y = \{y^0, y^1, \ldots, y^n\}$, with $m = n$, choosing an appropriate basis for the space $\mathcal{P}_n^1$ of linear polynomials of degree 1 in $\mathbb{R}^n$.

Simplices

Similarly, the convex hull of a given set $S \subset \mathbb{R}^n$ is the smallest convex set containing S (meaning that it is the intersection of all convex sets containing S). The convex hull of a set is always uniquely defined and consists of all convex combinations of elements of S, i.e., of all linear combinations of elements of S whose scalars are nonnegative and sum up to one (see [200]).

Definition 2.15. *Given an affinely independent set of points $Y = \{y^0, y^1, \ldots, y^m\}$, its convex hull is called a simplex of dimension m.*

A simplex of dimension 0 is a point, of dimension 1 is a closed line segment, of dimension 2 is a triangle, and of dimension 3 is a tetrahedron.

The vertices of a simplex are the elements of Y. A simplex in $\mathbb{R}^n$ cannot have more than $n + 1$ vertices. When there are $n + 1$ vertices, its dimension is n. In this case,

$$\left[y^1 - y^0 \cdots y^n - y^0 -(y^1 - y^0) \cdots -(y^n - y^0) \right] \qquad (2.10)$$

forms a (maximal) positive basis in $\mathbb{R}^n$. We illustrate this maximal positive basis in Figure 2.6.

The diameter of a simplex Y of vertices $y^0, y^1, \ldots, y^m$ is defined by

$$\text{diam}(Y) = \max_{0 \leq i < j \leq m} \| y^i - y^j \|.$$

One way of approximating $\text{diam}(Y)$ at y^0 is by computing the less expensive quantity

$$\Delta(Y) = \max_{1 \leq i \leq n} \| y^i - y^0 \|.$$

Clearly, we can write $\Delta(Y) \leq \text{diam}(Y) \leq 2\Delta(Y)$.

By the shape of a simplex we mean its equivalent class under similarity: the simplices of vertices Y and λY, $\lambda > 0$, share the same shape. The volume of a simplex of $n + 1$ vertices $Y = \{y^0, y^1, \ldots, y^n\}$ is defined by

$$\text{vol}(Y) = \frac{|\det(L)|}{n!},$$

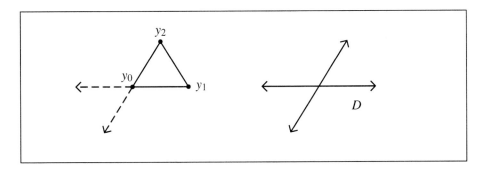

Figure 2.6. *How to compute a maximal positive basis from the vertices of a simplex.*

where

$$L = L(Y) = \left[y^1 - y^0 \cdots y^n - y^0 \right]^\top.$$

Since the vertices of a simplex form an affinely independent set, one clearly has that vol$(Y) > 0$. It is also left as an exercise to see that the choice of the centering point in L is irrelevant for the definition of the volume of a simplex.

The volume of a simplex is not a good measure of the quality of its geometry since it is not scaling independent. To see this let

$$Y_t = \left\{ \begin{bmatrix} 0 \\ 0 \end{bmatrix}, \begin{bmatrix} t \\ 0 \end{bmatrix}, \begin{bmatrix} 0 \\ t \end{bmatrix} \right\},$$

with $t > 0$. It is easy to see that vol$(Y_t) \to 0$ when $t \to 0$. However, the angles between the vectors formed by the vertices are the same for all positive values of t (or putting it differently all these simplices have the same shape).

A measure of the quality of a simplex geometry must be independent of the scale of the simplex, given by either $\Delta(Y)$ or diam(Y). One such measure is given by

$$\|[L(Y)/\Delta(Y)]^\dagger\|,$$

which reduces to

$$\|[L(Y)/\Delta(Y)]^{-1}\|$$

for simplices of $n+1$ vertices. One alternative when there are $n+1$ vertices is to work with a normalized volume

$$\text{von}(Y) = \text{vol}\left(\frac{1}{\text{diam}(Y)} Y \right) = \frac{|\det(L(Y))|}{n! \, \text{diam}(Y)^n}.$$

Poisedness and positive spanning

Another natural question that arises is the relationship between the concepts of poised interpolation and regression sets and positive spanning sets or positive bases. To study this relationship, let us assume that we have a positive spanning set D formed by nonzero vectors. Recall that consequently the cosine measure cm(D) is necessarily positive. For the sake of simplicity in this discussion let us assume that the elements in D have norm one. Then

$$\text{cm}(D) = \min_{\|v\|=1} \max_{d \in D} v^\top d \leq \min_{\|v\|=1} \max_{d \in D} |v^\top d| \leq \min_{\|v\|=1} \|D^\top v\|. \qquad (2.11)$$

We immediately conclude that D has full row rank.

Thus, given a point y^0 and a positive spanning set D for which $\text{cm}(D) > 0$, we know that the sample set $\{y^0\} \cup \{y^0 + d : d \in D\}$ is poised for linear regression. See Figure 2.7.

The contrary, however, is not true. Given a poised set $Y = \{y^0, y^1, \ldots, y^p\}$ for linear regression, the set of directions $\{y^1 - y^0, \ldots, y^p - y^0\}$ might not be a positive spanning set. It is trivial to construct a counterexample. For instance, let us take $n = 2$, $p = 3$, $y^0 = (0,0)$, and

$$\begin{bmatrix} y^1 & y^2 & y^3 \end{bmatrix} = \begin{bmatrix} 1 & 0 & -1 \\ 0 & 1 & 0 \end{bmatrix}.$$

See also Figure 2.7.

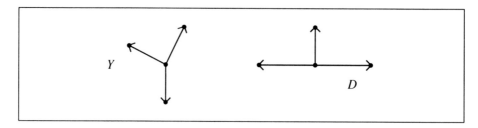

Figure 2.7. *For the positive spanning set on the left, the set Y marked by the bullets is poised for linear regression. However, given the poised set on the right, we see that the set of directions D marked by the arrows is not a positive spanning set.*

From the Courant–Fischer-type inequalities for singular values (see [131]), we conclude from (2.11) that

$$\text{cm}(D) \leq \min_{\|v\|=1} \|D^\top v\| = \sigma_{min}(D^\top),$$

where $\sigma_{min}(D^\top)$ represents the smallest singular value of $D^\top$. Hence,

$$\|(D^\top)^\dagger\| = \frac{1}{\sigma_{min}(D^\top)} \leq \frac{1}{\text{cm}(D)},$$

which shows that if the cosine measure of D is sufficiently away from zero, then the set $\{y^0\} \cup \{y^0 + d : d \in D\}$ is sufficiently "well poised."

2.6 Simplex gradients

Given a set $Y = \{y^0, y^1, \ldots, y^n\}$ with $n + 1$ sample points and poised for linear interpolation, the simplex gradient at y^0 is defined in the optimization literature (see Kelley [141]) by

$$\nabla_s f(y^0) = L^{-1} \delta f(Y),$$

where

$$L = \begin{bmatrix} y^1 - y^0 \cdots y^n - y^0 \end{bmatrix}^\top$$

and

$$\delta f(Y) = \begin{bmatrix} f(y^1) - f(y^0) \\ \vdots \\ f(y^n) - f(y^0) \end{bmatrix}.$$

However, it requires little effort to see that the simplex gradient is nothing else than the gradient of the linear interpolation model $m(x) = c + g^\top x$:

$$\nabla_s f(y^0) = g.$$

When the number of sample points exceeds $n + 1$, simplex gradients are defined in a regression sense as the least-squares solution of $L\nabla_s f(y^0) = \delta f(Y)$, where

$$L = \begin{bmatrix} y^1 - y^0 \cdots \cdots y^p - y^0 \end{bmatrix}^\top$$

and

$$\delta f(Y) = \begin{bmatrix} f(y^1) - f(y^0) \\ \vdots \\ \vdots \\ f(y^p) - f(y^0) \end{bmatrix}.$$

Again, one points out that a simplex gradient defined in this way is the gradient g of the linear regression model $m(x) = c + g^\top x$. We note that simplex gradients when $p > n$ are also referred to as stencil gradients (see the papers by C. T. Kelley). The set $\{y^1, \ldots, y^p\}$ is a stencil centered at y^0. For instance, the stencil could take the form $\{y^0 \pm he_i, i = 1, \ldots, n\}$, where h is the stencil radius, and e_i, $i = 1, \ldots, n$, are the coordinate vectors.

It is then obvious that under the assumptions stated for linear interpolation and linear regression the simplex gradient satisfies an error bound of the form

$$\|\nabla f(y^0) - \nabla_s f(y^0)\| \leq \kappa_{eg} \Delta,$$

where $\kappa_{cg} = p^{\frac{1}{2}} \nu \|\hat{L}^\dagger\|/2$ and $\hat{L} = L/\Delta$. In the case $p = n$, one has $\|\hat{L}^\dagger\| = \|\hat{L}^{-1}\|$.

2.7 Exercises

1. Prove that a set of nonzero vectors forms a positive basis for $\mathbb{R}^n$ if and only if their positive combinations span $\mathbb{R}^n$ and no proper subset exhibits the same property.

2. Show that $[I \ -I]$ is a (maximal) positive basis for $\mathbb{R}^n$ with cosine measure $1/\sqrt{n}$.

3. Prove Corollary 2.5.

4. Show that the value of a in (2.1) must be equal to $-1/n$.

5. Show that the cosine measure of a positive spanning set is necessarily positive.

6. In addition to the previous exercise, prove the reverse implication; that is, if the cosine measure of a set is zero, then the set cannot span $\mathbb{R}^n$ positively.

7. Show that the cosine measure of a minimal positive basis with uniform angles in $\mathbb{R}^n$ is $1/n$.

8. Prove, for linear interpolation and regression, that if Y is poised for some basis, then it is poised for any other basis in $\mathcal{P}_n^1$. Show that the definition of $m(x)$ in linear interpolation and regression is also independent of the basis chosen.

9. Show that (2.10) forms a (maximal) positive basis.

10. From (2.9), conclude the proof of Theorem 2.13.

11. Show that $\Delta(Y) \leq \mathrm{diam}(Y) \leq 2\Delta(Y)$ and $\mathrm{vol}(Y) > 0$ for any simplex of vertices Y.

Chapter 3

Interpolating nonlinear models

The sampling and linear models discussed in the previous chapter are appealing because they are simple to construct and to analyze. But as with any linear model they do not capture the curvature information of the function that they are approximating (unless it is flat). To achieve better local convergence rates in general it is essential to consider nonlinear models.

The quadratic polynomial model can be considered the simplest nonlinear model, yet it is often the most efficient. In this chapter we cover the properties of quadratic interpolation models. Quadratic regression models will be covered in a similar manner in Chapter 4. Nonpolynomial models, such as radial basis functions, can also be used effectively in derivative-free optimization. We will briefly cover these models in Chapter 12.

Everything discussed in this chapter applies to polynomial interpolation of any degree. But the quadratic case is of most interest to us, and hence all examples and pictures are limited to the quadratic case.

The goal of this chapter is to establish a theoretical foundation for using interpolant models as approximations of the true objective function in a derivative-free optimization algorithm. To ensure global convergence of an optimization algorithm that uses a model of the objective function it is typically necessary to guarantee a certain quality of this model. When a model is a truncated Taylor series expansion of first or second order, then the quality of the model is easily derived from the Taylor expansion error bounds. In the case of polynomial interpolation there exist similar bounds, but, unlike the Taylor expansion bounds, they depend not only on the center of the expansion and on the function that is being approximated but also on the set of interpolation points. (We have seen this in Chapter 2 for linear interpolation.) Thus, in order to maintain the quality of the interpolation model it is necessary to understand and maintain the quality of the interpolation set. In this chapter, we will examine several constants that characterize the quality of an interpolation set. We will study the relationship amongst these constants and the role they play in the error bound between a polynomial interpolant and the true function (and between their derivatives).

3.1   Basic concepts in interpolation

Polynomial bases

Let us consider $\mathcal{P}_n^d$, the space of polynomials of degree less than or equal to d in $\mathbb{R}^n$. Let $p_1 = p + 1$ be the dimension of this space. One knows that for $d = 1$, $p_1 = n + 1$ and that for $d = 2$, $p_1 = (n+1)(n+2)/2$.

A basis $\phi = \{\phi_0(x), \phi_1(x), \dots, \phi_p(x)\}$ of $\mathcal{P}_n^d$ is a set of p_1 polynomials of degree less than or equal to d that span $\mathcal{P}_n^d$. Since ϕ is a basis in $\mathcal{P}_n^d$, then any polynomial $m(x) \in \mathcal{P}_n^d$ can be written as $m(x) = \sum_{j=0}^{p} \alpha_j \phi_j(x)$, where the α_j's are some real coefficients.

There are a number of polynomial bases that are interesting to consider for various applications. We will focus only on those bases that are of interest in the context of this book. The simplest and the most common polynomial basis is the basis of monomials, or the *natural basis*.

The natural basis $\bar{\phi}$ can be conveniently described via the use of *multiindices*. Let a vector $\alpha^i = (\alpha_1^i, \dots, \alpha_n^i) \in \mathbb{N}_0^n$ be called a multiindex, and, for any $x \in \mathbb{R}^n$, let x^{α^i} be defined as

$$ x^{\alpha^i} = \prod_{j=1}^{n} x_j^{\alpha_j^i}. $$

Also, define

$$ |\alpha^i| = \sum_{j=1}^{n} \alpha_j^i \quad \text{and} \quad (\alpha^i)! = \prod_{j=1}^{n} (\alpha_j^i !). $$

Then the elements of the natural basis are

$$ \bar{\phi}_i(x) = \frac{1}{(\alpha^i)!} x^{\alpha^i}, \quad i = 0, \dots, p, \quad |\alpha^i| \le d. $$

The natural basis can be written out as follows:

$$ \bar{\phi} = \{1, x_1, x_2, \dots, x_n, x_1^2/2, x_1 x_2, \dots, x_{n-1}^{d-1} x_n/(d-1)!, x_n^d/d!\}. \tag{3.1} $$

For instance, when $n = 3$ and $d = 2$, we have

$$ \bar{\phi} = \{1, x_1, x_2, x_3, x_1^2/2, x_1 x_2, x_2^2/2, x_1 x_3, x_2 x_3, x_3^2/2\}. $$

The natural basis is the basis of polynomials as they appear in the Taylor expansion. For instance, assuming the appropriate smoothness, the Taylor model of order $d = 2$ in $\mathbb{R}^3$, centered at the point y, is the following polynomial in z_1, z_2, and z_3 (we write the elements of the natural basis within squared brackets for better identification):

$$ f(y)[1] + \frac{\partial f}{\partial x_1}(y)[z_1] + \frac{\partial f}{\partial x_2}(y)[z_2] + \frac{\partial f}{\partial x_3}(y)[z_3] $$
$$ + \frac{\partial^2 f}{\partial x_1^2}(y)[z_1^2/2] + \frac{\partial^2 f}{\partial x_1 x_2}(y)[z_1 z_2] + \frac{\partial^2 f}{\partial x_2^2}(y)[z_2^2/2] $$
$$ + \frac{\partial^2 f}{\partial x_1 x_3}(y)[z_1 z_3] + \frac{\partial^2 f}{\partial x_2 x_3}(y)[z_2 z_3] + \frac{\partial^2 f}{\partial x_3^2}(y)[z_3^2/2]. $$

When $d = 1$, we have $\bar{\phi} = \{1, x_1, \dots, x_n\}$, which appeared already in Chapter 2. There we have seen that the natural basis provides an immediate connection to geometrical concepts like affine independence and positive bases.

Other polynomial bases that are of interest in this chapter are the bases of Lagrange and Newton fundamental polynomials. We will discuss their definitions and properties later in this chapter.

Polynomial interpolation

We say that the polynomial $m(x)$ interpolates the function $f(x)$ at a given point y if $m(y) = f(y)$. Assume we are given a set $Y = \{y^0, y^1, \ldots, y^p\} \subset \mathbb{R}^n$ of interpolation points, and let $m(x)$ denote a polynomial of degree less than or equal to d that interpolates a given function $f(x)$ at the points in Y. By determining the coefficients $\alpha_0, \ldots, \alpha_p$ we determine the interpolating polynomial $m(x)$. The coefficients $\alpha_0, \ldots, \alpha_p$ can be determined from the interpolation conditions

$$\left(m(y^i) = \right) \quad \sum_{j=0}^{p} \alpha_j \phi_j(y^i) = f(y^i), \quad i = 0, \ldots, p. \tag{3.2}$$

Conditions (3.2) form a linear system in terms of the interpolation coefficients, which we will write in matrix form as

$$M(\phi, Y)\alpha_\phi = f(Y),$$

where

$$M(\phi, Y) = \begin{bmatrix} \phi_0(y^0) & \phi_1(y^0) & \cdots & \phi_p(y^0) \\ \phi_0(y^1) & \phi_1(y^1) & \cdots & \phi_p(y^1) \\ \vdots & \vdots & \vdots & \vdots \\ \phi_0(y^p) & \phi_1(y^p) & \cdots & \phi_p(y^p) \end{bmatrix},$$

$$\alpha_\phi = \begin{bmatrix} \alpha_0 \\ \alpha_1 \\ \vdots \\ \alpha_p \end{bmatrix}, \quad \text{and} \quad f(Y) = \begin{bmatrix} f(y^0) \\ f(y^1) \\ \vdots \\ f(y^p) \end{bmatrix}.$$

For instance, when $n = d = 2$ and $p = 5$, the matrix $M(\bar{\phi}, Y)$ becomes

$$\begin{bmatrix} 1 & y_1^0 & y_2^0 & (y_1^0)^2/2 & y_1^0 y_2^0 & (y_2^0)^2/2 \\ 1 & y_1^1 & y_2^1 & (y_1^1)^2/2 & y_1^1 y_2^1 & (y_2^1)^2/2 \\ 1 & y_1^2 & y_2^2 & (y_1^2)^2/2 & y_1^2 y_2^2 & (y_2^2)^2/2 \\ 1 & y_1^3 & y_2^3 & (y_1^3)^2/2 & y_1^3 y_2^3 & (y_2^3)^2/2 \\ 1 & y_1^4 & y_2^4 & (y_1^4)^2/2 & y_1^4 y_2^4 & (y_2^4)^2/2 \\ 1 & y_1^5 & y_2^5 & (y_1^5)^2/2 & y_1^5 y_2^5 & (y_2^5)^2/2 \end{bmatrix}.$$

For the above system to have a unique solution, the matrix $M(\phi, Y)$ has to be nonsingular.

Definition 3.1. *The set $Y = \{y^0, y^1, \ldots, y^p\}$ is poised for polynomial interpolation in $\mathbb{R}^n$ if the corresponding matrix $M(\phi, Y)$ is nonsingular for some basis ϕ in $\mathcal{P}_n^d$.*

Poisedness has other terminology in the literature (for example, the authors in [53] refer to a poised set as a *d-unisolvent* set). As we have seen in Section 2.5, the notion of poisedness in the linear case ($d = 1$) is the same as affine independence.

It is easy to see that if $M(\phi, Y)$ is nonsingular for some basis ϕ, then it is nonsingular for any basis of $\mathcal{P}_n^d$. Under these circumstances, one can also show that the interpolating polynomial $m(x)$ exists and is unique.

Given the polynomial space $\mathcal{P}_n^d$ and a basis ϕ, let

$$\phi(x) = [\phi_0(x), \phi_1(x), \ldots, \phi_p(x)]^\top$$

be a vector in $\mathbb{R}^{p_1}$ whose entries are the values of the elements of the polynomial basis at x (one can view ϕ as a mapping from $\mathbb{R}^n$ to $\mathbb{R}^{p_1}$).

Lemma 3.2. *Given a function $f : \mathbb{R}^n \to \mathbb{R}$ and a poised set $Y \in \mathbb{R}^n$, the interpolating polynomial $m(x)$ exists and is unique.*

Proof. It is obvious that $m(x)$ exists and is unique for a given basis $\phi(x)$, since Y is poised. What we now need to show is that $m(x)$ is independent of the choice of $\phi(x)$. Consider another basis $\psi(x) = P^\top \phi(x)$, where P is some nonsingular $p_1 \times p_1$ matrix. Clearly, we have that $M(\psi, Y) = M(\phi, Y)P$. Let α_ϕ be a solution of (3.2) for the basis ϕ, and let α_ψ be a solution of (3.2) for the basis ψ. Then, for any right-hand side $f(Y)$, we have that $\alpha_\psi = P^{-1}(M(\phi, Y))^{-1} f(Y)$ and

$$\alpha_\psi^\top \psi(x) = f(Y)^\top (M(\phi, Y))^{-\top} P^{-\top} P^\top \phi(x)$$
$$= f(Y)^\top (M(\phi, Y))^{-\top} \phi(x) = \alpha_\phi^\top \phi(x).$$

We conclude that $m(x) = \alpha_\psi^\top \psi(x) = \alpha_\phi^\top \phi(x)$ for any x.   □

The matrix $M(\phi, Y)$ is singular if and only if there exists $\gamma \in \mathbb{R}^{p_1}$ such that $\gamma \neq 0$ and $M(\phi, Y)\gamma = 0$ and that implies that there exists a polynomial, of degree at most d, expressed as

$$m(x) = \sum_{j=0}^{p} \gamma_j \phi_j(x),$$

such that $m(y) = 0$ for all $y \in Y$. In other words, $M(\phi, Y)$ is singular if and only if the points of Y lie on a "polynomial manifold" of degree d or less. For instance, six points on a second-order curve in $\mathbb{R}^2$, such as a curve, form a nonpoised set for quadratic interpolation.[4]

Now that we have established algebraic and geometric criteria for poisedness of an interpolation set it is natural to ask: can these conditions be extended to characterize a *well-poised* interpolation set. For instance, since nonsingularity of $M(\phi, Y)$ is the indicator of poisedness, will the *condition number* of $M(\phi, Y)$ be an indicator of *well poisedness*? The answer, in general, is "no," since the condition number of $M(\phi, Y)$ depends on the choice of ϕ. Moreover, for any given poised interpolation set Y, one can choose the basis ϕ so

---

[4]The fact that the points lie on a second-order curve trivially implies that columns of $M(\phi, Y)$ can be combined with some coefficients, not all of them zero, to obtain a zero vector.

that the condition number of $M(\phi, Y)$ can equal any number between 1 and $+\infty$. Also, for any fixed choice of ϕ, the condition number of $M(\phi, Y)$ depends on the scaling of Y. Hence, the condition number of $M(\phi, Y)$ is not considered to be a good characterization of the level of poisedness of a set of points. However, we will return to this issue and show that for a specific choice of ϕ, namely for the natural basis, and for $\hat{Y}$, a scaled version of Y, the condition number of $M(\bar{\phi}, \hat{Y})$ is a meaningful measure of well poisedness.

3.2 Lagrange polynomials

The most commonly used measure of well poisedness in the multivariate polynomial interpolation literature is based on Lagrange polynomials.

Definition 3.3. *Given a set of interpolation points* $Y = \{y^0, y^1, \ldots, y^p\}$, *a basis of* $p_1 = p + 1$ *polynomials* $\ell_j(x)$, $j = 0, \ldots, p$, *in* $\mathcal{P}_n^d$ *is called a basis of Lagrange polynomials if*

$$\ell_j(y^i) = \delta_{ij} = \begin{cases} 1 & \text{if } i = j, \\ 0 & \text{if } i \neq j. \end{cases}$$

If Y is poised, Lagrange polynomials exist, are unique, and have a number of useful properties.

Lemma 3.4. *If Y is poised, then the basis of Lagrange polynomials exists and is uniquely defined.*

Proof. The proof follows from the fact that each Lagrange polynomial $\lambda_j(x)$ is an interpolating polynomial (of a function that vanishes at all points in Y except at y^j where it is equal to 1) and from Lemma 3.2. □

The existence of Lagrange polynomials, in turn, implies poisedness of the sampling set. In fact, if Lagrange polynomials exist, then there exists a matrix A_ϕ (whose columns are the coefficients of these polynomials in the basis ϕ) such that $M(\phi, Y)A_\phi = I$ and, as a result, the matrix $M(\phi, Y)$ is nonsingular.

To show that the set of Lagrange polynomials as defined above forms a basis for $\mathcal{P}_n^d$, we need to show that any polynomial $m(x)$ in $\mathcal{P}_n^d$ is a linear combination of Lagrange polynomials. We know that any polynomial is uniquely defined by its values at the points in Y, since Y is poised. As the lemma below shows, these values are, in fact, the coefficients in the linear expression of $m(x)$ via Lagrange polynomials.

Lemma 3.5. *For any function $f : \mathbb{R}^n \to \mathbb{R}$ and any poised set $Y = \{y^0, y^1, \ldots, y^p\} \subset \mathbb{R}^n$, the unique polynomial $m(x)$ that interpolates $f(x)$ on Y can be expressed as*

$$m(x) = \sum_{i=0}^{p} f(y^i)\ell_i(x),$$

where $\{\ell_i(x), i = 0, \ldots, p\}$ is the basis of Lagrange polynomials for Y.

The proof of this last result is left for an exercise. To illustrate Lagrange polynomials in $\mathbb{R}^2$, consider interpolating the cubic function given by

$$f(x_1, x_2) = x_1 + x_2 + 2x_1^2 + 3x_2^3$$

at the six interpolating points $y^0 = (0,0)$, $y^1 = (1,0)$, $y^2 = (0,1)$, $y^3 = (2,0)$, $y^4 = (1,1)$, and $y^5 = (0,2)$. Clearly, $f(y^0) = 0$, $f(y^1) = 3$, $f(y^2) = 4$, $f(y^3) = 10$, $f(y^4) = 7$, and $f(y^5) = 26$. It is easy to see that the corresponding Lagrange polynomials $\ell_j(x_1, x_2)$, $j = 0, \ldots, 5$, are given by

$$
\begin{aligned}
\ell_0(x_1, x_2) &= 1 - \tfrac{3}{2}x_1 - \tfrac{3}{2}x_2 + \tfrac{1}{2}x_1^2 + \tfrac{1}{2}x_2^2 + x_1 x_2, \\
\ell_1(x_1, x_2) &= 2x_1 - x_1^2 - x_1 x_2, \\
\ell_2(x_1, x_2) &= 2x_2 - x_2^2 - x_1 x_2, \\
\ell_3(x_1, x_2) &= -\tfrac{1}{2}x_1 + \tfrac{1}{2}x_1^2, \\
\ell_4(x_1, x_2) &= x_1 x_2, \\
\ell_5(x_1, x_2) &= -\tfrac{1}{2}x_2 + \tfrac{1}{2}x_2^2.
\end{aligned}
$$

It is also easy to verify that

$$
\begin{aligned}
m(x_1, x_2) &= 0\ell_0(x_1, x_2) + 3\ell_1(x_1, x_2) + 4\ell_2(x_1, x_2) + 10\ell_3(x_1, x_2) \\
&\quad + 7\ell_4(x_1, x_2) + 26\ell_5(x_1, x_2)
\end{aligned}
$$

satisfies the interpolation conditions.

Computing the entire basis of Lagrange polynomials for a given set Y requires $\mathcal{O}(p^3)$ operations, which is of the same order of magnitude of the number of operations required to compute one arbitrary interpolating polynomial $m(x)$. If a set of Lagrange polynomials is available and the set Y is updated by one point, then is takes $\mathcal{O}(p^2)$ to update the set of Lagrange polynomials. We will discuss this further in Chapter 6.

One of the most useful features of Lagrange polynomials for the purposes of this book is that an upper bound on their absolute value in a region B is a classical measure of poisedness of Y in B. In particular in [53], it is shown that for any x in the convex hull of Y

$$\|\mathcal{D}^r f(x) - \mathcal{D}^r m(x)\| \leq \frac{1}{(d+1)!} \nu_d \sum_{i=0}^{p} \|y^i - x\|^{d+1} \|\mathcal{D}^r \ell_i(x)\|,$$

where $\mathcal{D}^r$ denotes the rth derivative of a function and ν_d is an upper bound on $\mathcal{D}^{d+1} f(x)$. Notice that this error bound requires $f(x)$ to have a bounded $(d+1)$st derivative. See [53] for a precise meaning of each of the elements in this bound. When $r = 0$, the bound on function values reduces to

$$|f(x) - m(x)| \leq \frac{1}{(d+1)!} p_1 \nu_d \Lambda_\ell \Delta^{d+1},$$

where

$$\Lambda_\ell = \max_{0 \leq i \leq p} \max_{x \in B(Y)} |\ell_i(x)|,$$

and Δ is the diameter of the smallest ball $B(Y)$ containing Y. See also [188] for a simple derivation of this bound. Λ_ℓ is closely related to the Lebesgue constant of Y (see [53]).

Equivalent definitions of Lagrange polynomials

Lagrange polynomials play an important role in the theory of uni- and multivariate polynomial interpolation. As seen from the properties described above, they can also be very useful in model-based derivative-free optimization. Since the interpolation model is trivially expressed through the basis of Lagrange polynomials, and such a basis can be computed and updated reasonably efficiently, it is practical to maintain the interpolation models by means of maintaining the Lagrange polynomial basis. From the interpolation error bound presented at the end of the last subsection it is evident that Lagrange polynomials are also useful for constructing criteria for selecting good interpolation sets. Hence, it is important for us to better understand the properties of Lagrange polynomials. Here we discuss two alternative ways to define Lagrange polynomials.

Given a poised set $Y = \{y^0, y^1, \ldots, y^p\} \subset \mathbb{R}^n$ and an $x \in \mathbb{R}^n$, because the matrix $M(\phi, Y)$ is nonsingular, we can express the vector $\phi(x)$ uniquely in terms of the vectors $\phi(y^i)$, $i = 0, \ldots, p$, as

$$\sum_{i=0}^{p} \lambda_i(x) \phi(y^i) = \phi(x) \tag{3.3}$$

or, in matrix form,

$$M(\phi, Y)^\top \lambda(x) = \phi(x), \quad \text{where } \lambda(x) = [\lambda_0(x), \ldots, \lambda_p(x)]^\top. \tag{3.4}$$

From (3.4) it is obvious that $\lambda(x)$ is a vector of polynomials in $\mathcal{P}_n^d$. We can also see directly from (3.3) that $\lambda_i(x)$ is the ith Lagrange polynomial for Y. Hence, the value of a Lagrange polynomial at a given point x is the weight of the contribution of the vector $\phi(y^i)$ in the linear combination that forms $\phi(x)$.

The second alternative way to define Lagrange polynomials is as follows. Given the set Y and a point x, consider the set $Y_i(x) = Y \setminus \{y^i\} \cup \{x\}$, $i = 0, \ldots, p$. From applying Cramer's rule to the system (3.4), we see that

$$\lambda_i(x) = \frac{\det(M(\phi, Y_i(x)))}{\det(M(\phi, Y))}. \tag{3.5}$$

From this expression it is clear that $\lambda(x)$ is a polynomial in $\mathcal{P}_n^d$ and that it satisfies Definition 3.3. Hence, $\{\lambda_i(x), i = 0, \ldots, p\}$ is exactly the set of Lagrange polynomials. It follows that $\lambda(x)$ does not, in fact, depend on the choice of ϕ as long as the polynomial space $\mathcal{P}_n^d$ is fixed.

To help further understand the meaning of (3.5), consider a set $\phi(Y) = \{\phi(y^i), i = 0, \ldots, p\}$ in $\mathbb{R}^{p_1}$. Let $\text{vol}(\phi(Y))$ be the volume of the simplex of vertices in $\phi(Y)$, given by

$$\text{vol}(\phi(Y)) = \frac{|\det(M(\phi, Y))|}{p_1!}.$$

(Such a simplex is the p_1-dimensional convex hull of $\phi(Y)$.) Then

$$|\lambda_i(x)| = \frac{\text{vol}(\phi(Y_i(x)))}{\text{vol}(\phi(Y))}. \tag{3.6}$$

In other words, the absolute value of the ith Lagrange polynomial at a given point x is the change in the volume of (the p_1-dimensional convex hull of) $\phi(Y)$ when y^i is replaced by x.

3.3   Λ-poisedness and other measures of well poisedness

So, how should we define a measure of well poisedness, after all? Given an interpolation set Y, a measure of poisedness should reflect how well this set "spans" the region where interpolation is of interest. Unlike the linear case, where a "good span" represents affine independence of the points of Y, in the nonlinear case we consider how well vectors $\phi(y^i)$, $i = 0, \ldots, p$, span the set of all $\phi(x)$ in a region of interest.

Clearly, such a measure will depend on Y and on the region considered. For example, in the case of linear interpolation, a set $Y = \{(0,0),(0,1),(1,0)\}$ is a well-poised set in $B(0;1)$—the ball of radius 1 centered at the origin—but is not a well-poised set in $B(0;10^6)$—the ball of radius 10^6 centered at the origin.

Well poisedness (and poisedness) of Y also has to depend on the polynomial space from which an interpolant is chosen. For instance, six points on a circle in $\mathbb{R}^2$ are not poised for interpolation by quadratic polynomials, but they are poised for interpolation, for example, in a subspace of cubic polynomials that does not have quadratic and constant terms.

Λ-poisedness

We will use the following equivalent definitions of well-poised sets.

Definition 3.6. *Let $\Lambda > 0$ and a set $B \in \mathbb{R}^n$ be given. Let $\phi = \{\phi_0(x), \phi_1(x), \ldots, \phi_p(x)\}$ be a basis in $\mathcal{P}_n^d$. A poised set $Y = \{y^0, y^1, \ldots, y^p\}$ is said to be Λ-poised in B (in the interpolation sense) if and only if*

1. *for the basis of Lagrange polynomials associated with Y*

$$\Lambda \geq \max_{0 \leq i \leq p} \max_{x \in B} |\ell_i(x)|,$$

   or, equivalently,

2. *for any $x \in B$ there exists $\lambda(x) \in \mathbb{R}^{p_1}$ such that*

$$\sum_{i=0}^{p} \lambda_i(x)\phi(y^i) = \phi(x) \qquad with \qquad \|\lambda(x)\|_\infty \leq \Lambda,$$

   or, equivalently,

3. *replacing any point in Y by any $x \in B$ can increase the volume of the set $\{\phi(y^0), \phi(y^1), \ldots, \phi(y^p)\}$ at most by a factor Λ.*

Note that this definition does not imply or require that the sample set Y be contained in the set B where the absolute values of the Lagrange polynomials are maximized. Indeed, the set Y can be arbitrarily far from B as long as all the Lagrange polynomials are bounded in absolute value on B. However, as we will see later, to guarantee the validity of algorithms that construct Λ-poised sets we have to generate points (up to the whole set Y) in B.

The second and the third definitions provide us with some geometric intuition of Λ-poisedness. As we mentioned earlier, we can consider a mapping $x \to \phi(x)$ and $\phi(Y)$ and

$\phi(B)$—the images of Y and B under this mapping. The second definition shows that Λ is a measure of how well $\phi(Y)$ spans $\phi(B)$. If $\phi(Y)$ spans $\phi(B)$ well, then any vector in $\phi(B)$ can be expressed as a linear combination of vectors in $\phi(Y)$ with reasonably small coefficients. The third definition has a similar interpretation, but it is expressed via volumes of simplices. If $\phi(Y)$ spans $\phi(B)$ well, then it is not possible to substantially increase the volume of $\phi(Y)$ by replacing one point in Y by any point in B.

Figures 3.1–3.4 show several sets of six points in B—the "squared" ball of radius $1/2$ around the point $(0.5, 0.5)$ in $\mathbb{R}^2$. The sets are listed under the figures, and for each of these sets the corresponding $\Lambda = \max_{0 \le i \le 5} \max_{x \in B} |\ell_i(x)|$ is listed. All the numbers are listed with sufficient accuracy to illustrate cases of well- and badly poised sets. For each interpolation set, we also show the model which interpolated the function $\cos(x_1) + \sin(x_2)$ on that set. It is evident from the pictures that the quality of the interpolation model noticeably deteriorates as Λ becomes larger.

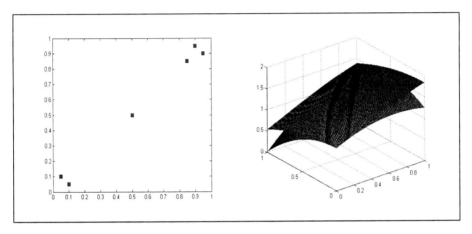

Figure 3.1. *A nonpoised set perturbed by about 0.1:* $Y = \{(0.05, 0.1),$ $(0.1, 0.05), (0.5, 0.5), (0.95, 0.9), (0.9, 0.95), (0.85, 0.85)\}$ *and* $\Lambda = 440$.

Λ-poisedness as the distance to linear independence

The constant Λ can be interpreted as an actual measure of distance to a nonpoised set. Given an interpolation set Y, let $B(y^0; \Delta(Y))$ be the smallest closed ball centered at the interpolation point y^0 and containing Y. Assume that, for a given Λ, the set Y is not Λ-poised in $B(y^0; \Delta(Y))$. Then there exists a $z \in B(y^0; \Delta(Y))$ such that

$$\sum_{i=0}^{p} \lambda_i(z)\phi(y^i) = \phi(z) \quad \text{and} \quad \|\lambda(z)\|_\infty > \Lambda,$$

and hence, without loss of generality, $\lambda_1(z) > \Lambda$. Then, dividing this expression by Λ, we have

$$\sum_{i=0}^{p} \frac{\lambda_i(z)}{\Lambda}\phi(y^i) = \sum_{i=0}^{p} \gamma_i(z)\phi(y^i) = \frac{\phi(z)}{\Lambda} \quad \text{and} \quad \gamma_1(z) > 1.$$

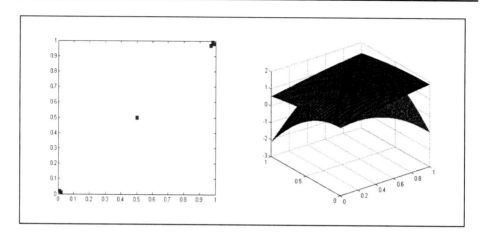

Figure 3.2. *A nonpoised set of points perturbed by about* 0.01: $Y = \{(0.01, 0.02), (0.02, 0.01), (0.5, 0.5), (0.99, 0.98), (0.98, 0.98), (0.97, 0.97)\}$ *and* $\Lambda = 21296$.

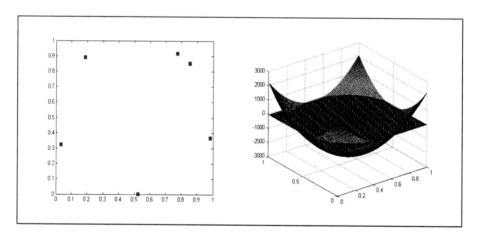

Figure 3.3. *A poised set, but the points nearly lie on a circle:* $Y = \{(0.524, 0.0006),\ (0.032, 0.323),\ (0.187, 0.890),\ (0.854, 0.853),\ (0.982, 0.368),\ (0.774, 0.918)\}$ *and* $\Lambda = 524982$.

Thus,

$$\left\| \sum_{i=0}^{p} \gamma_i(z) \phi(y^i) \right\|_\infty \leq \frac{\max_{x \in B(y^0; \Delta(Y))} \|\phi(x)\|_\infty}{\Lambda}.$$

If, for example, $\bar{\phi}$ is the natural basis, $y^0 = 0$, and the radius $\Delta(Y)$ of $B(y^0; \Delta(Y))$ is 1, then

$$\max_{x \in B(0;1)} \|\bar{\phi}(x)\|_\infty \leq 1$$

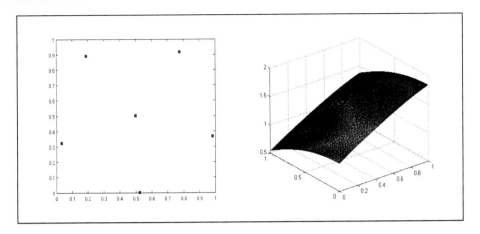

Figure 3.4. *An "ideal" set:* $Y = \{(0.524, 0.0006), (0.032, 0.323), (0.187, 0.890),$ $(0.5, 0.5), (0.982, 0.368), (0.774, 0.918)\}$ *and* $\Lambda = 1$.

and

$$\left\| \sum_{i=0}^{p} \gamma_i(z)\bar{\phi}(y^i) \right\|_{\infty} \leq \frac{1}{\Lambda} \quad \text{with} \quad \gamma_1(z) > 1.$$

It is now easy to see that $1/\Lambda$ bounds, in some sense, the distance to linear dependency of the vectors $\bar{\phi}(y^i)$, $i = 0, \ldots, p$. As Λ grows, the system represented by these vectors becomes increasingly linearly dependent. But the actual distance to singularity depends on the choice of ϕ.

Properties of Λ-poisedness

From now on we will omit saying that Y is Λ-poised with respect to some given $\mathcal{P}_n^d$. It will be assumed that the space $\mathcal{P}_n^d$ remains fixed according to a given context, unless stated otherwise.

We will state a few basic results that we hope serve to enhance understanding the poisedness constant Λ. The proofs are left as exercises.

Lemma 3.7.

1. *If B contains a point in Y and Y is Λ-poised in B, then $\Lambda \geq 1$.*

2. *If Y is Λ-poised in a given set B, then it is Λ-poised (with the same constant) in any subset of B.*

3. *If Y is Λ-poised in B for a given constant Λ, then it is $\tilde{\Lambda}$-poised in B for any $\tilde{\Lambda} \geq \Lambda$.*

4. *For any $x \in \mathbb{R}^n$, if $\lambda(x)$ is the solution of (3.3), then*

$$\sum_{i=0}^{p} \lambda_i(x) = 1.$$

We will now show that the poisedness constant Λ does not depend on the scaling of the sample set.

Lemma 3.8. *Let* $Y = \{y^0, y^1, \ldots, y^p\}$ *be an interpolation set and* $\lambda_i(x)$, $i = 0, \ldots, p$, *be the set of solutions to (3.3) for the given* Y. *Then, for any* $\Delta > 0$, $\lambda_i(x/\Delta)$, $i = 0, \ldots, p$, *is the set of solutions to (3.3) for* $\hat{Y}$, *where* $\hat{Y} = \{y^0/\Delta, y^1/\Delta, \ldots, y^p/\Delta\}$.

Proof. Since the solution $\lambda(x)$ of (3.3) does not depend on the choice of basis, we can consider $\bar{\phi}(x)$ (the natural basis). We know that $\lambda_i(x)$, $i = 0, \ldots, p$, satisfy

$$\sum_{i=0}^{p} \lambda_i \bar{\phi}(y^i) = \bar{\phi}(x).$$

If we scale each y^i and x by $1/\Delta$, this corresponds to scaling the above equations by different scalars $(1, 1/\Delta, 1/\Delta^2, \text{etc.})$. Clearly, $\lambda(x/\Delta)$ satisfies the scaled system of equations (3.3). $\quad\square$

An immediate corollary of this lemma is that if Y is Λ-poised in B, then $\hat{Y} = Y/\Delta$ is Λ-poised in $\hat{B} = B/\Delta$.

Our next step is to show that Λ-poisedness is invariant with respect to a shift of coordinates.

Lemma 3.9. *Let* $Y = \{y^0, y^1, \ldots, y^p\}$ *be an interpolation set and* $\lambda_i(x)$, $i = 0, \ldots, p$, *be the solution to (3.3) for a given* x. *Then, for any* $a \in \mathbb{R}^n$, *it is also the solution to (3.3) for* $Y_a = \{y^0 + a, y^1 + a, \ldots, y^p + a\}$ *and* $x_a = x + a$.

Proof. Since the solution $\lambda(x)$ of (3.3) does not depend on the choice of basis, we can consider $\bar{\phi}(x)$ (the natural basis). Let us recall the multiindex notation: $\bar{\phi}_j(x) = \frac{1}{(\alpha_j)!} x^{\alpha_j}$. Is it easy to show that, for any $j = 0, \ldots, p$,

$$\bar{\phi}_j(x + a) = \bar{\phi}_j(x) + \bar{\phi}_j(a) + \sum_{k:\, |\alpha_k| < |\alpha_j|} \gamma_k(a) \bar{\phi}_k(x), \tag{3.7}$$

where $\gamma_k(a)$ are some coefficients depending on the choice of a but independent of x. From (3.7) it is easy to see that $\bar{\phi}(x + a)$ defines a basis in $\mathcal{P}_n^d$. And, again, since the solution of (3.3) does not depend on the choice of basis,

$$\sum_{i=0}^{p} \lambda_i(x) \bar{\phi}(y^i + a) = \bar{\phi}(x + a). \quad\square$$

Newton fundamental polynomials

Newton fundamental polynomials (NFPs) are similar in nature to Lagrange polynomials. We briefly introduce them here, since, just as for Lagrange polynomials, they have been used in derivative-free optimization framework to maintain the quality of the interpolation models (see [59, 60]).

As for Lagrange polynomials, the NFPs form a basis and each member of the basis corresponds to one point in the interpolation set. The difference from the Lagrange polynomial basis is the following. The interpolation points and the corresponding NFPs are grouped into blocks, each block corresponding to a polynomial degree from 0 to d. Each block with degree ℓ contains precisely as many points and polynomials as there are multiindices α such that $|\alpha| = \ell$. The NFPs are constructed in such a way that the elements corresponding to the first ℓ blocks form the NFP basis for the space of polynomials with degree less than or equal to ℓ.

More specifically, the number of points in the ℓth block is

$$|Y^{[\ell]}| = \binom{\ell + n - 1}{\ell},$$

where the right-hand side is a binomial coefficient. Note that $|Y^{[0]}| = 1$. Corresponding to each point $y_i^{[\ell]} \in Y^{[\ell]}$ is a single NFP $n_i^{[\ell]}(x)$ of degree ℓ satisfying the conditions

$$n_i^{[\ell]}(y_j^{[m]}) = \delta_{ij}\delta_{\ell m} \quad \forall y_j^{[m]} \in Y^{[m]} \quad \text{with} \quad m \leq \ell. \tag{3.8}$$

The interpolating polynomial $m(x)$ is then given as

$$m(x) = \sum_{y_i^{[\ell]} \in Y} \gamma_i^{[\ell]}(Y, f) n_i^{[\ell]}(x),$$

where the coefficients $\gamma_i^{[\ell]}(Y, f)$ are generalized finite differences applied to f. We refer the readers to [205] for more details.

To exemplify NFPs we consider quadratic interpolation in $\mathbb{R}^2$ and the same set of points for which we computed the Lagrange polynomial basis in Section 3.2. The six interpolation points chosen are partitioned into three blocks ($d = 2$):

$$Y^{[0]} = \{(0,0)\}, \quad Y^{[1]} = \{(1,0),(0,1)\}, \quad \text{and} \quad Y^{[2]} = \{(2,0),(1,1),(0,2)\}.$$

The NFPs are given by

$$n_1^{[0]} = 1, \quad n_1^{[1]} = x_1, \quad n_2^{[1]} = x_2,$$

$$n_1^{[2]} = \frac{1}{2}(x_1^2 - x_1), \quad n_2^{[2]} = x_1 x_2, \quad \text{and} \quad n_3^{[2]} = \frac{1}{2}(x_2^2 - x_2),$$

for which the conditions (3.8) may easily be verified.

Just as in the case of the Lagrange polynomials, the NFPs exist if and only if the set Y is poised. The value

$$\Lambda_n = \max_{0 \leq i \leq p} \max_{x \in B(Y)} |n_i(x)|$$

serves as a measure of poisedness of the set Y (in $B(Y)$). The following bound is a simplification of the bound found in [205]:

$$|f(x) - m(x)| \leq \frac{2^d}{(d+1)!} p_1 \nu_d \Lambda_n \Delta^{d+1},$$

where ν_d is again an upper bound on $\mathcal{D}^{d+1} f(x)$ and Δ is the radius of $B(Y)$.

An advantage of the NFPs over the Lagrange polynomials lies in the fact that the polynomials associated with blocks of lower degree are independent of the higher-degree blocks. Notice that in the example above the first three NFPs provide a basis for linear interpolation and do not have quadratic terms, unlike the Lagrange polynomials for the same set of points discussed in Section 3.2.

The NFP basis is most advantageous when there are several blocks of moderate size, namely, when the dimension n is small and the degree d is relatively large. Such a situation rarely occurs in derivative-free optimization. However, in situations when Y does not contain enough points to construct a complete quadratic model it may be useful to select a block of sample points for linear interpolation and construct a linear basis based on that block. The remaining points can then serve to generate a set of second-order polynomials, which do not form a basis but nevertheless satisfy conditions (3.8). We will discuss this briefly in Chapter 5.

3.4 Condition number as a measure of well poisedness

As we concluded earlier, the condition number of the matrix $M(\phi, Y)$ of the interpolation conditions is a bad measure of poisedness of Y since it can be made arbitrarily large by changing ϕ or scaling Y.

We will now consider a specific choice of $\phi = \bar{\phi}$ (the natural basis). We will look at the condition number of $M(\bar{\phi}, \hat{Y})$, where $\hat{Y}$ is the shifted and scaled version of Y so that

$$\hat{Y} = \{0, \hat{y}^1, \ldots, \hat{y}^p\} \subset B(0; 1)$$

and

$$\max_{1 \le i \le p} \|\hat{y}^i - \hat{y}^0\| = \max_{1 \le i \le p} \|\hat{y}^i\| = 1.$$

Then we will establish how the condition number of $M(\bar{\phi}, \hat{Y})$ relates to the measure of poisedness. This will enable us to use the interpolation conditions and the matrix $M(\bar{\phi}, \hat{Y})$ directly in a derivative-free algorithm, without computing Lagrange polynomial or NFP bases, and still be able to maintain sufficient poisedness of the interpolation sets (and hence to guarantee a moderate bound on the interpolation error).

Shifting and scaling

For numerical reasons it is often useful to consider a set Y where one of its points (without loss of generality, y^0) is at the origin. If Y is not such a set, then a shift of coordinates is performed to move y^0 to the origin. In interpolation-based derivative-free optimization, the current best iterate is usually the center of the interpolation and it can always be viewed as the origin. As we have seen, neither the poisedness constant Λ of an interpolation set Y nor the quality of interpolation on Y changes under a shift of the coordinates.

We have established that Λ-poisedness is also independent of the scaling, although it does depend on the region B in which the poisedness is considered. The condition number of the matrix $M(\phi, Y)$ depends on the scaling of Y but is "independent" of any region B. To connect these two concepts we will fix the region to be $B(0; \Delta(Y))$—the smallest ball (centered at $y^0 = 0$) that contains Y. Then we scale $B = B(0; \Delta(Y))$ and Y so that the radius of $B(0; \Delta(Y))$ equals 1. We denote such a ball by $B(0; 1)$.

Thus, given any sample set written as

$$Y = \{y^0, y^1, \ldots, y^p\},$$

we first shift it by $-y^0$ to center the new set at the origin:

$$\{0, y^1 - y^0, \ldots, y^p - y^0\}.$$

Then we consider

$$\Delta = \Delta(Y) = \max_{1 \le i \le p} \|y^i - y^0\|$$

and scale the set by Δ:

$$\hat{Y} = \{0, \hat{y}^1, \ldots, \hat{y}^p\} = \{0, (y^1 - y^0)/\Delta, \ldots, (y^p - y^0)/\Delta\} \subset B(0;1).$$

The resulting sample set $\hat{Y}$ is contained in a ball of radius one centered at the origin and has at least one point on the ball boundary. The algorithms discussed in Chapter 6 which factorize $M(\bar{\phi}, Y)$ to achieve well poisedness will be stated and analyzed for sampling sets of this form, unless noted otherwise. If the original sample set Y does not meet this requirement, we first shift it and then scale it, so that $\hat{Y} \subset B(0;1)$.

Some auxiliary results

Let $\phi(x)$ be some basis of the space $\mathcal{P}_n^d$ of polynomials of degree less than or equal to d. We will show that any polynomial which can be expressed through $\phi(x)$ with coefficients that are not too small cannot vanish (or nearly vanish) on a unit ball.

Lemma 3.10. *There exists a number $\sigma_\infty > 0$ such that, for any choice of v satisfying $\|v\|_\infty = 1$, there exists a $y \in B(0;1)$ such that $|v^\top \phi(y)| \ge \sigma_\infty$.*

Proof. Consider

$$\psi(v) = \max_{x \in B(0;1)} |v^\top \phi(x)|. \tag{3.9}$$

It is easy to prove that $\psi(v)$ is a norm in the space of vectors v (which has the dimension of $\mathcal{P}_n^d$, given by $p_1 = p + 1$ in our notation). Since the ratio of any two norms in finite-dimensional spaces can be uniformly bounded by a constant, there exists a $\sigma_\infty > 0$ such that $\psi(v) \ge \sigma_\infty \|v\|_\infty$. The constant σ_∞ is defined as

$$\sigma_\infty = \min_{\|v\|_\infty = 1} \psi(v).$$

Thus, if v has ℓ_∞-norm one, then $\psi(v) \ge \sigma_\infty$ and there exists a $y \in B(0;1)$ such that $|v^\top \phi(y)| \ge \sigma_\infty$. $\square$

If we restrict our attention to the case of ϕ being the natural basis and $d \le 2$, then we can readily provide an explicit lower bound on σ_∞. We start by the case $d = 1$ where $\sigma_\infty \ge 1$.

Lemma 3.11. *Let $v^\top \bar{\phi}(x)$ be a linear polynomial, where $\|v\|_\infty = 1$ and $\bar{\phi}$ is the natural basis (defined by (3.1) when $d = 1$). Then*

$$\max_{x \in B(0;1)} |v^\top \bar{\phi}(x)| \geq 1.$$

The proof is given in Chapter 6. For $d = 2$ we have the following result. The proof is also postponed to Chapter 6 because it suggests an algorithm that can be part of the overall strategy to improve well poisedness of sample sets.

Lemma 3.12. *Let $v^\top \bar{\phi}(x)$ be a quadratic polynomial, where $\|v\|_\infty = 1$ and $\bar{\phi}$ is the natural basis (defined by (3.1) when $d = 2$). Then*

$$\max_{x \in B(0;1)} |v^\top \bar{\phi}(x)| \geq \frac{1}{4}.$$

Then, given any $\bar{v}$ such that $\|\bar{v}\| = 1$, we can scale $\bar{v}$ by at most $\sqrt{p_1}$ to $v = \alpha \bar{v}$, $0 < \alpha \leq \sqrt{p_1}$, such that $\|v\|_\infty = 1$. Thus,

$$\sigma_2 = \min_{\|\bar{v}\|=1} \max_{x \in B(0;1)} |\bar{v}^\top \bar{\phi}(x)| \geq \frac{1}{\sqrt{p_1}} \min_{\|v\|_\infty=1} \max_{x \in B(0;1)} |v^\top \bar{\phi}(x)| \geq \frac{1}{4\sqrt{p_1}}. \tag{3.10}$$

The last inequality is due to Lemma 3.12 applied to polynomials of the form $\hat{v}^\top \bar{\phi}(x)$. Specifying the bound on σ_∞ or σ_2 for polynomials of higher degree is also possible but is beyond the scope of this book.

Λ-poisedness and the condition number of $M(\bar{\phi}, \hat{Y})$

We will show how Λ-poisedness relates to the condition number of the following matrix:

$$\hat{M} = \begin{bmatrix} 1 & 0 & \cdots & 0 & 0 & 0 & \cdots & 0 & 0 \\ 1 & \hat{y}_1^1 & \cdots & \hat{y}_n^1 & \frac{1}{2}(\hat{y}_1^1)^2 & \hat{y}_1^1 \hat{y}_2^1 & \cdots & \frac{1}{(d-1)!}(\hat{y}_{n-1}^1)^{d-1}\hat{y}_n^1 & \frac{1}{d!}(\hat{y}_n^1)^d \\ \vdots & \vdots & & \vdots & \vdots & \vdots & & \vdots & \vdots \\ 1 & \hat{y}_1^p & \cdots & \hat{y}_n^p & \frac{1}{2}(\hat{y}_1^p)^2 & \hat{y}_1^p \hat{y}_2^p & \cdots & \frac{1}{(d-1)!}(\hat{y}_{n-1}^p)^{d-1}\hat{y}_n^p & \frac{1}{d!}(\hat{y}_n^p)^d \end{bmatrix}.$$

This matrix is exactly the matrix $M(\phi, \hat{Y})$ when ϕ is the natural basis $\bar{\phi}$.

The definition of Λ-poisedness in $B(y^0; \Delta(Y))$, when using the natural basis, implies, for all $x \in B(y^0; \Delta(Y))$, the existence of a $\lambda(x)$ such that

$$M^\top \lambda(x) = \bar{\phi}(x) \quad \text{with} \quad \|\lambda(x)\|_\infty \leq \Lambda.$$

From Lemmas 3.8 and 3.9, we know that this is equivalent to claiming, for all $x \in B(0;1)$, the existence of a $\lambda(x)$ such that

$$\hat{M}^\top \lambda(x) = \bar{\phi}(x) \quad \text{with} \quad \|\lambda(x)\|_\infty \leq \Lambda.$$

Also, since $x \in B(0; 1)$ and since at least one of the y^i's has norm 1 (recall that $B(0; 1)$ is the smallest enclosing ball centered at the origin), then the norm of this matrix is always bounded by

$$1 \leq \|\hat{M}\| \leq p_1^{\frac{3}{2}}. \tag{3.11}$$

The condition number of $\hat{M}$ is denoted by $\mathrm{cond}(\hat{M}) = \|\hat{M}\|\|\hat{M}^{-1}\|$. To bound $\mathrm{cond}(\hat{M})$ in terms of Λ it is then sufficient to bound $\|\hat{M}^{-1}\|$, and, conversely, to bound Λ in terms of $\mathrm{cond}(\hat{M})$ it is sufficient to bound it in terms of $\|\hat{M}^{-1}\|$. Theorem 3.14 below establishes this relationship. Its proof requires the following (easy to prove, see the exercises) auxiliary result.

Lemma 3.13. *Let w be a (normalized) right-singular vector of (a nonsingular square) matrix A corresponding to its largest singular value. Then, for any vector r of the appropriate size,*

$$\|Ar\| \geq |w^\top r| \|A\|.$$

Theorem 3.14. *If $\hat{M}$ is nonsingular and $\|\hat{M}^{-1}\| \leq \Lambda$, then the set $\hat{Y}$ is $\sqrt{p_1}\Lambda$-poised in the unit ball $B(0; 1)$ centered at 0. Conversely, if the set $\hat{Y}$ is Λ-poised in the unit ball $B(0; 1)$ centered at 0, then*

$$\|\hat{M}^{-1}\| \leq \theta p_1^{\frac{1}{2}} \Lambda, \tag{3.12}$$

where $\theta > 0$ is dependent on n and d but independent of $\hat{Y}$ and Λ.

Proof. Since the ℓ_2-norm is invariant under transposition, we can use $\hat{M}^\top$ in the proof. If $\hat{M}$ is nonsingular and $\|\hat{M}^{-1}\| \leq \Lambda$, then

$$\|\lambda(x)\|_\infty \leq \|\hat{M}^{-\top}\|_\infty \|\bar{\phi}(x)\|_\infty \leq p_1^{\frac{1}{2}} \|\hat{M}^{-\top}\| \|\bar{\phi}(x)\|_\infty \leq p_1^{\frac{1}{2}} \Lambda,$$

since $\max_{x \in B(0;1)} \|\bar{\phi}(x)\|_\infty \leq 1$.

To prove the reverse relation let $\bar{v}$ be a (normalized) right-singular vector of $\hat{M}^{-\top}$ corresponding to its largest singular value. Since $\|\bar{v}\| = 1$, we know from Lemma 3.10 that there exists a $y \in B(0; 1)$ such that

$$|\bar{v}^\top \bar{\phi}(y)| \geq \sigma_2 \geq \frac{\sigma_\infty}{\sqrt{p_1}}.$$

By applying Lemma 3.13 with $A = \hat{M}^{-\top}$, $w = \bar{v}$, and $r = \bar{\phi}(y)$,

$$\|\hat{M}^{-\top}\bar{\phi}(y)\| \geq |\bar{v}^\top \bar{\phi}(y)| \|\hat{M}^{-\top}\|,$$

and the result follows easily with $\theta = \sqrt{p_1}/\sigma_\infty$. $\qquad \square$

The constant θ can be estimated for specific values of d, for example $d = 1$ and $d = 2$. In fact, it is inversely related to σ_2, where

$$\sigma_2 = \min_{\|\bar{v}\|=1} \max_{x \in B(0;1)} |\bar{v}^\top \bar{\phi}(x)|.$$

Thus, as we have stated before, it is easy to see that $\theta \leq 1$ for $d = 1$. For $d = 2$ we can replace the constant θ of Theorem 3.14 by an upper bound $(\theta \leq 4\sqrt{p_1})$, which is easily derived using Lemma 3.12.

It is important to note that θ depends on the choice of ϕ. For example, if we scale every element of $\bar{\phi}$ by 2, then the appropriate θ will decrease by 2. Here we are interested in the condition number of a specific matrix $\hat{M} = M(\bar{\phi}, \hat{Y})$ arising in our algorithms and, hence, in a specific choice of ϕ.

The following corollary of Theorem 3.14 will be useful in Chapter 6. For the sake of clarity, we highlight the dependence of the constant involved on the natural basis $\bar{\phi}$.

Corollary 3.15. *If $\hat{Y}$ is Λ-poised in $B(0; 1)$, then $\mathrm{vol}(\bar{\phi}(\hat{Y})) \geq \Theta(p, \bar{\phi}) > 0$, where $\Theta(p, \bar{\phi})$ depends only on p (i.e., on n and d) and $\bar{\phi}$.*

Proof. Theorem 3.14 guarantees the existence of a constant θ dependent only on p and $\bar{\phi}$ such that $\|M(\bar{\phi}, \hat{Y})^{-1}\| \leq \theta\sqrt{p_1}\Lambda$. Since the absolute value of the determinant of a squared matrix is the product of its singular values, we obtain

$$\mathrm{vol}(\bar{\phi}(\hat{Y})) \;=\; \frac{1}{p_1! \, |\det(M(\bar{\phi}, \hat{Y})^{-1})|} \;\geq\; \frac{1}{p_1! (\theta\sqrt{p_1}\Lambda)^{p_1}}. \qquad \square$$

Derivation of error bounds in terms of the condition number

In the sections on Lagrange and Newton fundamental polynomial bases, we have seen the bounds on the error between the interpolating polynomial and the function being interpolated, between their gradients and, possibly, between their Hessians, expressed in terms of the bound on the maximum absolute value of these polynomials in the region of interest. We will now present similar bounds expressed via the condition number of $M(\bar{\phi}, \hat{Y})$. Since, as we have shown, the condition number of $M(\bar{\phi}, \hat{Y})$ is connected to Λ-poisedness, we can use the bound in Theorem 3.14 directly and express the error bound in terms of p, θ, and $\|M(\bar{\phi}, \hat{Y})^{-1}\|$ instead of Λ. However, the derivation of the bounds in [53] is quite complicated and requires knowledge of approximation theory. By contrast, the derivation of the bound presented below is very simple and requires no knowledge beyond calculus.

We consider interpolation of a function $f(y)$ by a polynomial $m(y)$, written using the natural basis:

$$m(y) \;=\; \sum_{k=0}^{p} \gamma_k \bar{\phi}_k(y).$$

The interpolation set satisfies $Y = \{y^0, y^1, \ldots, y^p\} \subset B(y^0; \Delta(Y))$, where $B(y^0; \Delta(Y))$ is the smallest closed ball, centered at y^0 and of radius $\Delta = \Delta(Y)$, containing Y. The coefficients γ_k, $k = 0, \ldots, p$, are defined by the linear system arising from the interpolation conditions (3.2).

We will consider the quadratic case ($d = 2$) in detail. The linear case ($d = 1$) has already been covered in Section 2.4. The error bounds are estimated in the quadratic case under the following assumptions.

Assumption 3.1. *We assume that $Y = \{y^0, y^1, \ldots, y^p\} \subset \mathbb{R}^n$, with $p_1 = p + 1 = (n+1)(n+2)/2$, is a poised set of sample points (in the quadratic interpolation sense, $d = 2$) contained in the ball $B(y^0; \Delta(Y))$ of radius $\Delta = \Delta(Y)$.*

Further, we assume that the function f is twice continuously differentiable in an open domain Ω containing $B(y^0; \Delta)$ and $\nabla^2 f$ is Lipschitz continuous in Ω with constant $\nu_2 > 0$.

Under these assumptions it is possible to build the quadratic interpolation model

$$m(y) = c + g^\top y + \frac{1}{2} y^\top H y = c + \sum_{1 \leq k \leq n} g_k y_k + \frac{1}{2} \sum_{1 \leq k, \ell \leq n} h_{k\ell} y_k y_\ell, \qquad (3.13)$$

where H is a symmetric matrix of order n. The unknown coefficients c, $g_1, \ldots, g_n$, and $h_{k\ell}, 1 \leq \ell \leq k \leq n$, are uniquely defined by the interpolation conditions (3.2) because the sample set is poised.

We will assume for the moment and without loss of generality that $y^0 = 0$. Later, we will lift this assumption and establish the interpolation bounds for any y^0.

We now consider a point y in the ball $B(0; \Delta(Y))$, for which we will try to estimate the error in the function value

$$m(y) = f(y) + e^f(y), \qquad (3.14)$$

in the gradient

$$\nabla m(y) = Hy + g = \nabla f(y) + e^g(y), \qquad (3.15)$$

and, since we are considering the quadratic case, also in the Hessian

$$H = \nabla^2 f(y) + E^H(y).$$

Since the Hessians of f and m are symmetric, we need only consider the error of the second-order derivatives in diagonal elements and in elements below the diagonal:

$$h_{k\ell} = \nabla^2_{k\ell} f(y) + E^H_{k\ell}(y), \quad 1 \leq \ell \leq k \leq n.$$

Using (3.13) and subtracting (3.14) from all the $p_1 = p + 1$ equalities in (3.2), we have that

$$(y^i - y)^\top g + \frac{1}{2}(y^i - y)^\top H(y^i - y) + (y^i - y)^\top Hy$$
$$= f(y^i) - f(y) - e^f(y), \quad i = 0, \ldots, p.$$

Substituting the expression for g from (3.15), regrouping terms, and applying the second-order Taylor expansion formula we get

$$(y^i - y)^\top e^g(y) + \frac{1}{2}(y^i - y)^\top [H - \nabla^2 f(y)](y^i - y)$$
$$= \mathcal{O}(\Delta^3) - e^f(y), \quad i = 0, \ldots, p. \qquad (3.16)$$

The next step is to subtract the first equation of (3.16) from the other equations, canceling $e^f(y)$ and obtaining (note that $y^0 = 0$)

$$(y^i)^\top (e^g(y) - E^H(y)y) + \frac{1}{2}(y^i)^\top [H - \nabla^2 f(y)](y^i) = \mathcal{O}(\Delta^3), \quad 1 \leq i \leq p.$$

Thus, the linear system that we need to analyze in this quadratic case can be written as

$$\sum_{1 \leq k \leq n} y^i_k t_k(y) + \frac{1}{2} \sum_{1 \leq k \leq n} (y^i_k)^2 E^H_{kk}(y) + \sum_{1 \leq \ell < k \leq n} [y^i_k y^i_\ell] E^H_{k\ell}(y)$$
$$= \mathcal{O}(\Delta^3), \quad 1 \leq i \leq p,$$

or, in matrix form, as

$$Q \begin{bmatrix} t(y) \\ e^H(y) \end{bmatrix} = \mathcal{O}(\Delta^3), \tag{3.17}$$

with

$$t(y) = e^g(y) - E^H(y)y = e^g(y) - [H - \nabla^2 f(y)]y. \tag{3.18}$$

Here $e^H(y)$ is a vector of dimension $n + n(n-1)/2$ storing the elements $E^H_{kk}(y)$, $k = 1, \ldots, n$, and $E^H_{k\ell}(y)$, $1 \le \ell < k \le n$.

Notice that the matrix Q is the same as the matrix formed by the last p rows and columns of $M(\bar{\phi}, Y)$ in the case when $d = 2$ (quadratic) and $y^0 = 0$, provided $e^H(y)$ is suitably ordered, which we will henceforth assume.

We will estimate an upper bound on the right-hand side vector in (3.17). Each element of this vector is the difference of two terms that can be bounded by $\nu_2 \|y^i - y\|^3/6$ and $\nu_2 \|y\|^3/6$, respectively, where ν_2 is the Lipschitz constant of $\nabla^2 f$ in Ω (see [76, Lemma 4.1.14]). Since $\|y^i - y\| \le 2\Delta$ and $\|y\| \le \Delta$, the difference can be bounded by $3\Delta^3/2$. Hence, the ℓ_∞-norm of the right-hand side can be bounded by that amount, and a bound on the ℓ_2-norm is

$$\left\| Q \begin{bmatrix} t(y) \\ e^H(y) \end{bmatrix} \right\| \le \frac{3}{2} p^{\frac{1}{2}} \nu_2 \Delta^3. \tag{3.19}$$

To eliminate the dependence of Q on Δ we need to consider the scaled matrix

$$\hat{Q} = Q \begin{bmatrix} D_\Delta^{-1} & 0 \\ 0 & D_{\Delta^2}^{-1} \end{bmatrix}, \tag{3.20}$$

where D_Δ is a diagonal matrix of dimension n with Δ in the diagonal entries and D_{Δ^2} is a diagonal matrix of dimension $p - n$ with Δ^2 in the diagonal entries. This scaled matrix is the same as the matrix Q corresponding to the scaled set $\hat{Y} = Y/\Delta \subset B(0; 1)$.

The next theorem states our error bounds in the quadratic case. As one might expect, the error bounds are linear in Δ for the second derivatives, quadratic in Δ for the first derivatives, and cubic in Δ for the function values, where $\Delta = \Delta(Y)$ is the radius of the smallest ball containing Y. The smaller $\|\hat{Q}^{-1}\|$ is, the better is the poisedness of Y and the better are the error bounds.

Theorem 3.16. *Let Assumption 3.1 hold. Then, for all points y in $B(y^0; \Delta(Y))$, we have that*

- *the error between the Hessian of the quadratic interpolation model and the Hessian of the function satisfies*

$$\|\nabla^2 f(y) - \nabla^2 m(y)\| \le \kappa_{eh} \Delta,$$

- *the error between the gradient of the quadratic interpolation model and the gradient of the function satisfies*

$$\|\nabla f(y) - \nabla m(y)\| \le \kappa_{eg} \Delta^2,$$

- *the error between the quadratic interpolation model and the function satisfies*

$$|f(y) - m(y)| \le \kappa_{ef} \Delta^3,$$

where κ_{eh}, κ_{eg}, and κ_{ef} are given by

$$\kappa_{eh} = 3\sqrt{2}p^{\frac{1}{2}}v_2\|\hat{Q}^{-1}\|/2,$$

$$\kappa_{eg} = 3(1+\sqrt{2})p^{\frac{1}{2}}v_2\|\hat{Q}^{-1}\|/2,$$

$$\kappa_{ef} = (6+9\sqrt{2})p^{\frac{1}{2}}v_2\|\hat{Q}^{-1}\|/4 + v_2/6.$$

Proof. Let us first assume that $y^0 = 0$ and write the left-hand side of the system (3.17) in the form

$$Q\begin{bmatrix} D_\Delta^{-1} & 0 \\ 0 & D_{\Delta^2}^{-1} \end{bmatrix}\begin{bmatrix} D_\Delta t(y) \\ D_{\Delta^2}e^H(y) \end{bmatrix}.$$

Then, using the bound (3.19) and the notation (3.20), we obtain

$$\left\|\begin{bmatrix} D_\Delta t(y) \\ D_{\Delta^2}e^H(y) \end{bmatrix}\right\| \leq \frac{3}{2}p^{\frac{1}{2}}v_2\|\hat{Q}^{-1}\|\Delta^3, \tag{3.21}$$

from which we get

$$\|D_{\Delta^2}e^H(y)\| \leq \frac{3}{2}p^{\frac{1}{2}}v_2\|\hat{Q}^{-1}\|\Delta^3,$$

yielding the bound $\|e^H(y)\| \leq (3/2)p^{\frac{1}{2}}v_2\|\hat{Q}^{-1}\|\Delta$. The error in the Hessian is therefore given by

$$\|E^H(y)\| \leq \|E^H(y)\|_F \leq \sqrt{2}\|e^H(y)\| \leq \frac{3\sqrt{2}}{2}p^{\frac{1}{2}}v_2\|\hat{Q}^{-1}\|\Delta,$$

where $\|\cdot\|_F$ denotes the Frobenius norm of a matrix given by the square root of the sum of all entries squared.

Now we would like to derive the bound on $\|e^g(y)\|$. From (3.21) we also have

$$\|D_\Delta t(y)\| \leq \frac{3}{2}p^{\frac{1}{2}}v_2\|\hat{Q}^{-1}\|\Delta^3$$

and

$$\|t(y)\| \leq \frac{3}{2}p^{\frac{1}{2}}v_2\|\hat{Q}^{-1}\|\Delta^2,$$

and therefore, from (3.18),

$$\begin{aligned}
\|e^g(y)\| &\leq \|t(y)\| + \|E^H(y)\|\|y\| \\
&\leq \frac{3}{2}p^{\frac{1}{2}}v_2\|\hat{Q}^{-1}\|\Delta^2 + \left(\frac{3\sqrt{2}}{2}p^{\frac{1}{2}}v_2\|\hat{Q}^{-1}\|\Delta\right)\Delta \\
&= \frac{3(1+\sqrt{2})}{2}p^{\frac{1}{2}}v_2\|\hat{Q}^{-1}\|\Delta^2.
\end{aligned}$$

Here we have used the fact that y is in the ball $B(0; \Delta(Y))$ centered at the origin.

Finally, from the detailed version of (3.16) for $i = 0$ and the bounds on $\|e^g(y)\|$ and $\|E^H(y)\|$ we have

$$
\begin{aligned}
|e^f(y)| &\leq \|e^g(y)\|\Delta + \tfrac{1}{2}\|E^H(y)\|\Delta^2 + \tfrac{\nu_2}{6}\Delta^3 \\
&\leq \tfrac{6+9\sqrt{2}}{4}p^{\frac{1}{2}}\nu_2\|\hat{Q}^{-1}\|\Delta^3 + \tfrac{\nu_2}{6}\Delta^3.
\end{aligned}
$$

If $y^0 \neq 0$, we would make a change of variables and consider a function $\hat{f}(z) = f(z + y^0)$ with $z = y - y^0$. We can then apply the results previously proved, since the shifted interpolation set $\{0, y^1 - y^0, \dots, y^p - y^0\}$ is now centered at the origin and $\hat{f}$ has the appropriate smoothness, and thus derive bounds for $\|\nabla^2 \hat{f}(z) - \nabla^2 \hat{m}(z)\|$, $\|\nabla \hat{f}(z) - \nabla \hat{m}(z)\|$, and $|\hat{f}(z) - \hat{m}(z)|$. Now we define $m(y) = \hat{m}(y^0 + z)$ and change the function and the polynomial model back to the original variables. It is a simple matter to see that we obtain the desired bounds since $\nabla^2 \hat{m}(z) = \nabla^2 m(y)$, $\nabla^2 \hat{f}(z) = \nabla^2 f(y)$, $\nabla \hat{m}(z) = \nabla m(y)$, and $\nabla \hat{f}(z) = \nabla f(y)$. $\square$

Since $\|\hat{Q}^{-1}\| \leq \|\hat{M}^{-1}\|$ (see the exercises), one can use Theorem 3.14 and express the error bounds given in the above theorem in terms of the constant Λ used in the definition of Λ-poisedness.

The error bounds given here for quadratic interpolation can be generalized for interpolation of any degree (for the generalization to the cubic case see [63]).

3.5   Exercises

1. Prove that if $M(\phi, Y)$ is nonsingular for some basis ϕ, then it is nonsingular for any basis of $\mathcal{P}_n^d$.

2. Show that for any given poised interpolation set Y, one can choose the basis ϕ so that the condition number of $M(\phi, Y)$ can equal any number between 1 and $+\infty$.

3. Construct a set of Lagrange polynomials in $\mathbb{R}^2$ for the set $y^0 = (0,0)$, $y^1 = (1,0)$, $y^2 = (0,1)$, $y^3 = (-1,0)$, $y^4 = (1,1)$, and $y^5 = (0,-1)$. Compare the Λ-poisedness constant for this set to the one relative to the set mentioned in Section 3.2.

4. Show that six points on a circle in $\mathbb{R}^2$ are not poised for interpolation by quadratic polynomials but are poised for interpolation in a subspace of cubic polynomials that does not have quadratic and constant terms.

5. Given $\mathcal{P}_n^d$, show that $\lambda(x)$ in (3.3) does not depend on the choice of ϕ.

6. Prove Lemma 3.5.

7. Prove Lemma 3.7.

8. Prove that $\psi(v)$ given by (3.9) is a norm in the space of vectors v.

9. Show that the bounds (3.11) for $\|\hat{M}\|$ are true.

10. Prove Lemma 3.13.

11. Show that $\|\hat{Q}^{-1}\| \leq \|\hat{M}^{-1}\|$.

Chapter 4

Regression nonlinear models

Let us again consider $\mathcal{P}_n^d$, the space of polynomials of degree less than or equal to d in $\mathbb{R}^n$. Let $q_1 = q + 1$ be the dimension of this space, and let $\phi = \{\phi_0(x), \phi_1(x), \ldots, \phi_q(x)\}$ be a basis for $\mathcal{P}_n^d$. Given a polynomial basis ϕ, we define $\phi(x) = [\phi_0(x), \phi_1(x), \ldots, \phi_q(x)]^\top$, as before, to be a vector in $\mathbb{R}^{q_1}$. As before, $Y = \{y^0, y^1, \ldots, y^p\} \subset \mathbb{R}^n$ denotes the set of $p_1 = p + 1$ sample points. Let $f(Y)$ denote the vector whose elements are $f(y^i)$, $i = 0, \ldots, p$. We will consider the context where the dimension q_1 of the polynomial space is fixed but the number p_1 of sample points can change.

It is known that a large portion of derivative-free optimization applications exhibits noisy functions. To some extent, most of the derivative-free methods are based on sampling the objective function at several points, sufficiently spaced apart. This approach by itself allows many of the derivative-free methods to be reasonably tolerant to noise. However, one may prefer to take the presence of noise into account when designing an algorithm.

The first and, perhaps, most natural approach to handling noisy data is to replace the interpolation of the objective function by least-squares regression. In this case, the interpolation conditions described in the previous chapter,

$$M(\phi, Y)\alpha = f(Y),$$

are solved in the least-squares sense, meaning that a solution α is found such that $\|M(\phi, Y)\alpha - f(Y)\|^2$ is minimized.

If function evaluations are relatively cheap, but still noisy, then it may be effective to sample the objective function at more local sample points (i.e., at points closer to the "center" of the model) than it would be necessary for complete interpolation. In that case, $p_1 > q_1$, which means that the number of rows in the matrix $M(\phi, Y)$ is larger than the number of columns and the interpolation system is overdetermined. As the number of sample points increases, the least-squares regression solution to the noisy problem converges (in some sense and under reasonable assumptions) to the least-squares regression of the underlying true function.

Specifically, let the noisy function $f(x) = f_{smooth}(x) + \varepsilon$, where $f_{smooth}(x)$ is the true, mostly likely smooth function which we are trying to optimize and ε is a random variable, independent of x and drawn from some distribution with zero mean. Assume that a polynomial basis ϕ is fixed, and consider a sample set Y^p whose size $|Y^p| = p + 1$ is variable. Let the random vector α^p be the least-squares solution to the system

$$M(\phi, Y^p)\alpha = f(Y^p)$$

and the real vector α^p_{smooth} be the least-squares solution to the system

$$M(\phi, Y^p)\alpha = f_{smooth}(Y^p).$$

Now assume that the size of the sample set Y^p is going to infinity and that the following condition holds:

$$\liminf_{p\to+\infty} \lambda_{min}\left(\frac{1}{p+1}M(\phi, Y^p)^\top M(\phi, Y^p)\right) > 0, \tag{4.1}$$

where λ_{min} denotes the minimum eigenvalue of a matrix. Then it is easy to prove the following consistency result (see the exercises at the end of this chapter and the books [116, 224] for further details).

Theorem 4.1. *If condition* (4.1) *holds and the sequence* $\{Y^p\}$ *is bounded, then*

$$\mathbf{E}(\alpha^p) = \alpha^p_{smooth} \quad \forall p \geq q$$

and

$$\lim_{p\to+\infty} \mathbf{Var}(\alpha^p - \alpha^p_{smooth}) = 0.$$

Condition (4.1) on the sequence of sets Y^p means that the data is sampled in a uniformly well-poised manner. We will return to this point when we study the poisedness properties of an overdetermined set Y.

When function evaluations are expensive, one rarely can afford to accumulate enough local sample points to exceed the number of points necessary for complete quadratic interpolation, namely $(n+1)(n+2)/2$. In this case least-squares regression and interpolation will produce identical results. But, even then, it may be useful to relax the interpolation conditions. If the objective function is very noisy, then the interpolation model may turn out to be unnecessarily "nonsmooth." One can then use regularized regression, which tries to trade off between optimizing the least-squares fit and the "smoothness" of the interpolation polynomial at the same time. We will explore these ideas further at the end of the chapter.

This chapter is dedicated entirely to the case when $p > q$. First, we will discuss the properties of the least-squares models and the corresponding well-poised sample sets.

4.1   Basic concepts in polynomial least-squares regression

Let $m(x)$ denote the polynomial of degree less than or equal to d that approximates a given function $f(x)$ at the points in Y via least-squares regression. Since ϕ is a basis in $\mathcal{P}_n^d$, then $m(x) = \sum_{k=0}^q \alpha_k \phi_k(x)$, where the α_k's are the unknown coefficients. The coefficients α can be determined from the least-squares regression conditions

$$M(\phi, Y)\alpha \overset{\ell.s.}{=} f(Y) \tag{4.2}$$

or, equivalently,

$$\min_{\alpha} \| M(\phi, Y)\alpha - f(Y) \|^2.$$

The above system has a unique solution if the matrix

$$M(\phi, Y) = \begin{bmatrix} \phi_0(y^0) & \phi_1(y^0) & \cdots & \phi_q(y^0) \\ \phi_0(y^1) & \phi_1(y^1) & \cdots & \phi_q(y^1) \\ \vdots & \vdots & \vdots & \vdots \\ \vdots & \vdots & \vdots & \vdots \\ \phi_0(y^P) & \phi_1(y^P) & \cdots & \phi_q(y^P) \end{bmatrix} \tag{4.3}$$

has full column rank.

It is easy to see that if $M(\phi, Y)$ is square and nonsingular, then the above problem becomes the interpolation problem. In many ways the properties of regression exactly or closely mimic those of interpolation. Just as in the case of interpolation, if $M(\phi, Y)$ has full column rank for some choice of ϕ, then it is so for any basis of $\mathcal{P}_n^d$. The following definition of poisedness is thus independent of the basis chosen.

Definition 4.2. *The set $Y = \{y^0, y^1, \ldots, y^P\}$ is poised for polynomial least-squares regression in $\mathbb{R}^n$ if the corresponding matrix $M(\phi, Y)$ has full column rank for some basis ϕ in $\mathcal{P}_n^d$.*

We now show that if Y is poised, then the least-squares regression polynomial is unique and independent of the choice of ϕ.

Lemma 4.3. *Given a function $f : \mathbb{R}^n \to \mathbb{R}$ and a poised set $Y \in \mathbb{R}^n$ with respect to polynomial least-squares regression, the least-squares regression polynomial $m(x)$ exists and is unique.*

Proof. Since the set is poised, it is obvious that the least-squares regression polynomial $m(x)$ exists and is unique for a given choice of basis. Now consider two different bases $\psi(x)$ and $\phi(x)$ related by $\psi(x) = P^\top \phi(x)$, where P is $q_1 \times q_1$ and nonsingular. Then $M(\psi, Y) = M(\phi, Y)P$.

Let α_ϕ (resp., α_ψ) be the vector of coefficients of the least-squares regression polynomial for the basis $\phi(x)$ (resp., $\psi(x)$). Since α_ϕ is the least-squares solution to the system $M(\phi, Y)\alpha = f(Y)$, then

$$\alpha_\phi = [M(\phi, Y)^\top M(\phi, Y)]^{-1} M(\phi, Y)^\top f(Y)$$

$$= [P^{-\top} M(\psi, Y)^\top M(\psi, Y) P^{-1}]^{-1} P^{-\top} M(\psi, Y)^\top f(Y)$$

$$= P[M(\psi, Y)^\top M(\psi, Y)]^{-1} M(\psi, Y)^\top f(Y) = P\alpha_\psi.$$

The last equality follows from the fact that α_ψ is the least-squares solution to the system $M(\psi, Y)\alpha = f(Y)$. Then, for any x,

$$\alpha_\psi^\top \psi(x) = \alpha_\psi^\top \left(P^\top \phi(x) \right) = \left(P\alpha_\psi \right)^\top \phi(x) = \alpha_\phi^\top \phi(x). \quad \square$$

4.2   Lagrange polynomials in the regression sense

The condition of poisedness and the existence of the regression polynomial is not sufficient in practical algorithms or in the derivation of "uniform" error bounds for use in the
convergence analysis of derivative-free algorithms. As in the case of interpolation, one
needs a condition of "well poisedness," characterized by a constant. We will extend the
notions of Λ-poisedness and Lagrange polynomials to the case of polynomial least-squares
regression.

Lagrange polynomials

Definition 4.4. *Given a set of sample points* $Y = \{y^0, y^1, \ldots, y^p\}$, *with* $p > q$, *a set of*
$p_1 = p + 1$ *polynomials* $\ell_j(x)$, $j = 0, \ldots, p$, *in* $\mathcal{P}_n^d$ *is called a set of regression Lagrange*
polynomials if

$$\ell_j(y^i) \stackrel{\ell.s.}{=} \delta_{ij} = \begin{cases} 1 & \text{if } i = j, \\ 0 & \text{if } i \neq j. \end{cases} \tag{4.4}$$

This set of polynomials is an extension of the traditional Lagrange polynomials to
the case $p > q$. Clearly, these polynomials are no longer linearly independent, since there
are too many of them. However, as we show below, many other properties of interpolation
Lagrange polynomials are preserved. Lemma 4.3 implies the following result.

Lemma 4.5. *If* Y *is poised for polynomial least-squares regression, then the set of regres-*
sion Lagrange polynomials exists and is uniquely defined.

Another essential property of Lagrange polynomials carries over to the regression
case.[5]

Lemma 4.6. *For any function* $f : \mathbb{R}^n \to \mathbb{R}$ *and any poised set* $Y = \{y^0, y^1, \ldots, y^p\} \subset \mathbb{R}^n$ *for*
polynomial least-squares regression, the unique polynomial $m(x)$ *that approximates* $f(x)$
via least-squares regression at the points in Y *can be expressed as*

$$m(x) = \sum_{i=0}^{p} f(y^i)\ell_i(x),$$

where $\{\ell_i(x), i = 0, \ldots, p\}$ *is the set of regression Lagrange polynomials for* Y.

Proof. The polynomial $m(x)$ can be written as

$$\sum_{i=0}^{q} \alpha_i \phi_i(x) = \alpha^\top \phi(x),$$

where α is the least-squares solution to the system

$$M(\phi, Y)\alpha = f(Y).$$

---

[5]It is interesting to note that the extension of Lagrange polynomials applies only to the case of least-
squares regression. In the case of ℓ_1-norm regression, for instance, the uniqueness property fails and $m(x) =$
$\sum_{i=0}^{p} f(y^i)\ell_i(x)$ may fail to hold for any choice of $\ell_i(x)$, $i = 0, \ldots, p$.

Since Y is poised, from the expression of the least-squares solution, we know that

$$\alpha = [M(\phi,Y)^\top M(\phi,Y)]^{-1} M(\phi,Y)^\top f(Y). \qquad (4.5)$$

We can write the jth Lagrange polynomial using the basis ϕ with some coefficient vector $a_\phi^j \in \mathbb{R}^{q_1}$:

$$\ell_j(x) = \phi(x)^\top a_\phi^j.$$

Then the Lagrange regression conditions (4.4), meant in the least-squares sense, can be written in the following matrix format:

$$M(\phi,Y)a_\phi^j \stackrel{\ell.s.}{=} e_{j+1}, \quad j = 0,\ldots,p,$$

where e_{j+1} is the $(j+1)$th column of the identity matrix of order $p+1$. In matrix notation, we have that

$$M(\phi,Y)A_\phi \stackrel{\ell.s.}{=} I,$$

where A_ϕ is the matrix whose columns are a_ϕ^j, $j = 0,\ldots,p$. From the solution of these least-squares problems, we have

$$A_\phi = [M(\phi,Y)^\top M(\phi,Y)]^{-1} M(\phi,Y)^\top. \qquad (4.6)$$

Hence, from the expression for α in (4.5),

$$\alpha^\top \phi(x) = f(Y)^\top A_\phi^\top \phi(x) = f(Y)^\top \ell(x),$$

where $\ell(x) = [\ell_0(x),\ldots,\ell_p(x)]^\top$. $\quad\square$

To illustrate the regression Lagrange polynomials, we consider the same six interpolation points $y^0 = (0,0)$, $y^1 = (1,0)$, $y^2 = (0,1)$, $y^3 = (2,0)$, $y^4 = (1,1)$, and $y^5 = (0,2)$ as we considered in the example in Section 3.2 and we add an extra point $y^6 = (0.5,0.5)$ to make interpolation overdetermined. It is difficult to verify the correctness of the regression Lagrange polynomials by hand, because their values at the corresponding points no longer have to be equal to 1 or 0. Below are the corresponding regression Lagrange polynomials $\ell_j(x_1,x_2), j = 0,\ldots,6$, whose coefficients were computed with the help of MATLAB® [1] software:

$$\ell_0(x_1,x_2) = 1 - \tfrac{3}{2}x_1 - \tfrac{3}{2}x_2 + \tfrac{1}{2}x_1^2 + \tfrac{1}{2}x_2^2 + x_1 x_2,$$

$$\ell_1(x_1,x_2) = \tfrac{5}{3}x_1 - \tfrac{1}{3}x_2 - 0.823x_1^2 + 0.176x_2^2 - 0.764x_1 x_2,$$

$$\ell_2(x_1,x_2) = -\tfrac{1}{3}x_1 + \tfrac{5}{3}x_2 + 0.176x_1^2 - 0.823x_2^2 - 0.764x_1 x_2,$$

$$\ell_3(x_1,x_2) = -\tfrac{5}{12}x_1 + \tfrac{1}{12}x_2 + 0.455x_1^2 - 0.044x_2^2 - 0.588x_1 x_2,$$

$$\ell_4(x_1,x_2) = -\tfrac{1}{6}x_1 - \tfrac{1}{6}x_2 + 0.088x_1^2 + 0.088x_2^2 + 1.117x_1 x_2,$$

$$\ell_5(x_1,x_2) = \tfrac{1}{12}x_1 - \tfrac{5}{12}x_2 - 0.044x_1^2 + 0.455x_2^2 - 0.588x_1 x_2,$$

$$\ell_6(x_1,x_2) = \tfrac{2}{3}x_1 + \tfrac{2}{3}x_2 - 0.352x_1^2 - 0.352x_2^2 - 0.47x_1 x_2.$$

The values of these regression Lagrange polynomials at their corresponding regression points are $\ell_0(y^0) = 1$, $\ell_1(y^1) = 0.84$, $\ell_2(y^2) = 0.84$, $\ell_3(y^3) = 0.99$, $\ell_4(y^4) = 0.96$, $\ell_5(y^5) = 0.99$, and $\ell_6(y^6) = 0.37$.

Equivalent definition of the Lagrange polynomials in the regression case

Recall the alternative definition given in (3.3), Chapter 3, for the (interpolation) Lagrange polynomials. We will show how this alternative definition extends to the regression Lagrange polynomials.

Given a poised set $Y = \{y^0, y^1, \ldots, y^p\} \subset \mathbb{R}^n$, with $p > q$ and $x \in \mathbb{R}^n$, we can express the vector $\phi(x)$ in terms of the vectors $\phi(y^i)$, $i = 0, \ldots, p$, as

$$\sum_{i=0}^{p} \lambda_i(x)\phi(y^i) = \phi(x) \tag{4.7}$$

or, equivalently,

$$M(\phi, Y)^\top \lambda(x) = \phi(x), \quad \text{where } \lambda(x) = [\lambda_0(x), \ldots, \lambda_p(x)]^\top.$$

Unlike the similar system in Chapter 3 this new system is underdetermined, and hence it might have multiple solutions. In order to establish uniqueness, we consider the minimum ℓ_2-norm solution. Thus, we are interested in the minimum-norm solution $\lambda(x)$ and in a bound on this norm. Just as in the interpolation case, where the $\lambda(x)$ that satisfy (3.3) are exactly the corresponding (interpolation) Lagrange polynomials, the same is true in the regression sense in that the $\lambda(x)$ that satisfy the minimum ℓ_2-norm solution of (4.7) are the corresponding (regression) Lagrange polynomials. In fact, when Y is poised, the matrix $M(\phi, Y)$ has full column rank and, from the definition of the minimum-norm solution of (4.7), we have that

$$\lambda(x) = M(\phi, Y)[M(\phi, Y)^\top M(\phi, Y)]^{-1}\phi(x),$$

and, as we have seen in the proof of Lemma 4.6, the regression Lagrange polynomials satisfy

$$\lambda(x) = M(\phi, Y)[M(\phi, Y)^\top M(\phi, Y)]^{-1}\phi(x) = \ell(x).$$

However, note that the alternative definition given by (3.6) which defines the value of Lagrange polynomials via a ratio of volumes of the sets Y and $Y_i(x) = Y \setminus \{y^i\} \cup \{x\}$ does not seem to extend in a straightforward manner to the regression case.

4.3   Λ-poisedness in the regression sense

As in the interpolation case, an upper bound on the absolute value of the Lagrange polynomials in a region B is the classical measure of poisedness of Y in B and is a constant present in the regression approximation error bounds. The results in [53] extend to the least-squares regression polynomials, and it is known that for any x in the convex hull of Y

$$|f(x) - m(x)| \leq \frac{1}{(d+1)!} \nu_d \sum_{i=0}^{p} \|y^i - x\|^{d+1} |\ell_i(x)|, \tag{4.8}$$

where ν_d is an upper bound on the derivative of order $d+1$ of f and $\ell_i(x)$, $i = 0, \ldots, p$, are the regression Lagrange polynomials. The bound on the error in the appropriate derivatives, as mentioned in Chapter 3 and presented in [53], also applies to the regression context. For the purposes of this chapter, we rewrite the bound (4.8) in two possible forms.

Λ-poisedness

First, as in Chapter 3, we have

$$|f(x) - m(x)| \leq \frac{1}{(d+1)!} v_d p_1 \Lambda_\ell \Delta^{d+1}, \tag{4.9}$$

where

$$\Lambda_\ell = \max_{0 \leq i \leq p} \max_{x \in B(Y)} |\ell_i(x)|,$$

and Δ is the diameter of the smallest ball $B(Y)$ containing Y.

This bound suggests that the measure of poisedness should involve the bound on Λ_ℓ, analogously to the interpolation case. Note, however, that in the case of regression, Λ_ℓ is not the only element of the bound that depends on the sample set. The number of points p_1 may change also. A simple solution would be to bound p_1; however, it is precisely by making p_1 grow large that we are able to achieve more accurate models for noisy problems, as we discussed in the beginning of this chapter. Hence, even if in reality having large sample sets is impractical, introducing artificial bounds on the number of sample points is perhaps also inappropriate. On the other hand, there are cases when such a bound occurs naturally. It is possible that, due to the cost of the function evaluations, only a few extra sample points are available in addition to a regular interpolation set. While one may want to use these points in the hopes of improving the accuracy of the model, one may not want to worry about the poisedness of these additional points. Hence, we will discuss two definitions of Λ-poisedness, one for the case when p is fixed and hence only the bound on Λ_ℓ is needed, and another for the case when p is allowed to grow. We start with the case of fixed p.

Definition 4.7. *Let $\Lambda > 0$ and a set $B \in \mathbb{R}^n$ be given. Let $\phi = \{\phi_0(x), \phi_1(x), \ldots, \phi_q(x)\}$, with $p > q$, be a basis in $\mathcal{P}_n^d$. A poised set $Y = \{y^0, y^1, \ldots, y^p\}$ is said to be Λ-poised in B (in the regression sense) if and only if*

1. *for the set of regression Lagrange polynomials associated with Y*

$$\Lambda \geq \max_{0 \leq i \leq p} \max_{x \in B} |\ell_i(x)|,$$

   or, equivalently,

2. *for any $x \in B$ the minimum ℓ_2-norm solution $\lambda(x) \in \mathbb{R}^{p_1}$ of*

$$\sum_{i=0}^{p} \lambda_i(x) \phi(y^i) = \phi(x)$$

   satisfies $\|\lambda(x)\|_\infty \leq \Lambda$.

Note that, as in the definition of Λ-poisedness for interpolation of Chapter 3, the set B is not required to contain the sample set Y. When $p = q$ the definitions of Λ-poisedness for interpolation and for regression coincide. When $p > q$ it is easy to see that if the set Y contains a subset of q_1 points that is Λ-poised, then the whole set Y is at least $\sqrt{q_1}\Lambda$-poised. The following theorem establishes the reverse relationship between a Λ-poised regression set Y and Λ-poisedness of a subset of Y.

Theorem 4.8. *Given a set* $Y = \{y^0, y^1, \ldots, y^p\}$, *which is* Λ-*poised in the regression sense, there is a subset of* $q_1 = q + 1$ *points in* Y *which is* $(p_1 - q_1 + 1)\Lambda$-*poised in the interpolation sense.*

The proof of Theorem 4.8 follows from the definitions for Λ-poisedness in the interpolation and regression senses and from Lemma 4.14 (given at the end of this chapter).

Most of the properties of Λ-poisedness in the interpolation sense extend to the regression case easily.

Lemma 4.9. *If* Y *is* Λ-*poised in a given set* B, *then*

    1. *it is* Λ-*poised (with the same constant) in any subset of* B,

    2. *it is* $\tilde{\Lambda}$-*poised in* B *for any* $\tilde{\Lambda} \geq \Lambda$,

    3. *for any* $x \in \mathbb{R}^n$, *if* $\lambda(x)$ *is a solution of* (4.7), *then*

$$\sum_{i=0}^{p} \lambda_i(x) = 1,$$

    4. $\hat{Y} = Y/\Delta$ *is* Λ-*poised in* $\hat{B} = B/\Delta$ *for any* $\Delta > 0$,

    5. $Y_a = \{y^0 + a, y^1 + a, \ldots, y^p + a\}$ *is* Λ-*poised in* $B_a = B + \{a\}$ *for any* $a \in \mathbb{R}^n$.

It is no longer true for the regression case, however, that if B contains a point in Y and Y is Λ-poised in B, then $\Lambda \geq 1$.

Strong Λ-poisedness

When the number of sample points for each regression model used by an optimization algorithm is bounded (and is moderate) the above Λ-poisedness condition on the sets is sufficient for establishing a Taylor-like error bound and, hence, convergence properties (see Chapter 10). But when p is allowed to grow arbitrarily large, for the error bound (4.9) to be useful, we need to have a uniform bound on the product $p_1 \Lambda_\ell$ for any sample set we wish to consider for accurate approximation. Such a bound is needed not only for theoretical purposes but also to serve as a practical condition to keep the error between the regression polynomial and the true function from growing.

Thus, the definition of Λ-poisedness needs to be strengthened to take the number of sample points into consideration when it becomes too large. Because the definition of the Lagrange polynomials, via $\lambda(x)$, involves the minimization of the ℓ_2-norm of $\lambda(x)$ rather than the ℓ_∞-norm, we find it more convenient to rewrite the error bound (4.8) in the form

$$|f(x) - m(x)| \leq \frac{1}{(d+1)!} v_d \sqrt{p_1} \Lambda_{\ell,2} \Delta^{d+1},$$

where

$$\Lambda_{\ell,2} = \max_{x \in B(Y)} \|\ell(x)\|.$$

Instead of bounding $p_1 \Lambda_\ell$, to keep the error bound in check, it is equivalent to bound $\sqrt{p_1} \Lambda_{\ell,2}$. If p grows, we need to consider sample sets whose ℓ_2-norm of the Lagrange polynomials reduces proportionally to $1/\sqrt{p_1}$. Let us first investigate if such sets exist.

For simplicity let us consider only sample sets whose number p_1 of points is an integer multiple of the dimension q_1 of the polynomial space. We can partition such a sample set Y into $l = p_1/q_1$ subsets of size q_1, say $Y_1, Y_2, \ldots, Y_l$. Each subset can be viewed as an interpolation set, since it has the right number of points. Assume now that each such set is Λ-poised in a given region B for some $\Lambda > 1$. This means that, for any $x \in B$ and for any $j = 1, \ldots, l$, we have

$$\sum_{i=0}^{q} \lambda_i^j(x) \phi(y_j^i) = \phi(x), \quad \|\lambda^j(x)\| \leq \sqrt{q_1} \|\lambda^j(x)\|_\infty \leq \sqrt{q_1} \Lambda,$$

where y_j^i is the ith point of the set Y_j and $\lambda_i^j(x)$ is the ith element of the Lagrange polynomial basis for the interpolation set Y_j (see Section 3.2). Now let us divide each of these expressions by l and add them together. What we obtain is

$$\sum_{j=1}^{l} \sum_{i=0}^{q} \frac{1}{l} \lambda_i^j(x) \phi(y_j^i) = \phi(x), \quad \|\lambda^j(x)/l\| \leq \frac{\sqrt{q_1}}{l} \Lambda.$$

Thus, for the entire sample set Y, there exists a solution $\bar{\lambda}_i(x), i = 0, \ldots, p_1$, to (4.7), which is obtained by concatenating the vectors $\lambda^j(x), j = 1, \ldots, l$, and dividing every component by l, such that

$$\|\bar{\lambda}(x)\| \leq \sqrt{l} \left(\frac{\sqrt{q_1}}{l} \Lambda \right) = \frac{q_1}{\sqrt{p_1}} \Lambda.$$

We are interested in the solution $\lambda(x)$ to (4.7) with the smallest ℓ_2-norm; hence, for such a solution

$$\|\lambda(x)\| \leq \|\bar{\lambda}(x)\| \leq \frac{q_1}{\sqrt{p_1}} \Lambda.$$

Since this solution is indeed the set of regression Lagrange polynomials for the set Y, that means that we have established that the product $\sqrt{p_1} \Lambda_{\ell,2}$ is bounded on B by $q_1 \Lambda$.

We derived this property by insisting that the set Y can be partitioned into $l = p_1/q_1$ subsets of size q_1 that are Λ-poised for interpolation. If p_1 is not an integer multiple of q_1, then we partition Y into $l = \lfloor p_1/q_1 \rfloor$ subsets (Λ-poised) and one more subset of less than q_1 points. We simply note that the case of fractional p_1/q_1 is a simple extension of the case where this number is an integer, and we do not consider it in detail here so as not to complicate the notation further.

We introduce the following definition of *strong* Λ-poisedness for a regression set Y.

Definition 4.10. *Let $\Lambda > 0$ and a set $B \in \mathbb{R}^n$ be given. Let $\phi = \{\phi_0(x), \phi_1(x), \ldots, \phi_q(x)\}$, with $p > q$, be a basis in $\mathcal{P}_n^d$. A poised set $Y = \{y^0, y^1, \ldots, y^p\}$ is said to be strongly Λ-poised in B (in the regression sense) if and only if*

1. *for the set of regression Lagrange polynomials associated with Y*

$$\frac{q_1}{\sqrt{p_1}} \Lambda \geq \max_{x \in B} \|\ell(x)\|,$$

or, equivalently,

2. *for any $x \in B$ the minimum ℓ_2-norm solution $\lambda(x) \in \mathbb{R}^{p_1}$ of*

$$\sum_{i=0}^{p} \lambda_i(x)\phi(y^i) = \phi(x)$$

satisfies $\|\lambda(x)\| \leq \frac{q_1}{\sqrt{p_1}}\Lambda$.

Consider $p_1 = q_1$ for a moment. If Y is Λ-poised in the interpolation sense, then $\|\lambda(x)\|_\infty \leq \Lambda$ and, since $\|\lambda(x)\| \leq \sqrt{q_1}\|\lambda(x)\|_\infty$, we have $\|\lambda(x)\| \leq \sqrt{q_1}\Lambda$. We can see that in the case when $p = q$ the definition of strong Λ-poisedness is not equivalent to the definition of Λ-poisedness for the interpolation but is implied by it. To induce the equivalence, it is possible to define Λ-poisedness via the ℓ_2-norm bound on the Lagrange polynomials in the case of interpolation in Chapter 3. This would not fundamentally alter any result related to approximation by polynomials presented in this book, but it would somewhat complicate the presentation and change several constants in various bounds related to polynomial interpolation.

Illustration of the Λ-poisedness definitions

With the help of the following two examples we try to illustrate the two definitions of Λ-poisedness and how they relate to each other and to the Λ-poisedness in the interpolation sense. We consider the following regression set:

$$Y_1 = \{(0,0),(1,0),(0,1),(2,0),(1,1),(0,2)\}$$

and let $Y_g = \{Y_1, Y_1, \ldots, Y_1\} + z$, which consists of 31 copies of Y_1 slightly perturbed by a random vector z of appropriate size with entries uniformly distributed between 0 and 0.05. It is illustrated on the left-hand side of Figure 4.1.

Let us consider one such typical random set which we generated and tested in MATLAB [1] software. The interpolation set Y_1 itself is Λ-poised (in the interpolation sense) with $\Lambda = 4.121$, while the largest Λ-poisedness constant among the 31 perturbed copies of Y_1 was 4.541. The entire regression set Y is Λ-poised (in the regression sense) with $\Lambda = 0.152$. First of all, we notice that the Λ-poisedness constant of the entire regression set is much smaller than the Λ-poisedness constant of any subset of the six points. Also notice that we can partition the set Y_g into 31 Λ-poised sets with $\Lambda \leq 4.541$. Although one might expect it, it is not true that the constant Λ_ℓ of Y_g is 31 times smaller than the maximum Λ_ℓ in any of these subsets, since $0.152 \times 31 = 4.721 > 4.541$. In our experiments, the Λ-poisedness constant, however, did reduce proportionally with the number of perturbed copies of Y_1 which we included in Y_g. Hence, we expect this set to be strongly Λ-poised.

To analyze strong Λ-poisedness, we need to estimate the strong Λ-poisedness constant of Y_g. This value is not simple to compute (in MATLAB [1] software) due to the need to maximize nonquadratic functions. However, instead of considering

$$\Lambda_{\ell,2} = \max_{x \in B(Y_g)} \|\ell(x)\|$$

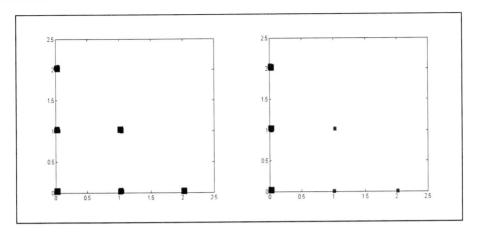

Figure 4.1. *A strongly well-poised regression set Y_g (left) and a well-poised (but not strongly poised) regression set Y_b (right). The thick squares are collections of slightly perturbed identical points. The thin squares are single points.*

we can easily compute the upper bound

$$\sqrt{\sum_{i=0}^{p}\left[\max_{x\in B(Y_g)}|\ell_i(x)|\right]^2},$$

which in the case of the set Y_g equals 1.122. Now we can see for this case that

$$\Lambda_{\ell,2} \leq 1.122 \leq \frac{q_1}{\sqrt{p_1}}4.541 = \frac{6}{13.64}4.541 = 1.998,$$

which means that the set Y_g is strongly Λ-poised with $\Lambda \leq 4.541$. Here we showed an example of a collection of subsets with roughly the same Λ-poisedness constants, which together make up a strongly Λ-poised regression set with a similar constant.

Now let us consider a different sample set Y_b which is composed of $Y_b = \{Y_1, Y_2, \ldots, Y_2\} + z$, where Y_1 is the same set as above and

$$Y_2 = \{(0,0),(0,1),(0,1),(0,0),(0,2),(0,2)\}.$$

This set Y_2 is obtained from Y_1 by repeating the first, third, and sixth points so that there is only a variation in the second argument. Thus the set Y_b contains one perturbed copy of a well-poised set Y_1 and 30 perturbed copies of a badly poised set Y_2 (note that z is again a random perturbation vector). The set Y_b is illustrated on the right-hand side of Figure 4.1. The Λ-poisedness constant for this Y_b is 3.046, which is smaller than the Λ-poisedness constant of Y_1, as expected, but is comparable to it, as opposed to being about 30 times smaller as in the first case. It is clear that the Λ-poisedness constant does not decrease significantly with p when badly poised subsets are used. Let us now show that the set Y_b is not strongly Λ-poised for Λ close to the constant of the poisedness of Y_1. To avoid computing $\Lambda_{\ell,2}$, we use the fact that the ℓ_2-norm of any vector is greater than or equal to its ℓ_∞-norm. It follows that $\Lambda_{\ell,2} \geq \Lambda_\ell$, and hence it is clear that Y_b is not strongly Λ-poised for any $\Lambda \le \frac{13.64}{6} 3.046 = 6.925$. Since $\Lambda_{\ell,2}$ could actually be up to $\sqrt{p_1}$ times larger than Λ_ℓ, the strong Λ-poisedness constant of the set Y_b may be as high as $\Lambda = 13.64 \times 6.925 \simeq 94$.

These two examples serve to illustrate the point that strong poisedness reflects how well the sample points are spread in space to form poised subsets. From the statistical point of view it is also clear that the strongly poised set Y of the first case is a better sample set to use in order to obtain an accurate model of a noisy function, since the second sample set will result in a model which puts strong bias on the variation of the second variable x_2.

4.4   Condition number as a measure of well poisedness

We recall that even though the condition number of the matrix $M(\phi, Y)$ is not a reliable measure of poisedness for arbitrary ϕ and Y, we have seen, in Chapter 3, that for the choice of $\phi = \bar{\phi}$ (the natural basis) and for the set $\hat{Y}$ (shifted and scaled from Y in such a way that the maximum distance between y^0 and any other points in Y is equal to 1) the Λ-poisedness constant and $\|M(\bar{\phi}, \hat{Y})^{-1}\|$ (and hence the condition number of $M(\bar{\phi}, \hat{Y})$) differ only by a constant factor.

In the regression case such a fact can be of particular importance since it supports the consistency of least-squares regression for noisy problems. Recall that for Theorem 4.1 to hold we needed condition (4.1) to be true. This means that for a sequence of sample sets Y^p, with $p \to +\infty$, the smallest eigenvalue of the matrix $M(\phi, Y^p)^\top M(\phi, Y^p)$ has to increase with a rate of at least p. Let us try to build some intuition behind this condition. As in the previous section, let us assume that the number of points p_1 equals lq_1, where l is an integer. Let us partition the set Y^p into l subsets, say $Y_1^p, Y_2^p, \ldots, Y_l^p$. Then the matrix $M(\phi, Y^p)^\top M(\phi, Y^p)$ can be written as $\sum_{i=1}^l M(\phi, Y_i^p)^\top M(\phi, Y_i^p)$. The smallest eigenvalue of $M(\phi, Y^p)^\top M(\phi, Y^p)$ is greater than or equal to the sum of the smallest eigenvalues of $M(\phi, Y_i^p)^\top M(\phi, Y_i^p)$, $i = 1, \ldots, l$. Hence, if the smallest eigenvalue of each of the $M(\phi, Y_i^p)^\top M(\phi, Y_i^p)$ is bounded away from zero by a constant $\sigma^2 > 0$ (or, equivalently, the smallest singular value of $M(\phi, Y^i)$ is bounded from below by $\sigma > 0$), then the smallest eigenvalue of the sum is bounded from below by $l\sigma^2$ (equivalently, the smallest singular value of the whole matrix is bounded from below by $\sqrt{l}\sigma$). From this argument and Theorem 3.14 we have the second part of the following result. The first part has already been shown in the previous section.

Theorem 4.11. *If a set Y^p of p_1 sample points can be partitioned into $l = p_1/q_1$ sets which are Λ-poised for interpolation, then the entire set is strongly Λ-poised for regression. If $Y^p = \hat{Y}^p \subset B(0; 1)$, then the inverse of the smallest singular value of $M(\bar{\phi}, Y^p)$ is bounded by $\sqrt{q_1/p_1}\Lambda$.*

Thus, we see that it is possible to ensure that the smallest eigenvalue of the matrix $M(\bar{\phi}, Y^p)^\top M(\bar{\phi}, Y^p)$ increases with a rate of at least p, which implies condition (4.1) required for the consistency of the least-squares regression model.

To connect strong Λ-poisedness in the regression sense directly with the bound on the condition number of $M(\bar{\phi}, \hat{Y})$ we need an analogue of Theorem 3.14. For this purpose, let us consider the reduced singular value decomposition of $\hat{M} = M(\bar{\phi}, \hat{Y}) = \hat{U} \hat{\Sigma} \hat{V}^\top$ (meaning that $\hat{U}$ is a $p_1 \times q_1$ matrix with orthonormal columns, $\hat{\Sigma}$ is a $q_1 \times q_1$ matrix of nonzero singular values, and $\hat{V}$ is a $q_1 \times q_1$ orthonormal matrix). Let $\hat{\sigma}_1$ (resp., $\hat{\sigma}_{q_1}$) denote the value of the largest (resp., smallest) singular value of $\hat{M}$. Then $\|\hat{M}\| = \|\hat{\Sigma}\| = \hat{\sigma}_1$ and

$\|\hat{\Sigma}^{-1}\| = 1/\hat{\sigma}_{q_1}$. Note that $\text{cond}(\hat{M}) = \hat{\sigma}_1/\hat{\sigma}_{q_1}$. To bound $\text{cond}(\hat{M})$ in terms of Λ it is then sufficient to bound $\|\hat{\Sigma}^{-1}\|$. The proof of the following theorem and related details can be found in [64].

Theorem 4.12. *If $\hat{\Sigma}$ is nonsingular and $\|\hat{\Sigma}^{-1}\| \leq \sqrt{q_1/p_1}\Lambda$, then the set $\hat{Y}$ is strongly Λ-poised in the unit ball $B(0; 1)$ centered at 0. Conversely, if the set $\hat{Y}$ is strongly Λ-poised in the unit ball $B(0; 1)$ centered at 0, then*

$$\|\hat{\Sigma}^{-1}\| \leq \frac{\theta q_1}{\sqrt{p_1}}\Lambda, \tag{4.10}$$

where $\theta > 0$ is dependent on n and d (i.e., on q_1) but independent of $\hat{Y}$ and Λ.

Error bounds in terms of the condition number

Now we turn our attention to the derivation of error bounds for polynomial least-squares regression. As in the polynomial interpolation case, we obtain Taylor-like bounds for linear and quadratic least-squares regression in terms of the poisedness constant Λ and in terms of the norm of $\hat{\Sigma}^{-1}$. The bound in terms of Λ was presented in Section 4.3. Due to the relation between $\hat{\Sigma}^{-1}$ and Λ this bound can also be expressed in terms of $\hat{\Sigma}^{-1}$. However, bounds using $\hat{\Sigma}^{-1}$ can be derived directly as was done in Chapter 3. Here we present the resulting bounds for the quadratic case. For the derivation and other details see [64]. Note that the linear case has already been covered in Section 2.4.

Assumption 4.1. *We assume that $Y = \{y^0, y^1, \ldots, y^p\} \subset \mathbb{R}^n$, with $p_1 = p + 1 > (n+1)(n+2)/2$, is a poised set of sample points (in the quadratic regression sense, $d = 2$) contained in the ball $B(y^0; \Delta(Y))$ of radius $\Delta = \Delta(Y)$. Further, we assume that the function f is twice continuously differentiable in an open domain Ω containing $B(y^0; \Delta)$ and $\nabla^2 f$ is Lipschitz continuous in Ω with constant $\nu_2 > 0$.*

Theorem 4.13. *Let Assumption 4.1 hold. Then, for all points y in $B(y^0; \Delta(Y))$, we have that*

- *the error between the Hessian of the quadratic regression model and the Hessian of the function satisfies*
$$\|\nabla^2 f(y) - \nabla^2 m(y)\| \leq \kappa_{eh}\,\Delta,$$

- *the error between the gradient of the quadratic regression model and the gradient of the function satisfies*
$$\|\nabla f(y) - \nabla m(y)\| \leq \kappa_{eg}\,\Delta^2,$$

- *the error between the quadratic regression model and the function satisfies*
$$|f(y) - m(y)| \leq \kappa_{ef}\,\Delta^3,$$

where κ_{eh}, κ_{eg}, and κ_{ef} are given by

$$
\begin{aligned}
\kappa_{eh} &= 3\sqrt{2}p^{\frac{1}{2}}\nu_2\|\hat{\Sigma}^{-1}\|/2, \\
\kappa_{eg} &= 3(1+\sqrt{2})p^{\frac{1}{2}}\nu_2\|\hat{\Sigma}^{-1}\|/2, \\
\kappa_{ef} &= (6+9\sqrt{2})p^{\frac{1}{2}}\nu_2\|\hat{\Sigma}^{-1}\|/4 + \nu_2/6.
\end{aligned}
$$

Note that from Theorem 4.12 if the set Y is strongly Λ-poised, then $p^{\frac{1}{2}}\|\hat{\Sigma}^{-1}\|$ is uniformly bounded, independently of p_1 and Y.

4.5   Notes and references

We have established in this chapter that the regression models enjoy similar properties to the interpolation models in terms of accuracy of approximation expressed via the bound on the corresponding Lagrange polynomials or on the norm of the inverse or the pseudoinverse of an appropriate matrix. These properties are useful for constructing provable convergent derivative-free algorithms that use least-squares regression to model the objective function.

Let us now turn our attention to other regression approximations which we mentioned in the introduction of this chapter. For instance, in the case of expensive but noisy function evaluations it may be useful to use regularized regression instead of interpolation. This will allow the approximating function to be more "smooth" by somewhat relaxing the interpolation conditions. Specifically, let α be the solution to our interpolation/regression conditions $M(\phi, Y)\alpha = f(Y)$, and let H_α be the resulting Hessian of the model

$$m(x) = \sum_{i=0}^{q} \alpha_i \phi_i(x).$$

When $\phi = \bar{\phi}$ is the natural basis, H_α is the symmetric matrix whose lower triangular part simply contains the elements of α corresponding to the monomials of degree 2. Now let us pick a regularization parameter ρ and consider the following problem:

$$\min_{\alpha} \rho\|H_\alpha\|^2 + \|M(\phi, Y)\alpha - f(Y)\|^2.$$

Assume that we are considering complete quadratic models with $p_1 \geq q_1$. We know that if $\rho \to 0$, then we recover the least-squares regression model which satisfies second-order Taylor-like error bounds for poised sample sets. If $\rho \to +\infty$, then, in this case, we recover linear least-squares regression, which also satisfies first-order Taylor-like error bounds. To preserve the second-order approximation we need to limit the size of ρ. In fact, it is possible to derive from the Taylor-like error bounds for least-squares regression models that, for small enough values of ρ, the solutions to the regularized regression problems also satisfy Taylor-like error bounds of second order. Here is a rough outline of the argument.

Let $m(x)$ be the regularized regression model and $m^{\ell s}(x)$ be the least-squares regression model. Let us consider a ball B of radius Δ and a strongly well-poised sample set $Y \subset B$. Let $m^{\ell s}(x) = (\alpha^{\ell s})^\top \phi(x)$ and $m(x) = \alpha^\top \phi(x)$. From the bound on the error of the least-squares regression, we know that for appropriately chosen Y

$$\|m^{\ell s}(Y) - f(Y)\| = \|M(\phi, Y)\alpha^{\ell s} - f(Y)\| = \mathcal{O}(\Delta^3).$$

This, in turn, means that by choosing ρ small enough (and if $\|[M(\phi, Y)^\top M(\phi, Y)]^{-1}\|$ is bounded) we can make the regularized regression model satisfy

$$\|M(\phi, Y)\alpha - f(Y)\| = \mathcal{O}(\Delta^3).$$

This implies that

$$\|M(\phi, Y)\alpha - M(\phi, Y)\alpha^{\ell s}\| = \mathcal{O}(\Delta^3).$$

If $\|[M(\phi,Y)^\top M(\phi,Y)]^{-1}\|$ is bounded, then we conclude that $\|\alpha - \alpha^{\ell s}\| = O(\Delta^3)$ and therefore $|m^{\ell s}(x) - m(x)| = \mathcal{O}(\Delta^3)$. Since for any $x \in B$ we know that $|m^{\ell s}(x) - f(x)| = \mathcal{O}(\Delta^3)$, we have $|m(x) - f(x)| = \mathcal{O}(\Delta^3)$; in other words, we have shown that if the least-squares regression model is accurate, then by keeping ρ sufficiently small we can ensure that the regularized regression model is also accurate. We conclude that the results that we discuss in this chapter are useful in applications to a variety of approximating functions, in addition to simple least-squares regression.

We end this chapter by proving the simple lemma which we used earlier to derive Theorem 4.8.

Lemma 4.14. *Consider a poised set* $Z = \{z^1, \dots, z^m\} \subset \mathbb{R}^n$, *with* $m > n$. *Let* $I \subset \{1, \dots, m\}$ *be a subset of indices with* $|I| = n$. *It is possible to choose* I *so that, for any* $x \in \mathbb{R}^n$ *such that*

$$x = \sum_{i=1}^m \lambda_i z^i, \quad |\lambda_i| \le \Lambda,$$

for some $\Lambda > 0$, *we can write*

$$x = \sum_{i \in I} \gamma_i z^i, \quad |\gamma_i| \le (m - n + 1)\Lambda.$$

Proof. Consider an $n \times n$ matrix A whose columns are the vectors z^i, $i \in I$. Among all possible sets I, choose the one that corresponds to the matrix A with the largest absolute value of the determinant. We will show that this I satisfies the statement of the lemma.

Let $\bar{I} = \{1, \dots, m\} \setminus I$, and let $Z_{\bar{I}}$ be the subset of Z containing points whose indices are in $\bar{I}$. First, we will show that for any z^j, $j \in \bar{I}$,

$$z^j = \sum_{i \in I} \alpha_i^j z^i, \quad |\alpha_i^j| \le 1.$$

By Cramer's rule $\alpha_i^j = \det(A_{z^j,i})/\det(A)$, where $A_{z^j,i}$ corresponds to the matrix A with its ith column replaced by the vector z^j. Since, by the selection of I, $|\det(A)| \ge |\det(A_{z^j,i})|$ for any $j \in \bar{I}$, we have $|\alpha_i^j| \le 1$.

Now consider any x such that

$$x = \sum_{i=1}^m \lambda_i z^i, \quad |\lambda_i| \le \Lambda.$$

We have

$$x = \sum_{i \in I} \lambda_i z^i + \sum_{j \in \bar{I}} \lambda_j \left(\sum_{i \in I} \alpha_i^j z^i \right) = \sum_{i \in I} \gamma_i z^i, \quad |\gamma_i| \le (m - n + 1)\Lambda, \, i \in I. \qquad \square$$

4.6   Exercises

1. Prove Theorem 4.1 from

$$\alpha^p - \alpha^p_{smooth} = \left[\frac{1}{p+1} M(\phi, Y^p)^\top M(\phi, Y^p) \right]^{-1} \frac{1}{p+1} M(\phi, Y^p)^\top \varepsilon^p,$$

   where ε^p is the appropriate vector of independent random variables drawn from the same distribution with mean 0.

2. Prove that if $M(\phi, Y)$ has full column rank for some basis ϕ, then it has full column rank for any basis of $\mathcal{P}_n^d$.

3. Show that $\lambda(x)$ in (4.7) does not depend on the choice of ϕ in $\mathcal{P}_n^d$.

4. Show that if a set Y with p_1 points $(p_1 > q_1)$ contains a subset of q_1 points that is Λ-poised, then it is $\sqrt{q_1}\Lambda$-poised.

5. Prove Lemma 4.9.

6. Prove Theorem 4.13.

Chapter 5

Underdetermined interpolating models

We will now consider the case when the number of interpolation points in Y is smaller than the number of elements in the polynomial basis ϕ ($p < q$, or $p_1 < q_1$ with $p_1 = p + 1$ and $q_1 = q + 1$). In this case, the matrix $M(\phi, Y)$ defining the interpolating conditions (see (4.3)) has more columns than rows and the interpolation polynomials defined by

$$m(y^i) = \sum_{k=0}^{q} \alpha_k \phi_k(y^i) = f(y^i), \quad i = 0, \ldots, p, \tag{5.1}$$

are no longer unique.

This situation is extremely frequent in model-based derivative-free optimization methods. Most of the derivative-free applications are based on objective functions that are costly to compute. The cost varies from application to application, and it is more or less relevant depending on the dimension of the problem. For instance, for a problem with 3 variables it takes 10 function evaluations to construct a determined quadratic polynomial interpolation model. If each such evaluation takes 1 hour, then even 10 evaluations might be too costly to insist upon. But if each evaluation takes 2 seconds, then it appears reasonable to construct accurate quadratic models whenever desired. Moreover, in the presence of noise it may be a good idea to use more than 10 points per model and apply regression, as was discussed in the previous chapter. For a 100 variable problem, however, to construct a determined quadratic polynomial interpolation model, we would need over 5000 points (see Table 1.1). Even with each function evaluation taking 2 seconds, evaluating the sample set would take hours. Of course, an efficient derivative-free method would not try to generate all these points from scratch for each new model, but rather it would try to reuse existing sample points whose values are known. For that to be useful, normally the sample points would have to be reasonably close to the current iterate. Unless the algorithm does not progress for a long time, which is clearly not desirable, there will typically occur situations where the number of useful sample points already evaluated is lower or much lower than the number of elements in the polynomial basis.

A simple way to conserve the cost of building a sample set is to use linear models; however, it is well known that convergence slows down significantly when no curvature is exploited. The answer to this dilemma is to use underdetermined quadratic interpolation models, using fewer points than are needed for a complete quadratic interpolation model but more points than those defining linear interpolation. By incorporating these extra points one hopes to capture some of the curvature of the function and observe an empirical superlinear rate of local convergence. The choice of the model and of the sample set plays a crucial role for such an approach. We will discuss these issues in this chapter.

5.1   The choice of an underdetermined model

First, we need to make sure that the model we are constructing is uniquely defined by the sample set. The simplest approach to restrict (5.1) so that it has a unique solution is to remove from the system the last $q - p$ columns of $M(\phi, Y)$. This causes the last $q - p$ elements of the solution α to be zero. Such an approach approximates some elements of α, while it sets others to zero based solely on the order of the elements in the basis ϕ. Clearly, this approach is not very desirable without any knowledge of, for instance, the sparsity structure of the gradient and the Hessian of the function f. There is also a more fundamental drawback: the first p_1 columns of $M(\phi, Y)$ may be linearly dependent. In this latter case, a natural conclusion would be that our sample points are not poised (in some sense) and we have to change them. However, if we had selected a different subset of p columns of $M(\phi, Y)$, it might have been well poised without changing any points. We will use a notion of *subbasis* of the basis ϕ to mean a subset of p_1 elements of the basis ϕ. Selecting p_1 columns of $M(\phi, Y)$ therefore corresponds to selecting the appropriate subbasis $\tilde{\phi}$. Let us consider the following example.

Let $\phi = \{1, x_1, x_2, \frac{1}{2}x_1^2, x_1x_2, \frac{1}{2}x_2^2\}$, $Y = \{y^0, y^1, y^2, y^3\}$, and let $y^0 = (0,0)$, $y^1 = (0,1)$, $y^2 = (0,-1)$, and $y^3 = (1,0)$. The matrix $M(\phi, Y)$ is given by

$$M(\phi, Y) = \begin{bmatrix} 1 & 0 & 0 & 0 & 0 & 0 \\ 1 & 0 & 1 & 0 & 0 & 0.5 \\ 1 & 0 & -1 & 0 & 0 & 0.5 \\ 1 & 1 & 0 & 0.5 & 0 & 0 \end{bmatrix}.$$

If we select the first four columns of $M(\phi, Y)$, then the system is still not well defined, since the matrix is singular. Hence, the set Y is not poised with respect to the subbasis $\tilde{\phi} = \{1, x_1, x_2, \frac{1}{2}x_1^2\}$, and a new set of sample points is needed. Notice that if another subbasis was selected, for instance, $\tilde{\phi} = \{1, x_1, x_2, \frac{1}{2}x_2^2\}$, then the set Y is well poised, the matrix consisting of the first, the second, the third, and the sixth columns of $M(\phi, Y)$ is well conditioned, and a unique solution exists. If the Hessian of f happens to look like

$$\begin{bmatrix} 0 & 0 \\ 0 & \frac{\partial^2 f}{\partial x_2^2}(x) \end{bmatrix},$$

then this reduced system actually produces the complete quadratic model of f.

If the sparsity structure of the derivatives of f is known in advance, then this advantage can be exploited by deleting appropriate columns from the system (5.1). A more sophisticated version of this idea is exploited in [56] for group partial separable functions and works well in practice when there is a known structure of the Hessian and gradient elements. If no such structure is known, then there is no reason to select one set of columns over another except for *geometry considerations*. Hence, it makes sense to select those columns that produce the best geometry. Given the sample set Y, this could be achieved by selecting the subbasis $\tilde{\phi}$ so that the poisedness constant Λ is minimized. The following example shows the possible disadvantages of this approach in computing a polynomial model.

Let us consider the purely linear case in $\mathbb{R}^3$ for simplicity. An example for a quadratic case can be constructed in a similar manner. Consider $\phi = \{1, x_1, x_2, x_3\}$ and $Y = \{y^0, y^1, y^2\}$, where, as usual, $y^0 = (0,0,0)$, and where $y^1 = (1,0,0)$, and $y^2 = (0,1,1-\epsilon)$. Assume $f(Y) = (0, b_1, b_2)$. The system (5.1) then becomes

$$
\begin{bmatrix}
1 & 0 & 0 & 0 \\
1 & 1 & 0 & 0 \\
1 & 0 & 1 & 1-\epsilon
\end{bmatrix}
\alpha =
\begin{bmatrix}
0 \\
b_1 \\
b_2
\end{bmatrix}.
$$

The best subbasis for Y is then $\tilde{\phi} = \{1, x_1, x_2\}$. If we select the appropriate columns of $M(\phi, Y)$ and solve the reduced system, we obtain the following solution for the coefficients of $m(x)$:

$$
\alpha =
\begin{bmatrix}
0 \\
b_1 \\
b_2 \\
0
\end{bmatrix}.
$$

Now, if we consider $y^2 = (0, 1-\epsilon, 1)$, then the best subbasis is $\tilde{\phi} = \{1, x_1, x_3\}$ and the solution that we will find with this approach is

$$
\alpha =
\begin{bmatrix}
0 \\
b_1 \\
0 \\
b_2
\end{bmatrix}.
$$

Notice that the two possible solutions are very different from each other, yet as ϵ goes to zero the two sets of sample points converge pointwise to each other. Hence, we see that the subbasis approach suffers from a lack of robustness with respect to small perturbations in the sample set. We also notice that in the first (second) case the fourth (third) element of the coefficient vector is set to zero and the third (fourth) element is set to b_2 (b_2). Hence, each solution is biased towards one of the basis components (x_2 or x_3) without using any actual information about the structure of f. A more suitable approach would be to treat all such components equally in some sense. This can be achieved by the minimum-norm solution of (5.1).

For this example, the minimum-norm solution in the first case is

$$
\alpha^{mn} = M(\phi, Y)^\top [M(\phi, Y) M(\phi, Y)^\top]^{-1} f(Y) =
\begin{bmatrix}
0 \\
b_1 \\
\frac{b_2}{2-2\epsilon+\epsilon^2} \\
\frac{(1-\epsilon)b_2}{2-2\epsilon+\epsilon^2}
\end{bmatrix}
$$

and in the second case is

$$\alpha^{mn} = \begin{bmatrix} 0 \\ b_1 \\ \frac{(1-\epsilon)b_2}{2-2\epsilon+\epsilon^2} \\ \frac{b_2}{2-2\epsilon+\epsilon^2} \end{bmatrix}.$$

These two solutions converge to $(0, b_1, b_2/2, b_2/2)$ as ϵ converges to zero. Hence, not only is the minimum-norm solution robust with respect to small perturbations of the data, but it also "evens out" the elements of the gradient over the x_2 and x_3 basis components.

For the reasons described above it is beneficial to consider the minimum-norm solution of the system (5.1). The minimum-norm solution is expressed as

$$M(\phi, Y)^\top [M(\phi, Y) M(\phi, Y)^\top]^{-1} f(Y)$$

and can be computed via the QR factorization or the reduced singular value decomposition of $M(\phi, Y)$. It is well known that a minimum-norm solution of an underdetermined system of linear equations is not invariant under linear transformations. In our case, this fact means that the minimum-norm solution depends on the choice of ϕ. It is easy to show that the resulting interpolation polynomial also depends on the choice of ϕ in the system (5.1).

Hence, depending on the choice of ϕ, we can obtain a better or a worse approximation to f by computing the minimum-norm interpolating polynomials. Ideally, we would like, for each set Y, to identify the "best" basis ϕ, which would generate the "best" minimum-norm interpolating polynomial. However, it is a nontrivial task to define such a basis. First of all, one needs to define the best interpolating polynomial. The natural choice is the polynomial that has the smallest approximation error with respect to the function f. But the definition of the best basis (and hence of the best polynomial) should depend only on Y. Moreover, the choice of the best subbasis model clearly depends on the initial choice of the basis as well. We are not aware of any work that answers these questions; hence, we do what seems most natural—we consider the minimum-norm underdetermined interpolant for the natural basis $\bar{\phi}$. As we will observe later, this turns out to be a reasonable choice.

5.2 Lagrange polynomials and Λ-poisedness for underdetermined interpolation

We will consider the natural basis $\bar{\phi}$ defined by (3.1) and the corresponding matrix $M(\bar{\phi}, Y)$. We will start by introducing a particular definition of the set of Lagrange polynomials for underdetermined interpolation.

Definition 5.1. *Given a set of interpolation points* $Y = \{y^0, y^1, \ldots, y^p\}$, *with* $p < q$, *where* $q_1 = q + 1$ *is the dimension of* $\mathcal{P}_n^d$, *a set of* $p_1 = p + 1$ *polynomials* $\ell_j(x) = \sum_{i=0}^q (\alpha_j)_i \bar{\phi}_i(x)$, $j = 0, \ldots, p$, *is called a set of minimum-norm Lagrange polynomials for the basis* $\bar{\phi}$ *if it is a minimum-norm solution of*

$$\ell_j(y^i) \overset{\text{m.n.}}{=} \delta_{ij} = \begin{cases} 1 & \text{if } i = j, \\ 0 & \text{if } i \neq j. \end{cases}$$

The minimum-norm Lagrange polynomials are thus given by the minimum-norm solution of

$$M(\bar{\phi},Y)\alpha_j \overset{m.n.}{=} e_{j+1}, \quad j = 0,\ldots,p,$$

where e_{j+1} is the $(j+1)$st column of the identity matrix of order $q+1$. This set of polynomials is an extension of the traditional Lagrange polynomials to the case when $p < q$. Clearly, these polynomials no longer compose a basis, since there are not enough of them. However, as in the regression case, many other properties of Lagrange interpolation polynomials are preserved.

The set of minimum-norm Lagrange polynomials exists and is unique if the matrix $M(\bar{\phi},Y)$ has full row rank. In this case, we will say that Y is poised. We again note that here the Lagrange polynomials generally depend on the choice of the basis ϕ, so the level of well poisedness does also, but it is easy to see that the poisedness of Y does not since the rank of $M(\bar{\phi},Y)$ does not depend on the choice of basis.

Just as in the case of standard Lagrange polynomials, the minimum-norm interpolating polynomial $m(x)$ in the underdetermined case has a simple representation in terms of the minimum-norm Lagrange polynomials.

Lemma 5.2. *For any function $f : \mathbb{R}^n \to \mathbb{R}$ and any poised set $Y = \{y^0, y^1, \ldots, y^p\} \subset \mathbb{R}^n$, the minimum-norm interpolating polynomial $m(x)$ (in terms of the basis $\bar{\phi}$) that interpolates $f(x)$ on Y can be expressed as*

$$m(x) = \sum_{i=0}^{p} f(y^i)\ell_i(x),$$

where $\{\ell_i(x), i = 0,\ldots,p\}$ is the set of minimum-norm Lagrange polynomials for Y.

We will now show that, as in the case of polynomial interpolation and regression, the alternative interpretations of the Lagrange polynomials can be easily derived. Thus, we will also have an analogous definition of Λ-poisedness.

Given a poised set $Y = \{y^0, y^1, \ldots, y^p\} \subset B \subset \mathbb{R}^n$ and $x \in B$, we attempt to express the vector $\bar{\phi}(x)$ in terms of the vectors $\bar{\phi}(y^i)$, $i = 0,\ldots,p$. Since the dimension of the vector $\bar{\phi}(x)$ is $q_1 > p_1$, it may no longer be possible to express it as a linear combination of the p_1 vectors $\bar{\phi}(y^i)$, $i = 0,\ldots,p$. Instead, we will be looking for the least-squares solution to the following system:

$$\sum_{i=0}^{p} \lambda_i(x)\bar{\phi}(y^i) \overset{\ell.s.}{=} \bar{\phi}(x). \tag{5.2}$$

We observe a duality in that in system (4.7) the minimum ℓ_2-norm solution $\lambda(x)$ to the system corresponded to the least-squares regression Lagrange polynomials, while in the current case the least-squares solution $\lambda(x)$ corresponds to the minimum ℓ_2-norm Lagrange polynomials.

To extend the definition given in (3.6) to the underdetermined case we use the same approach as in the full interpolation case. Given a set Y and a point x, we again define a set $Y_i(x) = Y \setminus \{y^i\} \cup \{x\}$, $i = 0,\ldots,p$. From Cramer's rule and from the definition of $\lambda_i(x)$ given in (5.2), we obtain that

$$\lambda_i(x) = \frac{\det(M(\bar{\phi},Y)M(\bar{\phi},Y_i(x)^\top))}{\det(M(\bar{\phi},Y)M(\bar{\phi},Y)^\top)}. \tag{5.3}$$

Let us denote $M(\bar{\phi}, Y)$ by $\bar{M}$ and consider the set $\phi_{\bar{M}}(Y) = \{\bar{M}\phi(y^i), i = 0, \ldots, p\}$ in $\mathbb{R}^{p_1}$. The ith element of this set is a vector $\phi(y^i) \in \mathbb{R}^{q_1}$ "projected" onto a p_1-dimensional space by computing its inner product with the rows of $\bar{M}$. Let $\text{vol}(\phi_{\bar{M}}(Y))$ be the volume of the simplex of vertices in $\phi_{\bar{M}}(Y)$, given by

$$\text{vol}(\phi_{\bar{M}}(Y)) = \frac{|\det(\bar{M}M(\bar{\phi}, Y)^{\top})|}{p_1!}.$$

(Such a simplex is the p_1-dimensional convex hull of $\phi_{\bar{M}}(Y)$.) Then

$$|\lambda_i(x)| = \frac{\text{vol}(\phi_{\bar{M}}(Y_i(x)))}{\text{vol}(\phi_{\bar{M}}(Y))}. \tag{5.4}$$

In other words, the absolute value of the ith Lagrange polynomial at a given point x is the change in the volume of (the p_1-dimensional convex hull of) $\phi_{\bar{M}}(Y)$ when y^i is replaced by x.

The following definition of well poisedness is analogous to Definitions 3.6 and 4.7.

Definition 5.3. *Let $\Lambda > 0$ and a set $B \in \mathbb{R}^n$ be given. Let $\bar{\phi}$ be the natural basis of monomials of $\mathcal{P}_n^d$. A poised set $Y = \{y^0, y^1, \ldots, y^p\}$, with $p < q$, where $q_1 = q + 1$ is the dimension of $\mathcal{P}_n^d$, is said to be Λ-poised in B (in the minimum-norm sense) if and only if*

1. *for the set of minimum-norm Lagrange polynomials associated with Y*

$$\Lambda \geq \max_{0 \leq i \leq p} \max_{x \in B} |\ell_i(x)|,$$

or, equivalently,

2. *for any $x \in B$ there exists $\lambda(x) \in \mathbb{R}^{p_1}$ such that*

$$\sum_{i=0}^{p} \lambda_i(x)\bar{\phi}(y^i) \stackrel{\ell.s.}{=} \bar{\phi}(x) \qquad with \qquad \|\lambda(x)\|_{\infty} \leq \Lambda,$$

or, equivalently,

3. *replacing any point in Y by any $x \in B$ can increase the volume of the set $\{\bar{M}\phi(y^0), \bar{M}\phi(y^1), \ldots, \bar{M}\phi(y^p)\}$ at most by a factor Λ.*

It is proved in [64] that if a set is well poised in the minimum-norm sense, then it is well poised in the best subbasis (interpolation) sense and vice versa.

Derivation of error bounds in terms of the condition number

As far as Taylor-like error bounds are concerned, the underdetermined models do not offer in general any accuracy at all. First of all, the bounds from [53] do not apply since they cover only the cases when the model of a polynomial itself is the polynomial. This is not generally the case for underdetermined interpolation, since the polynomial is not usually defined uniquely by the interpolation conditions.

One can prove for minimum-norm models in the underdetermined case that a projection of the errors onto an appropriate linear subspace obeys similar error bounds as for interpolation or regression models, but such a result is clearly insufficient (see [64]).

As we will see in Theorem 5.4 below, when the sample set is linearly poised, underdetermined quadratic interpolation can offer an accuracy similar to linear interpolation or regression. However, these error bounds depend not only on the linear poisedness constant but also on the poisedness constant for the whole interpolation set. In fact, in Theorem 5.4, we will see that the error bounds depend on the norm of the Hessian of the model. The norm of the Hessian can be bounded, in turn, by the overall poisedness constant (in the quadratic underdetermined sense, or in the minimum Frobenius norm sense as discussed in the next section). It is important to note that the derivation of these error bounds is done for any underdetermined quadratic interpolating model based on a set poised for linear interpolation or regression.

We first present the derivation of these bounds. We begin, as usual, by stating our assumptions.

Assumption 5.1. *We assume that* $Y = \{y^0, y^1, \ldots, y^p\} \subset \mathbb{R}^n$ *is a set of sample points poised in the linear interpolation sense (or in the linear regression sense if $p > n$) contained in the ball $B(y^0; \Delta(Y))$ of radius $\Delta = \Delta(Y)$.*

Further, we assume that the function f is continuously differentiable in an open domain Ω containing $B(y^0; \Delta)$ and ∇f is Lipschitz continuous in Ω with constant $\nu > 0$.

We recall here, from Chapter 2, the definitions of

$$\hat{L} = \frac{1}{\Delta} L = \frac{1}{\Delta} \left[y^1 - y^0 \cdots y^p - y^0 \right]^\top$$

and $\hat{L}^\dagger = (\hat{L}^\top \hat{L})^{-1} \hat{L}^\top$. The error bounds are stated in terms of the conditioning of this scaled version of L.

Theorem 5.4. *Let Assumption 5.1 hold. Then, for all points y in $B(y^0; \Delta(Y))$, we have that*

- *the error between the gradient of a quadratic underdetermined interpolation model and the gradient of the function satisfies*

$$\|\nabla f(y) - \nabla m(y)\| \leq \frac{5\sqrt{p}}{2} \|\hat{L}^\dagger\| (\nu + \|H\|) \Delta,$$

- *the error between a quadratic underdetermined interpolation model and the function satisfies*

$$|f(y) - m(y)| \leq \frac{5\sqrt{p}}{2} \|\hat{L}^\dagger\| (\nu + \|H\|) \Delta^2 + \frac{1}{2} (\nu + \|H\|) \Delta^2,$$

where H is the Hessian of the model.

Proof. Let us write our quadratic underdetermined interpolation model as in (3.13) and denote the errors in function values and in the gradient, $e^f(x)$ and $e^g(x)$, as in (3.14) and (3.15), respectively.

Similarly to what we have done in the proof of Theorem 3.16, we start by subtracting (3.14) from the interpolation conditions and obtain

$$(y^i - y)^\top g + \frac{1}{2}(y^i - y)^\top H(y^i - y) + (y^i - y)^\top Hy$$
$$= f(y^i) - f(y) - e^f(y), \quad i = 0, \ldots, p.$$

Now, instead of using a second-order Taylor expansion as we did in Theorem 3.16, we use a first-order argument and write

$$(y^i - y)^\top e^g(y) + \frac{1}{2}(y^i - y)^\top H(y^i - y)$$
$$= \mathcal{O}(\Delta^2) - e^f(y), \quad i = 0, \ldots, p. \tag{5.5}$$

The next step is to subtract the first of these equations from the others, canceling the term $e^f(y)$ and obtaining

$$(y^i - y^0)^\top e^g(y) + \frac{1}{2}(y^i - y)^\top H(y^i - y) - \frac{1}{2}(y^0 - y)^\top H(y^0 - y)$$
$$= \mathcal{O}(\Delta^2), \quad i = 1, \ldots, p.$$

The bound for the error in the gradient can now be easily derived using Assumption 5.1 and the matrix form of these equalities.

Finally from (5.5) with $y = y^0$ and from the bound on the error in the gradient we get the bound on the error in the function values.   □

5.3   Minimum Frobenius norm models

It is important to note that not any set of $p_1 \geq n + 1$ points that is well poised for minimum norm or best subbasis interpolation is poised for linear interpolation. A simple example is the set $Y = \{(0,0,0),(1,0,0),(0,1,0),(1,1,0)\}$ which is poised for the subbasis $\{1, x_1, x_2, x_1 x_2\}$ but not poised for the subbasis $\{1, x_1, x_2, x_3\}$.

Typically, in a derivative-free optimization framework which uses incomplete interpolation it is desirable to construct accurate linear models and then enhance them with curvature information, hoping that the actual accuracy of the model is better than that of a purely linear model. Hence, it is important to construct sample sets that are poised for linear interpolation or regression. Those sets will give us at least linear accuracy as we have seen from Theorem 5.4. We also know from this result that it is relevant to build models for which the norm of the Hessian is moderate.

Let us consider underdetermined quadratic interpolation ($q = (n + 1)(n + 2)/2$), and let us split the natural basis $\bar{\phi}$ into linear and quadratic parts: $\bar{\phi}_L = \{1, x_1, \ldots, x_n\}$ and $\bar{\phi}_Q = \{\frac{1}{2}x_1^2, x_1 x_2, \ldots, \frac{1}{2}x_n^2\}$. The interpolation model can thus be written as

$$m(x) = \alpha_L^\top \bar{\phi}_L(x) + \alpha_Q^\top \bar{\phi}_Q(x),$$

where α_L and α_Q are the appropriate parts of the coefficient vector α. Let us define the *minimum Frobenius norm* solution α^{mfn} as a solution to the following optimization problem in α_L and α_Q:

$$
\begin{aligned}
\min \quad & \tfrac{1}{2}\|\alpha_Q\|^2 \\
\text{s.t.} \quad & M(\bar{\phi}_L,Y)\alpha_L + M(\bar{\phi}_Q,Y)\alpha_Q = f(Y).
\end{aligned}
\tag{5.6}
$$

We call such a solution a minimum Frobenius norm solution because, due to our specific choice of $\bar{\phi}(x)$ and the separation $\alpha = (\alpha_L, \alpha_Q)$, minimizing the norm of α_Q is equivalent to minimizing the Frobenius norm[6] of the Hessian of $m(x)$.

The condition for the existence and uniqueness of the minimum Frobenius norm model is that the following matrix is nonsingular (see the exercises):

$$
F(\bar{\phi},Y) = \begin{bmatrix} M(\bar{\phi}_Q,Y)M(\bar{\phi}_Q,Y)^\top & M(\bar{\phi}_L,Y) \\ M(\bar{\phi}_L,Y)^\top & 0 \end{bmatrix}.
\tag{5.7}
$$

So, we say that a sample set Y is poised in the minimum Frobenius norm sense if the matrix (5.7) is nonsingular. Poisedness in the minimum Frobenius norm sense implies poisedness in the linear interpolation or regression senses and, as a result, poisedness for quadratic underdetermined interpolation in the minimum-norm sense. Note that the matrix (5.7) is nonsingular if and only if $M(\bar{\phi}_L,Y)$ has full column rank and $M(\bar{\phi}_Q,Y)M(\bar{\phi}_Q,Y)^\top$ is positive definite in the null space of $M(\bar{\phi}_L,Y)^\top$ (and that this latter condition can be guaranteed, for instance, if $M(\bar{\phi}_Q,Y)$ has full row rank). Note also that the Hessian H of the minimum Frobenius norm model can be easily calculated as (see the exercises)

$$
H = \sum_{i=0}^{p} \mu_i \left((y^i)(y^i)^\top - \tfrac{1}{2}D^i \right),
$$

where D^i is the diagonal matrix with diagonal entries $[y_j^i]^2$, $j = 1,\ldots,n$, and μ_i, $i = 0,\ldots,p$, are the Lagrange multipliers associated with the equality constraints in (5.6).

We can define minimum Frobenius norm Lagrange polynomials as follows.

Definition 5.5. *Given a set of interpolation points $Y = \{y^0, y^1, \ldots, y^p\}$, with $n < p < q$, where $q+1$ is the dimension of $\mathcal{P}_n^d$ with $d = 2$, a set of $p_1 = p+1$ polynomials $\ell_i(x) = \sum_{j=0}^{q}(\alpha_i)_j\bar{\phi}_j(x)$, $i = 0,\ldots,p$, is called a set of minimum Frobenius norm Lagrange polynomials for the basis $\bar{\phi}$ if the ith Lagrange polynomial is a solution of (5.6) with the function f equal to the indicator function for y^i, for each $i = 0,\ldots,p$.*

There is also an alternative definition for such minimum Frobenius norm Lagrange polynomials consisting of defining $\lambda(x)$ as the solution of the problem

$$
\begin{aligned}
\min \quad & \|M(\bar{\phi}_Q,Y)^\top\lambda(x) - \bar{\phi}_Q(x)\|^2 \\
\text{s.t.} \quad & M(\bar{\phi}_L,Y)^\top\lambda(x) = \bar{\phi}_L(x).
\end{aligned}
\tag{5.8}
$$

For a set Y poised in the minimum Frobenius norm, $\lambda(x)$ satisfies

$$
F(\bar{\phi},Y)\begin{bmatrix} \lambda(x) \\ \mu(x) \end{bmatrix} = \begin{bmatrix} M(\bar{\phi}_Q,Y)\bar{\phi}_Q(x) \\ \bar{\phi}_L(x) \end{bmatrix}
\tag{5.9}
$$

---

[6]The Frobenius matrix norm is defined for squared matrices by $\|A\|_F^2 = \sum_{1 \le i,j \le n} a_{ij}^2$. The Frobenius norm is the norm defined by the trace matrix inner product $\|A\|_F^2 = \langle A,A \rangle_{\text{tr}} = \text{tr}(A^\top A)$ (recall from linear algebra that the trace of a squared matrix is the sum of its diagonal entries).

for some vector (of multipliers) $\mu(x)$. Each of the elements of the solution vector $\lambda(x)$ for this problem is the corresponding minimum Frobenius norm Lagrange polynomial, i.e., $\lambda(x) = \ell(x)$. It is also true that the minimum Frobenius norm model is expressible via the minimum Frobenius norm Lagrange polynomials:

$$m^{\mathrm{mfn}}(x) = \sum_{i=0}^{p} f(y^i)\ell_i(x). \tag{5.10}$$

The proof of these facts are left as exercises.

One can also extend the definition given in (3.6) to the minimum Frobenius norm case. Given a set Y and a point x, we again consider $Y_i(x) = Y \setminus \{y^i\} \cup \{x\}$, $i = 0, \ldots, p$. From Cramer's rule and from the definition of $\lambda_i(x)$ given in (5.9), we obtain that

$$\lambda_i(x) = \frac{\det(F_{\bar{M}}(\bar{\phi}, Y_i(x)))}{\det(F_{\bar{M}}(\bar{\phi}, Y))},$$

where (using $\bar{M} = M(\bar{\phi}, Y) = [\bar{M}_L \ \bar{M}_Q]$)

$$F_{\bar{M}}(\bar{\phi}, Y_i(x)) = \begin{bmatrix} \bar{M}_Q M(\bar{\phi}_Q, Y_i(x))^\top & \bar{M}_L \\ M(\bar{\phi}_L, Y_i(x))^\top & 0 \end{bmatrix}.$$

Thus, we can also regard the absolute value of the Lagrange polynomial $\lambda_i(x)$ as a ratio of two volumes, in the sense that

$$|\lambda_i(x)| = \frac{\mathrm{vol}(\bar{\phi}_{\bar{M}}(Y_i(x)))}{\mathrm{vol}(\bar{\phi}_{\bar{M}}(Y))}$$

with the volume of Y (in the minimum Frobenius norm sense) defined as

$$\mathrm{vol}(\bar{\phi}_{\bar{M}}(Y)) = \frac{|\det(F_{\bar{M}}(\bar{\phi}, Y))|}{(p_1 + n + 1)!}.$$

We can now state a definition of Λ-poisedness in the spirit of this book.

Definition 5.6. *Let $\Lambda > 0$ and a set $B \in \mathbb{R}^n$ be given. Let $\bar{\phi}$ be the natural basis of monomials of $\mathcal{P}_n^d$ with $d = 2$. A poised set $Y = \{y^0, y^1, \ldots, y^p\}$, with $p < q$, where $q_1 = q + 1$ is the dimension of $\mathcal{P}_n^d$, is said to be Λ-poised in B (in the minimum Frobenius norm sense) if and only if*

1. *for the set of minimum Frobenius norm Lagrange polynomials associated with Y*

$$\Lambda \geq \max_{0 \leq i \leq p} \max_{x \in B} |\ell_i(x)|,$$

or, equivalently,

2. *for any $x \in B$ the solution $\lambda(x) \in \mathbb{R}^{p_1}$ of (5.8) is such that*

$$\|\lambda(x)\|_\infty \leq \Lambda,$$

or, equivalently,

3. *replacing any point in Y by any $x \in B$ can increase the volume* $\mathrm{vol}(\bar{\phi}_{\bar{M}}(Y))$ *at most by a factor* Λ.

This definition of Λ-poisedness in the minimum Frobenius norm sense automatically implies the definition of $\sqrt{p_1}\Lambda$-poisedness for linear regression. One can see this easily since $\lambda_r(x)$ for linear regression is the minimum-norm solution of the constraints in (5.8) and thus $\|\lambda_r(x)\|_\infty \leq \|\lambda_r(x)\| \leq \|\lambda(x)\| \leq \sqrt{p_1}\|\lambda(x)\|_\infty \leq \sqrt{p_1}\Lambda$. In particular, since $\theta \leq 1$ in the linear case (see the discussion after Theorem 3.14), we get $\|M(\bar{\phi}_L, Y)^\dagger\| \leq \sqrt{p_1}\Lambda$, which can then be used in Theorem 5.4.

To establish that the minimum Frobenius norm models based on Λ-poised sets in the minimum Frobenius norm sense yield first-order Taylor-like errors of the type given in Theorem 5.4, it remains to show that the Hessians of the models are bounded.

Theorem 5.7. *Let Assumption 5.1 hold, and assume further that Y is Λ-poised in the minimum Frobenius norm sense. Given an upper bound Δ_{max} on Δ, we have that the Hessian H of the minimum Frobenius norm model satisfies*

$$\|H\| \leq \frac{4p_1\sqrt{q_1}\nu\Lambda}{c(\Delta_{max})},$$

where $c(\Delta_{max}) = \min\{1, 1/\Delta_{max}, 1/\Delta_{max}^2\}$.

Proof. Note that we can assume without loss of generality that $y^0 = 0$. From Lemma 3.10, we know that there exists a constant $\sigma_\infty > 0$, independent of the set of Lagrange polynomials, such that, for any $i \in \{0, \ldots, p\}$, one has $\ell_i(y) = a_i^\top \bar{\phi}(y)$ and (with $x = y/\Delta$ and $a_i(\Delta)$—the vector obtained from a_i by dividing the terms corresponding to linear components in $\bar{\phi}$ by Δ and the terms corresponding to quadratic components in $\bar{\phi}$ by Δ^2)

$$
\begin{aligned}
\max_{y \in B(0;\Delta)} |a_i^\top \bar{\phi}(y)| &= \|a_i(\Delta)\|_\infty \max_{x \in B(0;1)} |(a_i(\Delta)/\|a_i(\Delta)\|_\infty)^\top \bar{\phi}(x)| \\
&\geq \|a_i(\Delta)\|_\infty \sigma_\infty \\
&\geq \|a_i\|_\infty c(\Delta_{max})\Delta^2 \sigma_\infty.
\end{aligned}
$$

The general definition of σ_∞ is given in Lemma 3.10, and we know, from Lemma 3.12, that $\sigma_\infty \geq 1/4$ in the quadratic case. Thus, from the Λ-poisedness assumption we get

$$\|a_i\|_\infty \leq \frac{4\Lambda}{c(\Delta_{max})\Delta^2}.$$

On the other hand, we also know that

$$
\begin{aligned}
\|\nabla^2 \ell_i(y)\| &\leq \|\nabla^2 \ell_i(y)\|_F \leq \sqrt{2}\|a_i^Q\| \leq \sqrt{2}(\|a_i^L\| + \|a_i^Q\|) \\
&\leq 2\|a_i\| \leq 2\sqrt{q_1}\|a_i\|_\infty,
\end{aligned}
$$

where $\ell_i(y) = a_i^\top \bar{\phi}(y) = (a_i^L)^\top \bar{\phi}_L(y) + (a_i^Q)^\top \bar{\phi}_Q(y)$.

Without loss of generality we can also assume that $f(y^0) = 0$ and $\nabla f(y^0) = 0$. Such an assumption requires a subtraction of a linear polynomial from f. If we now subtract the same polynomial from m^{mfn}, this new model will remain the minimum Frobenius norm model of the transformed f, and the Hessian of m^{mfn} remains unchanged. But then we can conclude, expanding about y^0 and using the properties of f stated in Assumption 5.1, that

$$\max_{y \in B(y^0; \Delta)} |f(y)| \leq \frac{\nu}{2} \Delta^2.$$

Finally, from (5.10) and the above inequalities, we obtain that

$$\|H\| = \|\nabla^2 m^{\text{mfn}}(y)\| \leq \sum_{i=0}^{p} |f(y^i)| \|\nabla^2 \ell_i(y)\| \leq \frac{4 p_1 \sqrt{q_1} \nu \Lambda}{c(\Delta_{max})}. \qquad \square \qquad (5.11)$$

Note that the poisedness constant Λ appears twice in the error bounds of Theorem 5.4, once as the bound on the linear regression and once in the bound on the Hessian of the model.

As in Theorem 3.14, we can prove that the condition number of $F(\bar{\phi}, Y)$ and the Λ-poisedness constant are related to each other for poised sets in the minimum Frobenius norm sense. To establish such a result we consider a set $\hat{Y}$, obtained by shifting and scaling Y, so that it is contained tightly in $B(0; 1)$ (meaning that $\hat{Y} \subset B(0; 1)$ and at least one of the points in $\hat{Y}$ lies on the boundary of $B(0; 1)$) and work with $\hat{F} = F(\bar{\phi}, \hat{Y})$.

Theorem 5.8. *If $\hat{F}$ is nonsingular and $\|\hat{F}^{-1}\| \leq \Lambda$, then the set $\hat{Y}$ is $\kappa \Lambda$-poised in the unit ball $B(0; 1)$ centered at 0, where κ depends only on n and p. Conversely, if the set $\hat{Y}$ is Λ-poised in the unit ball $B(0; 1)$ centered at 0, then*

$$\|\hat{F}^{-1}\| \leq \max\{\theta_0, \theta_2(\Lambda), \theta_4(\Lambda)\}, \qquad (5.12)$$

where θ_0, $\theta_2(\Lambda)$, and $\theta_4(\Lambda)$ are polynomials in Λ of degrees 0, 2, and 4, respectively, with nonnegative coefficients dependent only on n and p.

Proof. Let $\psi(x) = (\hat{M}_Q \bar{\phi}_Q(x), \bar{\phi}_L(x))$ and $p_F = p_1 + n + 1$. If $\hat{F}$ is nonsingular and $\|\hat{F}^{-1}\| \leq \Lambda$, then, from (5.9) and $\max_{x \in B(0;1)} \|\bar{\phi}(x)\|_\infty \leq 1$,

$$\|\lambda(x)\|_\infty \leq \|\hat{F}^{-1}\|_\infty \|\psi(x)\|_\infty \leq p_F^{\frac{1}{2}} \|\hat{F}^{-1}\| \|\psi(x)\|_\infty \leq \kappa \Lambda,$$

with

$$\kappa = p_F^{\frac{1}{2}} \left\| \begin{bmatrix} \hat{M}_Q & 0 \\ 0 & I \end{bmatrix} \right\|.$$

To prove the reverse relation let $\bar{v}$ be a normalized eigenvector of $\hat{F}^{-1}$ corresponding to its largest eigenvalue in absolute value. By applying Lemma 3.13 with $A = \hat{F}^{-1}$, $w = \bar{v}$, and $r = \psi(y)$ for some $y \in B(0; 1)$, we obtain

$$\|\hat{F}^{-1} \psi(y)\| \geq |\bar{v}^\top \psi(y)| \|\hat{F}^{-1}\|.$$

Now note that $\bar{v}^\top \psi(y) = v_F^\top \bar{\phi}(y)$ with $v_F = (\hat{M}_Q^\top \bar{v}_\lambda, \bar{v}_\mu)$ and $\bar{v} = (\bar{v}_\lambda, \bar{v}_\mu)$. Thus,

$$\|\hat{F}^{-1} \psi(y)\| \geq |v_F^\top \bar{\phi}(y)| \|\hat{F}^{-1}\|. \qquad (5.13)$$

We know that $\bar{v}$ is also an eigenvector of $\hat{F}$ associated with its smallest eigenvalue e_{min} in absolute value (note that $|e_{min}| = 1/\|\hat{F}^{-1}\|$). Thus, using $\hat{F}\bar{v} = e_{min}\bar{v}$ we can derive

$$\|\hat{M}_Q^\top \bar{v}_\lambda\|^2 + e_{min}\|\bar{v}_\mu\|^2 = e_{min}\|\bar{v}_\lambda\|^2, \tag{5.14}$$

which then implies

$$\|v_F\|^2 = e_{min}\|\bar{v}_\lambda\|^2 + (1 - e_{min})\|\bar{v}_\mu\|^2. \tag{5.15}$$

Now we will prove that either $|e_{min}| \geq 1/2$ (in which case $\|\hat{F}^{-1}\| \leq 2 = \theta_0$) or it happens that v_F is such that $\|v_F\|^2 \geq 1/4$ or $\|v_F\|^2 \geq |e_{min}|$. In fact, if $|e_{min}| < 1/2$, then, from (5.15),

$$\|v_F\|^2 \geq e_{min}\|\bar{v}_\lambda\|^2 + |e_{min}|\|\bar{v}_\mu\|^2.$$

Thus, when $e_{min} > 0$, since $\|\bar{v}\| = 1$,

$$\|v_F\|^2 \geq |e_{min}|\left(\|\bar{v}_\lambda\|^2 + \|\bar{v}_\mu\|^2\right) = |e_{min}|.$$

If $e_{min} < 0$, then we know from (5.14) that it must be the case where $\|\bar{v}_\lambda\|^2 \leq \|\bar{v}_\mu\|^2$. Now note that since $\|\bar{v}\| = 1$, either $\|\bar{v}_\mu\|^2 \geq 1/4$ or $\|\bar{v}_\lambda\|^2 \geq 1/4$; in the latter case one also gets $\|\bar{v}_\mu\|^2 \geq \|\bar{v}_\lambda\|^2 \geq 1/4$. Then, as a result of the definition of v_F, we obtain $\|v_F\|^2 \geq 1/4$.

From Lemma 3.10, we can choose $y \in B(0;1)$ so that, for some appropriate $\sigma_2 > 0$,

$$|v_F^\top \bar{\phi}(y)| \geq \sigma_2 \|v_F\|.$$

Thus, either

$$|v_F^\top \bar{\phi}(y)| \geq \frac{\sigma_2}{2},$$

in which case, from (5.13), we obtain $\|\hat{F}^{-1}\psi(y)\| \geq (\sigma_2/2)\|\hat{F}^{-1}\|$, or

$$|v_F^\top \bar{\phi}(y)| \geq \sigma_2|e_{min}|^{\frac{1}{2}} = \sigma_2\frac{1}{\|\hat{F}^{-1}\|^{\frac{1}{2}}},$$

in which case we have $\|\hat{F}^{-1}\psi(y)\| \geq \sigma_2\|\hat{F}^{-1}\|^{\frac{1}{2}}$, again using (5.13). Thus, using (5.9),

$$\left\|\begin{bmatrix} \lambda(y) \\ \mu(y) \end{bmatrix}\right\| = \|\hat{F}^{-1}\psi(y)\| \geq \sigma_2\min\left\{\frac{\|\hat{F}^{-1}\|}{2}, \|\hat{F}^{-1}\|^{\frac{1}{2}}\right\}. \tag{5.16}$$

Finally, we use (5.9) to express $\mu(y)$ in terms of $\lambda(y)$:

$$\hat{M}_L\mu(y) = \hat{M}_Q\bar{\phi}_Q(y) - \hat{M}_Q\hat{M}_Q^\top\lambda(y).$$

As we have seen immediately after Definition 5.6, Λ-poisedness in the minimum Frobenius norm sense implies Λ-poisedness in the linear (interpolation or regression) sense. Thus, using that we consider a unit ball, we can bound $\|\mu(y)\|$ in terms of a quadratic polynomial in Λ:

$$\|\mu(y)\| \leq \sqrt{p_1}\|\hat{M}_Q\|\Lambda + p_1\|\hat{M}_Q\hat{M}_Q^\top\|\Lambda^2.$$

Then, from $\|\lambda(y)\|_\infty \leq \Lambda$, we see that

$$\left\|\begin{bmatrix} \lambda(y) \\ \mu(y) \end{bmatrix}\right\| \leq \|\lambda(y)\| + \|\mu(y)\| \leq \sqrt{p_1}(\|\hat{M}_Q\| + 1)\Lambda + p_1\|\hat{M}_Q\hat{M}_Q^\top\|\Lambda^2.$$

As a result of this bound and using (5.16), we see that either

$$\|\hat{F}^{-1}\| \le \frac{2\sqrt{p_1}}{\sigma_2}(\|\hat{M}_Q\|+1)\Lambda + \frac{2p_1}{\sigma_2}\|\hat{M}_Q\hat{M}_Q^\top\|\Lambda^2 = \theta_2(\Lambda)$$

or

$$\|\hat{F}^{-1}\| \le \left(\frac{\sqrt{p_1}}{\sigma_2}(\|\hat{M}_Q\|+1)\Lambda + \frac{p_1}{\sigma_2}\|\hat{M}_Q\hat{M}_Q^\top\|\Lambda^2\right)^2 = \theta_4(\Lambda),$$

and the proof is concluded.   □

The following corollary of Theorem 5.8 will be useful in Chapter 6. For the sake of clarity, we highlight the dependence of the constant involved on the natural basis $\bar{\phi}$.

Corollary 5.9. *If $\hat{Y}$ is Λ-poised in $B(0;1)$ in the minimum Frobenius norm sense, then $|\det(F(\bar{\phi},\hat{Y}))| \ge \Theta(n,p,\bar{\phi}) > 0$, where $\Theta(n,p,\bar{\phi})$ depends only on n, p, and $\bar{\phi}$.*

Proof. Theorem 5.8 guarantees the existence of a constant $\max\{\theta_0,\theta_2(\Lambda),\theta_4(\Lambda)\}$ dependent only on n, p, and $\bar{\phi}$ such that $\|F(\bar{\phi},\hat{Y})^{-1}\| \le \max\{\theta_0,\theta_2(\Lambda),\theta_4(\Lambda)\}$. Since the absolute value of the determinant of a symmetric matrix is the product of its eigenvalues, we obtain

$$|\det(F(\bar{\phi},\hat{Y})| = \frac{1}{|\det(F(\bar{\phi},\hat{Y})^{-1})|} \ge \frac{1}{(\max\{\theta_0,\theta_2(\Lambda),\theta_4(\Lambda)\})^{p_1+n+1}}.\quad □$$

Least Frobenius norm updating of quadratic models

Powell [191] suggested choosing the solution to the underdetermined interpolation system (5.1) which provides the Hessian model H closest, in the Frobenius norm sense, to a previously calculated Hessian model H^{old}. In the notation of this chapter, such a model is the solution of

$$\begin{aligned}
\min \quad & \tfrac{1}{2}\|\alpha_Q - \alpha_Q^{old}\|^2 \\
\text{s.t.} \quad & M(\bar{\phi}_L,Y)\alpha_L + M(\bar{\phi}_Q,Y)\alpha_Q = f(Y).
\end{aligned} \quad (5.17)$$

Rather than solving (5.17) directly, one can solve a shifted problem on $\alpha_Q^{dif} = \alpha_Q - \alpha_Q^{old}$ and then compute α_Q as $\alpha_Q^{dif} + \alpha_Q^{old}$ (and the same for α_L). The shifted problem is posed as follows:

$$\begin{aligned}
\min \quad & \tfrac{1}{2}\|\alpha_Q^{dif}\|^2 \\
\text{s.t.} \quad & M(\bar{\phi}_L,Y)\alpha_L^{dif} + M(\bar{\phi}_Q,Y)\alpha_Q^{dif} = f^{dif}(Y),
\end{aligned} \quad (5.18)$$

where $f^{dif}(Y) = f(Y) - m^{old}(Y)$. Problem (5.18) is then of the type given in (5.6).

It is possible to show (see [191] or the exercise below) that if f itself is a quadratic function, then

$$\|H - \nabla^2 f\| \le \|H^{old} - \nabla^2 f\|,\quad (5.19)$$

where $\nabla^2 f$ is the constant Hessian of f. This property is an indication of the good behavior of Powell's least Frobenius update (as we will discuss in Chapter 11).

5.4   Notes and references

Another effective idea for constructing a "quasi-Newton" model with guaranteed linear accuracy is to employ NFPs, which we discussed in Chapter 3. First, a subset of $n +$ 1 affinely independent points of Y is selected and the set of linear NFPs is constructed. Then for the remaining points the set of NFPs is underdetermined. As described above for Lagrange polynomials, we can consider the minimum Frobenius norm NFPs for the quadratic block. NFPs can then serve as a measure of poisedness of the sample set, a criterion for selecting sample points, and to construct the interpolation model. Further references on the use of NFPs can be found in [59, 205].

Minimum Frobenius norm models have so far proven to be the most successful second-order models in interpolation-based trust-region methods. They are used in at least two software implementations. In the DFO code (see Chapter 11 and the appendix) they are used to construct a model for any $n + 1 < p < q$ with the smallest Frobenius norm of the Hessian and where the sample set selection is based on the NFPs. As we have mentioned above, the models in [191, 192] are constructed to minimize the Frobenius norm not of the Hessian itself but of the change in the Hessian of the model from one iteration to the next. This clever idea works very successfully with $p_1 = 2n + 1$ and is an attempt to recreate the second-order information in the style of quasi-Newton methods. The sample set selection is based on minimum Frobenius norm Lagrange polynomials. We will discuss both of these approaches in more detail in Chapter 11.

5.5   Exercises

1. Show that the set of minimum-norm Lagrange polynomials (associated with $\bar{\phi}$) exists and is unique if the matrix $M(\bar{\phi}, Y)$ has full row rank.

2. Given a poised set Y, show that the functions $\lambda_i(x), i = 0, \ldots, p$, defined by the least-squares solution of (5.2), form the set of minimum-norm Lagrange polynomials for Y given in Definition 5.1.

3. Show that the minimum Frobenius norm polynomial exists and is unique if and only if the matrix (5.7) is nonsingular. (State the first-order necessary and sufficient conditions for problem (5.6), eliminate α_Q, and write these conditions in terms of the Lagrange multipliers and α_L.)

4. Show that the Hessian H of the minimum Frobenius norm model is

$$ H = \sum_{i=0}^{p} \mu_i \left((y^i)(y^i)^\top - \frac{1}{2} D^i \right), $$

where D^i is the diagonal matrix with diagonal entries $[y_j^i]^2, j = 1, \ldots, n$, and $\mu_i, i = 0, \ldots, p$, are the Lagrange multipliers associated with the equality constraints in (5.6). (Given the previous exercise, it is easy to see that $[\alpha_Q]_1 = \sum_{i=0}^{p} \frac{1}{2} \mu_i [y_1^i]^2, [\alpha_Q]_2 = \sum_{i=0}^{p} \mu_i y_1^i y_2^i, \ldots, [\alpha_Q]_{q-n} = \sum_{i=0}^{p} \frac{1}{2} \mu_i [y_n^i]^2$.)

5. Assume that Y is poised in the minimum Frobenius norm sense. Show that the solution of (5.8) is the (unique) set of minimum Frobenius norm Lagrange polynomials given in Definition 5.5. (State the first-order necessary and sufficient conditions

of (5.6) for the $(i+1)$st indicator function associated with y^i and express it as a function of the ith column of the inverse of (5.7), and do the same for the ith component of $\lambda(x)$.)

6. Assume that Y is poised in the minimum Frobenius norm sense. Using the derivation of the previous exercise, show (5.10).

7. Assume that Y is poised in the minimum Frobenius norm sense. Show that the minimum Frobenius norm Lagrange polynomials given in Definition 5.5 are the Lagrange multipliers associated with the equality constraints of the following problem:

$$\min \quad \tfrac{1}{2}\|\alpha_Q\|^2 + (\bar{\phi}_L(x))^\top \alpha_L$$
$$\text{s.t.} \quad M(\bar{\phi}_L, Y)\alpha_L + M(\bar{\phi}_Q, Y)\alpha_Q = M(\bar{\phi}_Q, Y)\bar{\phi}_Q(x).$$

8. Prove (5.19) when f is quadratic by taking the following steps. First, consider the function

$$h(\theta) = \|(H - H^{old}) + \theta(\nabla^2 f - H)\|$$

and show that it attains its minimizer at $\theta = 0$. Then, write down $h'(0) = 0$ and see that $H - H^{old}$ and $\nabla^2 f - H$ are orthogonal in the trace matrix inner product. Basically you have shown that H is the projection of H^{old} onto the affine set of quadratic functions defined by the interpolating conditions, applied using the Frobenius norm in H.

Chapter 6

Ensuring well poisedness and suitable derivative-free models

In Chapters 3–5 the sample set Y was typically considered given and fixed, and we were exploring the properties of the polynomial bases and models associated with such a set. In contrast, in this chapter we will consider modifying a set Y to improve the quality of the models. From Chapters 3–5 we know that the quality of the model is connected with the poisedness of the sample set. To motivate the material of this chapter we first put together the conditions and error bounds on derivative-free models required by model-based derivative-free algorithms, such as the trust-region methods of Chapter 10. We abstract from the specifics of the model functions and state general requirements for these models. For polynomial interpolation and regression models, in particular, these requirements imply that the sample sets Y have to be Λ-poised with a fixed upper bound on Λ for all sample sets used by a derivative-free algorithm. It turns out that this property does not actually have to hold for *all* models used by a derivative-free algorithm, but it has to be checked and enforced whenever necessary. To enforce this we need to have algorithms which can perform the checking and enforcing and which can be used efficiently within a derivative-free framework (such as, for instance, the one described in Chapter 10).

We will discuss algorithms which construct poised interpolation or regression sets, and those which improve and maintain the poisedness constants of these sets. Recall that the definition of Λ-poisedness requires the maximum absolute value of the Lagrange polynomials associated with the sample set to be no greater than Λ. The algorithms that we discuss here are based on constructing Lagrange or other (similar) polynomial bases and using those as a guide for the modification of the sample set Y.

We will discuss two main algorithmic approaches for updating the set Y. One approach consists of directly controlling the absolute values of the Lagrange polynomials. It is discussed in Section 6.2. The other approach controls Λ-poisedness "indirectly" by applying pivotal algorithms to control the conditioning of the scaled version $M(\bar{\phi}, \hat{Y})$ of the matrix $M(\phi, Y)$ (see Chapters 3–5), which defines the system of Lagrange polynomials. In Section 6.3 we will describe such a pivotal algorithm.

We shall always refer to the sample set as Y in this chapter, and it should be understood that the set is being updated as an algorithm progresses. We will not index different interpolation sets by iteration number or algorithmic step, in order to keep the notation simple. The meaning of each use of the Y notation should be clear from the context. When it is not so, we employ some additional notation. We also would like to mention that in this chapter the order of the points in Y is important, since the points in Y may be considered, accepted, rejected, or replaced by each algorithm in a certain order. Naturally, once the set Y has been fixed, the order of the points has no influence on the interpolation or regression model. We will consider an arbitrary set Y and the set, which we denote by $\hat{Y}$, which is obtained from Y by shifting and scaling it so that the smallest superscribing ball containing $\hat{Y}$ is $B(0;1)$—the ball of radius 1, centered at the origin.

6.1   Fully linear and fully quadratic models

In this section we will abstract from the specifics of the models that we use. We will impose only those requirements on the models that are essential to reproduce the properties of the Taylor models needed to guarantee global convergence of interpolation-based derivative-free methods (such as the trust-region methods of Chapter 10). We will show that polynomial interpolation and regression models, in particular, can satisfy such requirements.

In a derivative-free algorithmic framework (such as the one described in Chapter 10), it is essential to guarantee that whenever necessary a model of the objective function with uniformly good local accuracy can be constructed. In the case of linear approximation, we will say that such a model has to belong to a *fully linear* class. This concept requires the following assumption.

Assumption 6.1. *Suppose that a set S and a radius Δ_{max} are given. Assume that f is continuously differentiable with Lipschitz continuous gradient in an appropriate open domain containing the Δ_{max} neighborhood $\bigcup_{x \in S} B(x; \Delta_{max})$ of the set S.*

In the algorithmic context of model-based methods, one is typically given a point x_0 and the smoothness conditions on the function f are imposed in a level set of the form $S = L(x_0) = \{x \in \mathbb{R}^n : f(x) \le f(x_0)\}$. It is also typical to impose an upper limit Δ_{max} on the size of the balls where the model is considered. The definition of a *fully linear* class of models is given below.

Definition 6.1. *Let a function $f : \mathbb{R}^n \to \mathbb{R}$, that satisfies Assumption 6.1, be given. A set of model functions $\mathcal{M} = \{m : \mathbb{R}^n \to \mathbb{R}, m \in C^1\}$ is called a fully linear class of models if the following hold:*

1. *There exist positive constants κ_{ef}, κ_{eg}, and ν_1^m such that for any $x \in S$ and $\Delta \in (0, \Delta_{max}]$ there exists a model function $m(y)$ in $\mathcal{M}$, with Lipschitz continuous gradient and corresponding Lipschitz constant bounded by ν_1^m, and such that*

   - *the error between the gradient of the model and the gradient of the function satisfies*

$$\|\nabla f(y) - \nabla m(y)\| \le \kappa_{eg}\,\Delta \quad \forall y \in B(x; \Delta), \tag{6.1}$$

   and

   - *the error between the model and the function satisfies*

$$|f(y) - m(y)| \le \kappa_{ef}\,\Delta^2 \quad \forall y \in B(x; \Delta). \tag{6.2}$$

Such a model m is called fully linear on $B(x;\Delta)$.

2. *For this class $\mathcal{M}$ there exists an algorithm, which we will call a "model-improvement" algorithm, that in a finite, uniformly bounded (with respect to x and Δ) number of steps can*

   - *either establish that a given model $m \in \mathcal{M}$ is fully linear on $B(x;\Delta)$ (we will say that a certificate has been provided and the model is certifiably fully linear),*

   - *or find a model $\tilde{m} \in \mathcal{M}$ that is fully linear on $B(x;\Delta)$.*

In the case of quadratic approximation, we will ask the models to belong to a *fully quadratic* class, and for that purpose we require one more degree of smoothness.

Assumption 6.2. *Suppose that a set S and a radius Δ_{max} are given. Assume that f is twice continuously differentiable with Lipschitz continuous Hessian in an appropriate open domain containing the Δ_{max} neighborhood $\bigcup_{x \in S} B(x;\Delta_{max})$ of the set S.*

The fully quadratic class of models defined below has uniformly good local second-order accuracy.

Definition 6.2. *Let a function f, that satisfies Assumption 6.2, be given. A set of model functions $\mathcal{M} = \{m : \mathbb{R}^n \to \mathbb{R}, m \in C^2\}$ is called a fully quadratic class of models if the following hold:*

1. *There exist positive constants κ_{ef}, κ_{eg}, κ_{eh}, and ν_2^m, such that for any $x \in S$ and $\Delta \in (0, \Delta_{max}]$ there exists a model function $m(y)$ in $\mathcal{M}$, with Lipschitz continuous Hessian and corresponding Lipschitz constant bounded by ν_2^m, and such that*

   - *the error between the Hessian of the model and the Hessian of the function satisfies*

   $$\|\nabla^2 f(y) - \nabla^2 m(y)\| \leq \kappa_{eh} \Delta \quad \forall y \in B(x;\Delta), \tag{6.3}$$

   - *the error between the gradient of the model and the gradient of the function satisfies*

   $$\|\nabla f(y) - \nabla m(y)\| \leq \kappa_{eg} \Delta^2 \quad \forall y \in B(x;\Delta), \tag{6.4}$$

   and

   - *the error between the model and the function satisfies*

   $$|f(y) - m(y)| \leq \kappa_{ef} \Delta^3 \quad \forall y \in B(x;\Delta). \tag{6.5}$$

Such a model m is called fully quadratic on $B(x;\Delta)$.

2. *For this class $\mathcal{M}$ there exists an algorithm, which we will call a "model-improvement" algorithm, that in a finite, uniformly bounded (with respect to x and Δ) number of steps can*

   - *either establish that a given model $m \in \mathcal{M}$ is fully quadratic on $B(x;\Delta)$ (we will say that a certificate has been provided and the model is certifiably fully quadratic),*

- *or find a model $\tilde{m} \in \mathcal{M}$ that is fully quadratic on $B(x; \Delta)$.*

Note that the definition of a fully quadratic (resp., fully linear) class does not imply that *all* models contained in that class satisfy (6.3)–(6.5) (resp., (6.1)–(6.2)) for given Δ and x. This has to be the case to make the definition of the class independent of the choice of Δ and x. It is required only that for any specific choice of Δ and x there exists a member of the class which satisfies (6.3)–(6.5) (resp., (6.1)–(6.2)).

As a consequence of the flexibility of Definitions 6.1 and 6.2, some of the model classes that fit in their framework are usually of no interest for a practical algorithm. For instance, consider $\mathcal{M} = \{f\}$—a class consisting of the function f itself. Clearly, by Definition 6.2 such an $\mathcal{M}$ is a fully quadratic class of models, since f is a fully quadratic model of itself for any x and Δ and since the algorithm for verifying that f is fully quadratic is trivial.

Another source of impractical fully linear or fully quadratic classes is the flexibility in the choice of a model-improvement algorithm. The definition in the fully linear case requires the existence of a finite procedure which either certifies that a model is fully linear or produces such a model. For example, Taylor models based on suitably chosen finite-difference gradient evaluations are a fully linear class of models, but a model-improvement algorithm needs to build such models "from scratch" for each new x and Δ. In a derivative-free algorithm with expensive (and often noisy) function evaluations this approach is typically impractical. However, the framework of fully linear models still supports such an approach and guarantees its convergence, provided that all necessary assumptions are satisfied.

The purpose of the abstraction of fully linear and fully quadratic models is, however, to allow for the use of models possibly different from polynomial interpolation and regression, as long as these models fit Definitions 6.1 and 6.2. The abstraction highlights, in our opinion, the fundamental requirements for obtaining the appropriate global convergence results for interpolation-based trust-region algorithms or other sampling-based algorithms where the models are expected to reproduce the local uniform behavior of Taylor models.

The case of polynomial models

One can immediately observe that linear and quadratic, interpolation and regression, polynomial models can be chosen to satisfy the error bounds of Definitions 6.1 and 6.2. In fact, given $\Lambda > 1$, any linear (resp., quadratic) polynomial interpolation model built on a Λ-poised set or a polynomial regression model built on a strongly Λ-poised set is an element of the same fully linear (resp., fully quadratic) class, for which the constants κ_{ef} and κ_{eg} (resp., κ_{ef}, κ_{eg}, and κ_{eh}) depend only on Λ, the cardinality p_1 of the sample set, the dimension q_1 of the polynomial space, and the Lipschitz constant of ∇f (resp., $\nabla^2 f$) in Assumption 6.1 (resp., Assumption 6.2). From the discussion in Section 5.3 we also know that the minimum Frobenius norm models based on a Λ-poised set (in the minimum Frobenius norm sense) are fully linear with κ_{ef} and κ_{eg} depending only on Λ, p_1, and q_1 of the polynomial space and on the Lipschitz constant of ∇f. Note that since the gradient of a linear polynomial model and the Hessian of a quadratic polynomial model are constant, they are trivially Lipschitz continuous for any positive Lipschitz constant.

One can also guarantee under some conditions (see the exercises) that the gradient of a quadratic polynomial model based on a Λ-poised set is Lipschitz continuous—and thus that (interpolation or regression) quadratic polynomial models also satisfy the bounds needed in the definition of a fully linear class.

We will show in the remainder of this chapter how polynomial interpolation and regression models can provide fully linear and fully quadratic model classes. Toward that end we need to address the issue of the existence of finite model-improvement algorithms as required by Definitions 6.1 and 6.2.

6.2 Ensuring well poisedness using Lagrange polynomials

We know from Chapter 3 that, given a set of interpolation points, the maximum of the absolute values of the Lagrange polynomials $\ell(x)$ (see Definition 3.3) in the region of interest defines the Λ-poisedness of the interpolation set (see Definition 3.6) in that region. It is natural, therefore, to directly monitor and control the size of $\ell(x)$ as a way of maintaining well poisedness in interpolation sets. The cases of regression sets (see Chapter 4) and underdetermined interpolation sets (see Chapter 5) are somewhat different; therefore, for the moment, we will focus on complete determined interpolation.

Before we discuss the method of maintaining the Λ-poisedness of the interpolation set via Lagrange polynomials, let us discuss, given an interpolation set, how to compute and update Lagrange polynomials.

Computing and updating Lagrange polynomials

Given an interpolation set $Y = \{y^0, y^1, \ldots, y^p\}$, whose cardinality is equal to the dimension of the polynomial space $\mathcal{P}_n^d$, the following procedure generates the basis of Lagrange polynomials.

Algorithm 6.1 (Computing Lagrange polynomials).

Initialization: Start by choosing an initial approximation to the Lagrange polynomial basis, e.g., given by the monomial basis (3.1), $\ell_i(x) = \bar{\phi}_i(x)$, $i = 0, \ldots, p$.

For $i = 0, \ldots, p$

1. **Point selection:** Find $j_i = \text{argmax}_{i \leq j \leq p} |\ell_i(y^j)|$. If $\ell_i(y^{j_i}) = 0$, then stop (the set Y is not poised). Otherwise, swap points y^i and y^{j_i} in set Y.

2. **Normalization:**
$$\ell_i(x) \leftarrow \ell_i(x)/\ell_i(y^i). \tag{6.6}$$

3. **Orthogonalization:** For $j = 0, \ldots, p, \ j \neq i$,
$$\ell_j(x) \leftarrow \ell_j(x) - \ell_j(y^i)\ell_i(x). \tag{6.7}$$

It is easy to verify that this algorithm indeed generates the basis of Lagrange polynomials. After the first iteration we have that

$$\ell_0(y^0) = 1 \quad \text{and} \quad \ell_j(y^0) = 0, \ j = 1, \ldots, p. \tag{6.8}$$

After the second iteration we get that

$$\ell_1(y^1) = 1 \quad \text{and} \quad \ell_j(y^1) = 0, \; j = 0, \ldots, p, \; j \neq 1.$$

However, this second iteration does not destroy (6.8) since $\ell_1(y^0) \leftarrow \ell_1(y^0)/\ell_1(y^1)$ and $\ell_j(y^0) \leftarrow \ell_j(y^0) - \ell_j(y^1)\ell_1(y^0)$, $j = 0, \ldots, p$, $j \neq 1$. The rest of the proof is left as an exercise.

Each Lagrange polynomial can have up to $p_1 = p + 1$ coefficients; hence evaluating and updating each polynomial in (6.6) and (6.7) takes $\mathcal{O}(p)$ operations. There are $p + 1$ such polynomials, and at each iteration of the algorithm up to $p + 1$ of these polynomials get evaluated and updated. Since the algorithm always makes $p + 1$ steps it requires $\mathcal{O}(p^3)$ operations to compute the set of Lagrange polynomials from scratch.

Now assume that the basis of Lagrange polynomials is already at hand. The question is how will it change if one of the points, say y^k, in the interpolation set $Y = \{y^0, y^1, \ldots, y^p\}$ is replaced by a new one, say y_*^k. The new interpolation set becomes

$$Y^* = Y \cup \{y_*^k\} \setminus \{y^k\}.$$

Denote by $\ell_j^*(x)$, $j = 0, \ldots, p$, the new Lagrange polynomials corresponding to Y^*. If $\ell_k(y_*^k) \neq 0$, then we simply perform the normalization and orthogonalization steps of the above algorithm for $i = k$. In particular,

$$\ell_k^*(x) = \ell_k(x)/\ell_k(y_*^k), \tag{6.9}$$

$$\ell_j^*(x) = \ell_j(x) - \ell_j(y_*^k)\ell_k^*(x) \quad \forall j \in \{0, \ldots, p\}, \; j \neq k. \tag{6.10}$$

If it happens that $\ell_k(y_*^k) = 0$, then we have a nontrivial polynomial that vanished at all points of Y^*, and hence the resulting set Y^* is not poised and the above steps are not valid.

It is easy to see now from (6.9) and (6.10) that the effort of updating the Lagrange polynomials is $\mathcal{O}(p^2)$, since it is simply the same as performing one iteration of Algorithm 6.1. In the case of linear interpolation, $p = \mathcal{O}(n)$ and the cost of computing the whole set of Lagrange polynomials is not very high. On the other hand, in the case of quadratic interpolation, $p = \mathcal{O}(n^2)$ and the cost of computing the whole set of of Lagrange polynomials is $\mathcal{O}(n^6)$, which can be prohibitive for anything but small n if used often. Thus, restricting the changes of interpolation set to simple one-point updates can imply significant savings with respect to Lagrange polynomial computations. Now we can recall that the interpolation model can be expressed as a simple linear combination of the Lagrange polynomials with known coefficients. Hence, the computation of the interpolation model reduces to computing the set of Lagrange polynomials and, therefore, has the same complexity. In particular, if a new set of Lagrange polynomials can be recomputed in $\mathcal{O}(n^2)$ steps, then so can the new model.

Geometry checking algorithm

In order to make the Λ-poisedness requirement practical we need to have a procedure which presented with an interpolation set Y can either verify that the set is Λ-poised in a given closed ball B or, in case it fails to do so, is able to make improvement steps in such a way that after a finite number of these steps the set Y is guaranteed to become Λ-poised.

There are two cases to consider. The first case is when Y is either not poised or it contains less than p_1 points. This case is detected by the point selection step of Algorithm 6.1 (in practice with the zero test replaced with a suitable tolerance on the absolute value). To improve the set Y in this situation we consider the following modification of Algorithm 6.1. (Later we will consider a second case when the given set Y contains at least p_1 points and it is poised.)

Algorithm 6.2 (Completing the nonpoised set Y via Lagrange polynomials).

Initialization: Start by choosing an initial approximation to the Lagrange polynomial basis, e.g., given by the monomial basis (3.1), $\ell_i(x) = \bar{\phi}_i(x)$, $i = 0, \ldots, p$. Let $p_{ini} + 1$ be the number of given points in the initial set Y.

For $i = 0, \ldots, p$

1. **Point selection:** Find $j_i = \operatorname{argmax}_{i \leq j \leq p_{ini}} |\ell_i(y^j)|$. If $|\ell_i(y^{j_i})| > 0$ and $i \leq p_{ini}$, then swap points y^i and y^{j_i} in set Y. Otherwise, compute (or recompute if $i \leq p_{ini}$) y^i as

$$y^i \in \operatorname*{argmax}_{x \in B} |\ell_i(x)|.$$

2. **Normalization:**

$$\ell_i(x) \leftarrow \ell_i(x)/\ell_i(y^i).$$

3. **Orthogonalization:** For $j = 0, \ldots, p$, $j \neq i$,

$$\ell_j(x) \leftarrow \ell_j(x) - \ell_j(y^i)\ell_i(x).$$

This algorithm will first include those points in Y which make it a poised set. Once it runs out of such points, it will discard any remaining points (those which make the set nonpoised). Then it will generate a point at a time, by maximizing the next suitable Lagrange polynomial until Y contains p_1 points. The complexity of this algorithm depends on the complexity of performing global optimization of Lagrange polynomials in B.

Now let us focus on the case where Y is poised and has exactly p_1 points (it can have more, but we will simply ignore the unused points). We want an algorithm that will check if Y is Λ-poised, for a given $\Lambda > 1$, and, if it is not, then it will make it so. To check whether Y is Λ-poised, we can compute the maximum absolute value of the Lagrange polynomials on B. If such a maximum absolute value is below a given Λ, then Y is Λ-poised. If it is not below Λ, then the interpolation point corresponding to the Lagrange polynomial with the largest maximum absolute value is replaced by a point that maximizes this Lagrange polynomial on the ball.

Algorithm 6.3 (Improving well poisedness via Lagrange polynomials).

Initialization: Choose some constant $\Lambda > 1$. Assume that a poised interpolation set Y is given with cardinality p_1 (if not, apply Algorithm 6.2 to generate such a set). If not already available, compute the Lagrange polynomials $\ell_i(x)$, $i = 0, \ldots, p$, associated with $Y(= Y_0)$.

For $k = 1, 2, \ldots$

1. Estimate

$$\Lambda_{k-1} = \max_{0 \leq i \leq p} \max_{x \in B} |\ell_i(x)|.$$

Λ_{k-1} can be computed exactly by maximizing the absolute value of all Lagrange polynomials on B or estimated via upper and lower bounds. Ensure that Λ_{k-1} is known with enough accuracy to guarantee either that $\Lambda_{k-1} > \Lambda$ or $\Lambda_{k-1} \leq \Lambda$.

2. If $\Lambda_{k-1} > \Lambda$, then let $i_k \in \{0, \ldots, p\}$ be an index for which

$$\max_{x \in B} |\ell_{i_k}(x)| > \Lambda,$$

and let $y_*^{i_k} \in B$ be a point that maximizes $|\ell_{i_k}(x)|$ in B (approximately or exactly depending on the optimization performed in Step 1).

Update $Y(= Y_k)$ by performing the point exchange

$$Y \leftarrow Y \cup \{y_*^{i_k}\} \setminus \{y^{i_k}\}.$$

Otherwise (i.e., $\Lambda_{k-1} \leq \Lambda$), Y is Λ-poised and stop.

3. Update all Lagrange polynomial coefficients.

Assuming that the global maximization of the quadratic polynomials on B can be performed (for instance, as in [172]), let us show that for any $\Lambda > 1$ the above algorithm will stop after a finite, uniformly bounded number of steps.

First of all, let us recall the properties of Lagrange polynomials with respect to shifting and scaling. Hence, given a set Y in a ball B we observe that Algorithm 6.3 performs identically to the case when it is applied to the shifted and scaled version of Y which lies tightly in $B(0; 1)$ (meaning that it is a subset of $B(0; 1)$ and at least one of its points lies on the boundary of $B(0; 1)$). By identical behavior we mean that the algorithm will replace the same sequence of points and that the resulting new sample sets will be identical up to shifting and scaling.

Theorem 6.3. *For any given $\Lambda > 1$, a closed ball B, and a fixed polynomial basis ϕ, Algorithm 6.3 terminates with a Λ-poised set Y after at most $N = N(\Lambda, \phi)$ iterations, where N is a constant which depends on Λ and ϕ.*

Proof. We divide the proof of the theorem into three parts.

Part 1. Recall that $\ell(x) = \lambda(x)$, (3.6) for Lagrange polynomials, and the definition of $\phi(Y) = M(\phi, Y)$ for some fixed polynomial basis ϕ. Equation (3.6) states that $|\ell_{i_k}(y_*^{i_k})|$ is equal to the ratio of the volume of the simplex of vertices in $\phi(Y^*)$, where $Y^* = Y \cup \{y_*^{i_k}\} \setminus \{y^{i_k}\}$, to the volume of the simplex of vertices in $\phi(Y)$:

$$|\ell_{i_k}(y_*^{i_k})| = \frac{\text{vol}(\phi(Y^*))}{\text{vol}(\phi(Y))} > \Lambda > 1.$$

This means that each time a point y^{i_k} is replaced by a new point $y_*^{i_k}$ such that $|\ell_{i_k}(y_*^{i_k})| > \Lambda > 1$, the volume of the appropriate simplex increases by at least a factor of Λ.

All the points are considered within the closed ball B, and hence $\phi(B)$ is a compact set in $\mathbb{R}^{p_1}$. This means that the volume of the simplex defined by $\phi(Y)$ cannot increase infinitely many times by a constant factor $\Lambda > 1$. Let $V_{max} = V(\phi, Y_{max})$ be the maximum volume of all simplices formed by $\phi(Y)$ for $Y \subset B$, and let $V(\phi, Y_0)$ be the volume of the simplex defined by the vertices in $\phi(Y_0)$ for the initial set Y_0. The number of steps of Algorithm 6.3 cannot exceed $\lfloor \log_\Lambda(V_{max}/V(\phi, Y_0)) \rfloor$. Note that this quantity is scale independent.

Part 2. We will now show that, after at most p_1 steps, Algorithm 6.3 computes a set Y such that the simplex volume satisfies $V(\phi, Y) \geq \Theta(p, \phi) > 0$. First, we will show that after at most p_1 iterations the algorithm obtains a set Y which is 2^p-poised.

Assume that one iteration of Algorithm 6.3 requiring an update is performed, the point y^{i_1} is replaced, and the Lagrange polynomials get updated. After the update, since ℓ_{i_1} is divided by its maximum value, we have $\max_{x \in B} |\ell_{i_1}(x)| \leq 1$. Then, on the next step requiring a replacement point, $i_2 \neq i_1$, since we choose a polynomial whose value is larger than 1 on B. After the second step, we have, again, $\max_{x \in B} |\ell_{i_2}(x)| \leq 1$, and also $\max_{x \in B} |\ell_{i_1}(x)| \leq 2$, since $\ell_{i_1}(x) \leftarrow \ell_{i_1}(x) - \ell_{i_1}(y^{i_2})\ell_{i_2}(x)$. At the third step we either have a 2-poised set or we have $i_3 \notin \{i_1, i_2\}$. By the same logic as used above after the third step we have $\max_{x \in B} |\ell_{i_3}(x)| \leq 1$, $\max_{x \in B} |\ell_{i_2}(x)| \leq 2$, and $\max_{x \in B} |\ell_{i_1}(x)| \leq 4$. By extending this argument and applying mathematical induction we easily achieve that after p_1 steps we will have a 2^p-poised set Y.

Part 3. From the remark before this theorem we know that if the algorithm runs for p_1 steps, then it will produce the same set (up to shifting and scaling) as if it were applied to the shifted and scaled $\hat{Y}$ in $B(0; 1)$. This means that we can apply Corollary 3.15, which states that the volume $V(\phi, Y)$ of the simplex of vertices Y obtained after p_1 steps is uniformly bounded away from zero: $V(\phi, \hat{Y}) \geq \Theta(p, \phi) > 0$, where $\Theta(p, \phi)$ depends only on p and ϕ. (Corollary 3.15 is stated for the natural basis $\bar{\phi}$ but holds for any ϕ with an according redefinition of θ.) Hence, $\lfloor \log_\Lambda(V_{max}/V(\phi, Y)) \rfloor$ is uniformly bounded from above by a constant N_1 that depends only on Λ, p, and ϕ. Letting $N = N_1 + p_1$ concludes the proof. $\square$

Note that the first part of the proof, where the number of steps is shown to be finite (but not necessarily uniformly bounded), still holds if we use only the fact that at each step the absolute value of the Lagrange polynomial at the new point is greater than Λ. However, the proof that the number of steps is not only finite but also uniformly bounded required a uniform lower bound on the volume of the initial set Y_0. This was accomplished in the second and third parts of the proof. Note that in the second part of the proof, to show that we can obtain a 2^p-poised set after p_1 steps, we needed the fact that the global maximum absolute value of a Lagrange polynomial on B can be found at each iteration. Such a requirement could be circumvented if one finds a different way to ensure a uniform lower bound on the volume of the initial set Y_0.

We now give an outline on how one can modify Algorithm 6.3 to avoid global optimization of all polynomials at each step. Each improving step of Algorithm 6.3 performs an update of Lagrange polynomials, which we know requires $\mathcal{O}(p^2)$ steps. Hence, the global optimization of all Lagrange polynomials on B at each step can dominate the overall computational effort. Such optimization can be necessary to find out if there is any polynomial whose maximum absolute value exceeds Λ. It is possible to reduce the computational effort by looking at the coefficients of the Lagrange polynomials, rather than at their actual maximum absolute values. This will remove the necessity of optimizing all polynomials at each step.

By knowing B and the polynomial coefficients it is possible to estimate an upper bound on its maximum absolute value. For example, a polynomial in $\mathbb{R}^2$ of degree 2 whose coefficients do not exceed 100 cannot have an absolute value of more that 600 in a ball of radius 1 around the origin, because its value is the sum of six monomials, each of which cannot exceed 100 in absolute value. An upper bound on the value of a polynomial in an arbitrary ball can be derived similarly: from the value of the polynomial at the center of the ball and the radius of the ball, one can estimate the maximum change in the value of the polynomial within the ball. One can use these upper bounds to estimate an upper bound on the Λ-poisedness constant and also to select the Lagrange polynomials that need to be optimized.

To avoid the global optimization of the polynomials whenever possible, we also would need to use specific lower bounds on the maximum absolute values of the Lagrange polynomials. This can be done, for instance, by the enumeration method outlined in the proof of Lemma 6.7 below. The procedure in the proof is outlined for polynomials with at least one of the coefficients having an absolute value of at least 1. Clearly, the same procedure can be applied to any positive lower bound on the maximum absolute value of the coefficients. For instance, in the case of quadratic polynomials, if $b > 0$ is such a bound, then the lower bound on the maximum absolute value of the polynomial itself is guaranteed to be at least $b/4$ (instead of $1/4$). It is important to remember, however, that the application of Theorem 6.3 is not straightforward when the global optimization of the Lagrange polynomials is replaced by procedures to construct upper and lower bounds.

In the examples of the performance of Algorithm 6.3 that we present in this chapter we perform the global optimization of all the Lagrange polynomials and select as i_k the index of the polynomial which has the largest absolute value. In one algorithm presented in Chapter 11 (see Section 11.3), we consider selecting i_k based on the distance of the interpolation point y^{i_k} from the center of B. If the maximum absolute value of the chosen Lagrange polynomial does not exceed Λ, then the index of the next furthest point can be considered. Such an approach fits into the framework of Algorithm 6.3, and it has the advantages of economizing on the Lagrange polynomial global optimization and allowing flexibility in the point selection process.

An example

To illustrate the step-by-step outcome of Algorithm 6.3 we present an example where the algorithm is applied to the poorly poised set in Figure 3.2 with $\Lambda = 21296$. To simplify presentation and computations we shifted and scaled the set here so that it lies in or around the unit ball centered at the origin. This involves scaling the points by a factor of two; hence the Λ constant is reduced by a factor of 4 to $\Lambda = 5324$. We are computing the poisedness within the unit ball around the origin, i.e., $B = B(0; 1)$ (despite the fact that Y is not a subset of B). Below we list the initial set Y_0 and the sets $Y_1, \ldots, Y_5$ computed by the first five iterations of Algorithm 6.3 which are also illustrated in Figure 6.1. At each iteration, one point in the set is being replaced by a better point and the poisedness improves.

$$
Y_0 = \begin{bmatrix} -0.98 & -0.96 \\ -0.96 & -0.98 \\ 0 & 0 \\ 0.98 & 0.96 \\ 0.96 & 0.98 \\ 0.94 & 0.94 \end{bmatrix}, \quad
Y_1 = \begin{bmatrix} -0.98 & -0.96 \\ -0.96 & -0.98 \\ 0 \\ 0.98 & 0.96 \\ 0.96 & 0.98 \\ 0.707 & -0.707 \end{bmatrix},
$$

$$
Y_2 = \begin{bmatrix} -0.848 & 0.528 \\ -0.96 & -0.98 \\ 0 \\ 0.98 & 0.96 \\ -0.96 & -0.98 \\ 0.707 & -0.707 \end{bmatrix}, \quad
Y_3 = \begin{bmatrix} 0.848 & 0.528 \\ -0.96 & -0.98 \\ 0 & 0 \\ 0.98 & 0.96 \\ -0.89 & 0.996 \\ 0.707 & -0.707 \end{bmatrix}, \qquad (6.11)
$$

$$
Y_4 = \begin{bmatrix} -0.967 & 0.254 \\ -0.96 & -0.98 \\ 0 & 0 \\ 0.98 & 0.96 \\ -0.089 & 0.996 \\ 0.707 & -0.707 \end{bmatrix}, \quad
Y_5 = \begin{bmatrix} -0.967 & 0.254 \\ -0.96 & -0.98 \\ 0 & 0 \\ 0.98 & 0.96 \\ -0.199 & 0.979 \\ 0.707 & -0.707 \end{bmatrix}.
$$

As we can see, the algorithm achieves almost optimal poisedness in five steps. These steps forced the algorithm to replace the interpolation set almost entirely. However, we notice that near-optimal poisedness is already achieved after three steps, which exactly reflects the number of points that need to be replaced in this case to remedy near-affine dependence. If instead we consider the badly poised set from Figure 3.3, where the six points almost lie on a circle, we see in Figure 6.2 that just by performing one step of Algorithm 6.3 we achieve optimal poisedness.

6.3 Ensuring well poisedness using pivotal algorithms

We recall that in Chapter 3 we have seen that the poisedness of the interpolation set can be expressed through the condition number of the matrix $M(\bar{\phi}, \hat{Y})$, where $\hat{Y}$ is the scaled and shifted version of Y. It turns out that an algorithm equivalent to Algorithm 6.2 can be stated, which considers this condition number rather than the bound on the value of Lagrange polynomials explicitly. Recall that Algorithm 6.2 was introduced to check if a set Y is poised in a given region B and modify/complete Y if necessary to make a poised set. The algorithm below achieves the same result by applying Gaussian elimination with row pivoting to the matrix $M(\bar{\phi}, Y)$ and monitoring the pivots. The ith pivot during the Gaussian elimination is selected from the ith column of the modified $M(\bar{\phi}, Y)$ after the factorization is applied to the first $i - 1$ columns. The elements of the ith column of the modified $M(\bar{\phi}, Y)$ are in fact the values of a certain polynomial, evaluated at the sample

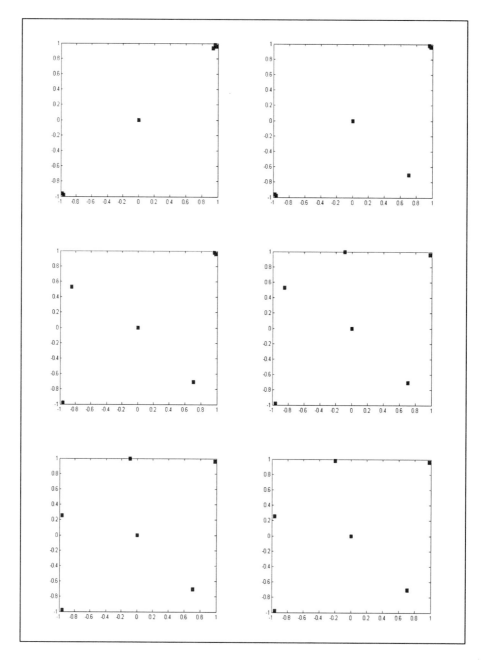

Figure 6.1. *Result of the application of Algorithm 6.3 starting from Y_0 in* (6.11). *The corresponding poisedness constants are the following:* $\Lambda_0 = 5324$, $\Lambda_1 = 36.88$, $\Lambda_2 = 15.66$, $\Lambda_3 = 1.11$, $\Lambda_4 = 1.01$, *and* $\Lambda_5 = 1.001$.

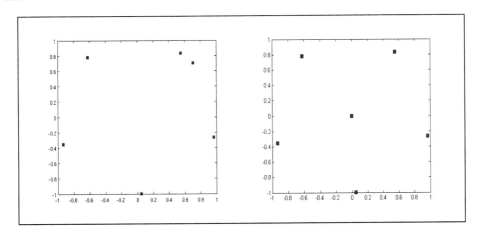

Figure 6.2. *Result of the application of Algorithm 6.3 starting from a set of points that nearly lie on a circle. Changing only one point reduces the poisedness constant $\Lambda_0 = 524982$ to $\Lambda_1 = 1$.*

points in Y. We will call these polynomials *pivot polynomials* and denote them by $u_i(x)$, $i = 0, \ldots, p$. As we will see later, the pivot polynomials are closely related to Lagrange polynomials.

Algorithm 6.4 (Completing the nonpoised set Y via LU factorization).

Initialization: Start by choosing an initial approximation to the pivot polynomial basis, e.g., given by the monomial basis (3.1), $u_i(x) = \bar{\phi}_i(x)$, $i = 0, \ldots, p$. Let $p_{ini} + 1$ be the number of given points in the initial set Y.

For $i = 0, \ldots, p$

    1. **Point selection:** Find $j_i = \text{argmax}_{i \leq j \leq p_{ini}} |u_i(y^j)|$. If $|u_i(y^{j_i})| > 0$ and $i \leq p_{ini}$, then swap points y^i and y^{j_i} in set Y. Otherwise, compute (or recompute if $i \leq p_{ini}$) y^i as

$$y^i \in \underset{x \in B}{\text{argmax}} |u_i(x)|.$$

    2. **Gaussian elimination:** For $j = i + 1, \ldots, p$

$$u_j(x) \leftarrow u_j(x) - \frac{u_j(y^i)}{u_i(y^i)} u_i(x).$$

Let us observe the differences between Algorithms 6.2 and 6.4. The latter algorithm does not normalize the pivot polynomials. In fact, one can easily see that after the algorithm terminates we have a set of pivot polynomials $u = \{u_0(x), u_1(x), \ldots, u_p(x)\}$ and the corresponding matrix $M(u, Y)$ is nothing else but the upper triangular factor of the LU factorization of $M(\phi, Y)$.

Apart from these differences we observe that the two algorithms have the same complexity and produce identical sets of points.

Lemma 6.4. *Given the same, possibly not poised, starting set Y, the same region B, and the same initial basis $\bar{\phi}$, Algorithms 6.2 and 6.4 produce identical poised sets Y.*

Proof. We will use the induction to prove this statement. It is clear that during the zeroth step ($i = 0$) of both algorithms the point selection step will produce the same point for y^0. Let us now assume that the first k steps of both algorithms resulted in an identical set $\{y^0, \ldots, y^{k-1}\}$. Indeed, at the kth step of Algorithm 6.4 (we start from the zeroth step) the kth pivot polynomial has the following properties: it is a linear combination of the first $k + 1$ elements of the basis $\bar{\phi}$ and its value is zero at the first k points $y^0, \ldots, y^{k-1}$ of Y (the points that have already been included by the pivotal algorithm). The same is true for the kth Lagrange polynomial $\ell_k(x)$ at the beginning of step k of Algorithm 6.2. Since by the induction assumption the first k points produced by the two algorithms are the same, this means that at the beginning of step k of each of the two algorithms we are considering two polynomials which (i) lie in the same $(k + 1)$-dimensional subspace of polynomials (the subspace spanned by $\phi_0(x), \phi_1(x), \ldots, \phi_k(x)$) and (ii) have value zero at a given poised subset of k points. Hence, the two polynomials have to coincide up to a constant factor. It immediately follows that when $i = k$, the point selection step of Algorithms 6.2 and 6.4 produce the same point y^k. By induction the lemma follows. $\square$

Let us now consider possible algorithms for the improvement of poisedness of a poised set Y using the pivot polynomials. It is natural to introduce the following modification to Algorithm 6.4. We apply a threshold ξ for accepting the next pivot during the point selection step.

Algorithm 6.5 (Improving poisedness of Y via LU factorization).

Initialization: Initialize the pivot polynomial basis with some basis, e.g., the monomial basis (3.1), $u_i(x) = \bar{\phi}_i(x), i = 0, \ldots, p$. Select pivot threshold $\xi > 0$.

For $i = 0, \ldots, p$

1. **Point selection:** If possible, choose $j_i \in \{i, \ldots, |Y| - 1\}$ such that $|u_i(y^{j_i})| \geq \xi$. If such j_i is found, then swap the positions of points y^i and y^{j_i} in set Y. Otherwise, recompute y^i as

$$y^i \in \operatorname*{argmax}_{x \in B} |u_i(x)|$$

and stop if $|u_i(y^i)| < \xi$ (the pivot threshold is too big).

2. **Gaussian elimination:** For $j = i + 1, \ldots, p$

$$u_j(x) \leftarrow u_j(x) - \frac{u_j(y^i)}{u_i(y^i)} u_i(x).$$

The result of the modified algorithm is a poised set Y such that when Gaussian elimination is applied to $M(\bar{\phi}, Y)$ all pivots have absolute values not smaller than ξ. In this case, it is possible to prove (see the exercises) that the norm of the inverse of $M(\bar{\phi}, Y)$ is bounded by

$$\|M(\bar{\phi}, Y)^{-1}\| \leq \frac{\sqrt{p_1}\, \varepsilon_{growth}}{\xi}, \tag{6.12}$$

where ε_{growth} can be seen as an estimate of the growth factor (see, e.g., [127, Sections 9.3 and 9.4]) that occurs during the factorization, and ξ is the lower bound on the absolute value of the pivots imposed by the algorithm. Notice that there is some freedom in how the index j_i is chosen, as long as the corresponding pivot is large enough. To keep the growth factor down it is desirable to choose the index which provides the largest value of a pivot (this corresponds to partial pivoting in Gaussian elimination). On the other hand, in an optimization algorithm, one may give preference to points according to their proximity to the current iterate. In a practical algorithm a balance of these two policies usually provides the best approach.

The above bound holds for any matrix $M(\bar{\phi}, Y)$ and for any set Y which is generated by Algorithm 6.5. However, to have a simple pivoting strategy and a meaningful bound on $\|M(\bar{\phi}, Y)^{-1}\|$ we need to shift and scale $Y = \{y^0, y^1, \ldots, y^p\}$ to get

$$\hat{Y} = \{0, \hat{y}^1, \ldots, \hat{y}^p\} = \{0, (y^1 - y^0)/\Delta, \ldots, (y^p - y^0)/\Delta\} \subset B(0; 1),$$

where

$$\Delta = \Delta(Y) = \max_{1 \leq i \leq p} \|y^i - y^0\|.$$

It can be seen that $\hat{Y}$ lies tightly in a ball of radius one centered at the origin. Algorithm 6.5 is then applied to the shifted and scaled set $\hat{Y}$. Unlike Algorithm 6.3, where the Lagrange polynomials may be available beforehand and hence shifting and scaling of the set Y may result in increased computational complexity, Algorithm 6.5 always builds the set of pivot polynomials from scratch, and hence shifting and scaling of Y do not cause significant additional work.

From Theorem 3.14 we know that if $\|M(\bar{\phi}, \hat{Y})^{-1}\|$ is bounded, then the set Y is Λ-poised, with Λ being within a constant factor of $\|M(\bar{\phi}, \hat{Y})^{-1}\|$. We conclude that to verify or guarantee Λ-poisedness of a given interpolation set Y, we apply scaling and shifting to obtain an appropriate set $\hat{Y}$. We then apply Algorithm 6.5, which computes a factorization of $M(\bar{\phi}, \hat{Y})$ by computing the pivot polynomials and assigning them to interpolation points. If small pivots are encountered during the factorization, then the "unacceptable" points are replaced by "acceptable" ones. After the factorization is completed the resulting set $\hat{Y}$ can be transformed back to Y, which is Λ-poised for some constant Λ independent of the original interpolation set Y.

The threshold ξ is used to trade off between maintaining better poised sets and not having to discard many of the existing points in Y. If ξ is chosen to be very small, then it is easy for the pivot values to pass this threshold and hence for the points to be accepted into the interpolation set. But since the resulting poisedness constant is inversely proportional to ξ as (6.12) implies, the resulting set may not be well poised. On the other hand, if the value of ξ is chosen too high, it may be necessary to replace almost all points in Y to satisfy the threshold condition. If one is not careful, it may even be impossible to find a point which can provide large enough pivot value. Fortunately, there exists a reasonably large upper bound on ξ, for which we guarantee that the algorithm will always be successful. We know that the $(k+1)$st pivot polynomial $u_k(x)$ can be expressed as $(v^k)^\top \bar{\phi}(x)$, where $v^k = (v_0, \dots, v_{k-1}, 1, 0, \dots, 0)$, and hence $\|v^k\|_\infty \geq 1$. By Lemma 6.7 below, there exists a $\theta > 0$ such that if the pivot threshold is chosen to be any number between 0 and θ, then Algorithm 6.5 will always be successful. Note that θ can be set to 1 when $d = 1$ and to $1/4$ when $d = 2$. Thus, we can state the following result.

Theorem 6.5. *For any given $\xi \in (0,1)$ and $p_1 = n+1$ in the linear case, and any given $\xi \in (0,1/4)$ and $p_1 = (n+1)(n+2)/2$ in the quadratic case, Algorithm 6.5 computes a set $\hat{Y}$ of $p_1 = p + 1$ points in the unit ball $B(0;1)$ for which the pivots of the Gaussian elimination of $\hat{M} = M(\bar{\phi}, \hat{Y})$ satisfy*

$$\left| u_i(y^i) \right| \geq \xi, \quad i = 0, \dots, p.$$

To guarantee Λ-poisedness, Algorithm 6.5 relies on the bound (6.12). Since the growth factor is usually unknown and theoretically can be very large (but finite), then, in theory, the outcome of a pivot algorithm can be an interpolation set with a very large Λ-poisedness constant. For theoretical purposes what matters is that in this case Λ is bounded by a constant, however large. From the practical point of view, the growth factor is known to be moderate in almost all cases and grows exponentially large only for artificial examples.

The effort required by one run of Gaussian elimination in the quadratic case is of the order of $\mathcal{O}(n^6)$ floating point operations. The algorithm does not require, however, the global maximization of the absolute value of the pivot polynomials in a ball. Although such optimization will help and should be used if possible, strictly speaking we need only guarantee the determination of a point that generates an absolute value for the pivot polynomial greater than or equal to ξ. This can be done by using the arguments used in the proof of Lemma 6.7 for scaled and shifted sets. It shows that by simple enumeration we can find such a point for any value of ξ such that $\theta > \xi > 0$ for some positive constant θ which is equal to $1/4$ in the case of quadratic interpolation. The complexity of such an enumeration is $\mathcal{O}(p)$, and it has to be performed at most $\mathcal{O}(p)$ times; hence it does not have a significant contribution to the overall complexity, which is $\mathcal{O}(p^3)$.

Incremental improvements via pivotal algorithms

Algorithm 6.3 improves Λ-poisedness in a gradual manner, replacing one point at a time and reducing the upper bound on the Lagrange polynomial absolute values. Algorithm 6.5, on the other hand, selects the threshold a priori and replaces as many points as necessary to satisfy the threshold condition. One can apply Algorithm 6.5 repetitively with decreasing threshold values to encourage gradual improvement of the poisedness. However, picking a good sequence of threshold values is a nontrivial task. One may happen to decrease the value of ξ too slowly and not obtain any improvement after several applications of the algorithm, or one may decrease the threshold too quickly and end up having to replace too many points. There are cases where the gradual updates of the threshold value make algorithmic sense, for example, when the pivot values from the last application of Algorithm 6.5 are remembered and used as a guideline. Such considerations should be addressed during the development of a specific derivative-free optimization algorithm (see Chapter 11).

The example in the following subsection applies the threshold reduction idea to show the possible progress of the poisedness of the sets produced by Algorithm 6.5. We will see that in this case Algorithm 6.5 works similarly to Algorithm 6.3, but it may replace more than one point per run. In a practical derivative-free optimization framework, this property may be advantageous when a few points actually need to be replaced to obtain good geometry—Algorithm 6.5 may be able to identify them all during one run. On the other hand, it may replace some of the points unnecessarily, obtaining marginal improvement of the poisedness but wasting valuable function evaluations.

If we want to use the pivoting approach but replace only one point per each run, the scheme described next can be used. Essentially, we run Algorithm 6.5 without a threshold (i.e., the threshold value is set to zero). Assuming that the initial set Y is poised and has p_1 points, the algorithm will run to the last iteration, where we can then optimize the last pivot polynomial and replace the last point in Y.

Algorithm 6.6 (Incremental improvement of poisedness of Y via LU factorization).

Initialization: Start by choosing an initial approximation to the pivot polynomial basis, e.g., given by the monomial basis (3.1), $u_i(x) = \bar{\phi}_i(x)$, $i = 0, \ldots, p$. Assume Y contains p_1 poised points.

For $i = 0, \ldots, p - 1$

1. **Point selection:** Find $j_i = \mathrm{argmax}_{i \leq j \leq |Y|} |u_i(y^j)|$ and swap points y^i and y^{j_i} in set Y.

2. **Gaussian elimination:** For $j = i+1, \ldots, p$

$$u_j(x) \leftarrow u_j(x) - \frac{u_j(y^i)}{u_i(y^i)} u_i(x).$$

Improvement step:

$$y^p_{\text{new}} \in \mathrm{argmax}_{x \in B} |u_p(x)|.$$

This algorithm attempts to identify the best candidate point for replacement. Since the LU procedure pivots by rows (that is, by points) the points which result in bad pivot values are pushed to the bottom rows. Hence, by replacing the very last point we hopefully obtain the best improvement of the poisedness.

Note that the new point generated by Algorithm 6.6 is the same as a point generated at an iteration of Algorithm 6.3 if the sets under consideration are the same and the last point in the LU factorization of Algorithm 6.6 corresponds to the Lagrange polynomial of maximum absolute value in Algorithm 6.3—since in this case the pth pivot polynomial is the same (up to a constant factor) as the Lagrange polynomial just mentioned (see the proof of Lemma 6.4). The workload involved in the application of Algorithm 6.6 is $\mathcal{O}(p^3)$, but it does not need to optimize all Lagrange polynomials.

Algorithm 6.6 constitutes one iteration of the improvement procedure. To identify the next point to be replaced, Algorithm 6.6 should be repeated again from the beginning. If after applying Algorithm 6.6 the last point is already the optimal point for the corresponding pivot polynomial, then no further improvement can be obtained by this strategy. There is no guarantee that the sequence of points which is replaced by Algorithm 6.3 is identical to the sequence of points replaced by a strategy consisting of a sequential application of Algorithm 6.6. However, as long as these sequences are identical, namely, if the point which produces the largest absolute value of the Lagrange polynomials and the point which produces the smallest (hence the last) pivot are the same on each iteration, then the resulting interpolation sets are identical (since, recall, the pth pivotal polynomial is, up to a constant, the corresponding Lagrange polynomial).

An example

Let us again turn to the example we used to illustrate Algorithm 6.3, with $\Lambda = 5324$, to show the effects of Algorithm 6.5. We want to show that Algorithm 6.5 can produce interpolation sets of comparable quality to the ones produced by Algorithm 6.3. To that end we will apply the following adjustments of ξ: we run Algorithm 6.5 three times, applying thresholds $\xi = 0.01, 0.1, 0.5$.

Below we list the initial set and the results of the three runs of Algorithm 6.5 which are also illustrated in Figure 6.3:

$$
Y_0 = \begin{bmatrix} -0.98 & -0.96 \\ -0.96 & -0.98 \\ 0 & 0 \\ 0.98 & 0.96 \\ 0.96 & 0.98 \\ 0.94 & 0.94 \end{bmatrix}, \quad
Y_1 = \begin{bmatrix} -0.98 & -0.96 \\ -0.96 & -0.98 \\ 0 & 0 \\ 0.98 & 0.96 \\ 0.96 & 0.98 \\ 0.707 & -0.707 \end{bmatrix},
$$

$$
(6.13)
$$

$$
Y_2 = \begin{bmatrix} -0.98 & -0.96 \\ -0.135 & -0.99 \\ -0.0512 & 0.0161 \\ 0.98 & 0.96 \\ -0.14 & 0.99 \\ 6.99 & -7.14 \end{bmatrix}, \quad
Y_3 = \begin{bmatrix} -0.98 & -0.96 \\ 0.943 & -0.331 \\ 0.0281 & 0.0235 \\ 0.98 & 0.96 \\ -0.14 & 0.99 \\ 0.699 & -0.714 \end{bmatrix}.
$$

At each step, some of the points in the set are replaced by better points and the poisedness improves. Moreover, the poisedness progress during the three applications of Algorithm 6.5 is slightly better than the poisedness progress of the first three iterations of Algorithm 6.3 on the same example (see Figure 6.1). However, on average 2 points per run are replaced in the former case, versus one point per iteration in the latter. The replacement of the origin on the second step by the point $(-0.0512, 0.0161)$ is, in fact, an example of wasting a sample point to obtain a relatively negligible improvement of the poisedness. Notice that the final sets are very similar for both cases. We used global optimization of the pivot polynomials to illustrate that their behavior is similar to using Lagrange polynomials.

If we apply Algorithm 6.6 to the same set, then it will produce exactly the same sequence of sets as the first four iterations of Algorithm 6.3; that is, it will produce the first four sets in the example of Lagrange polynomials given in Section 3.3 and illustrated in Figure 6.1. After four iterations it will stop, unable to improve set Y any further.

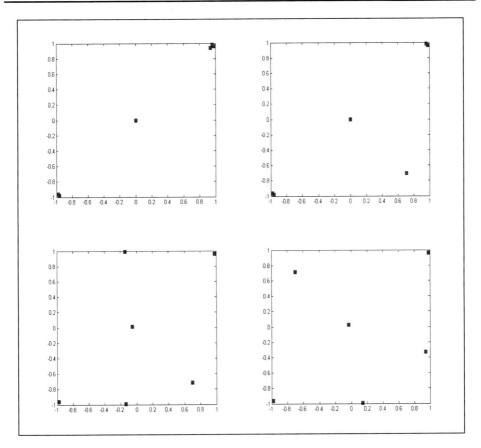

Figure 6.3. *Result of applying Algorithm 6.6 four times starting from Y_0 in (6.13). The corresponding poisedness constants are the following:* $\Lambda_0 = 5324$, $\Lambda_1 = 36.88$, $\Lambda_2 = 1.699$, *and* $\Lambda_3 = 1.01$.

6.4 Practical considerations of geometry improvement algorithms

Let us discuss whether the proposed procedures are "practical." Derivative-free optimization problems typically address functions whose evaluation is expensive; hence a practical approach should attempt to economize on function evaluations. The first question about the pivotal algorithms is whether too many interpolation points need to be replaced at each iteration of a model-based derivative-free optimization algorithm. The secondary consideration is the linear algebra cost of a model-improvement algorithm.

Several situations arise during the course of a model-based derivative-free algorithm. First, as the iterates progress, some of the points in the sample set may become too distant from the current best iterate. In this case they no longer belong to a region B where the poisedness is considered. These points simply need to be replaced by points closer to the current iterate. These are the cases when Algorithms 6.2 and 6.4 are useful. If the basis of Lagrange polynomials is available, then the Lagrange polynomial that corresponds to the point which is being removed is optimized over B. The optimal point then replaces the point that is being removed. The rest of the polynomials are updated. The cost of this procedure is $\mathcal{O}(p^2)$ plus the cost of optimizing the Lagrange polynomial. If a point which produces a large enough value (greater than 1) of the absolute value of the Lagrange polynomial is found, then the global optimization is not necessary, since we know that the poisedness improves. If the set of Lagrange polynomials is not available, or if a pivot approach is used, then the complexity is $\mathcal{O}(p^3)$.

Another situation when an interpolation point needs to be replaced is when a new optimization iterate is found and needs to be included in the interpolation set. In this case, the Lagrange polynomial algorithm will identify the polynomial whose value at the new point is the largest and replace the point in Y corresponding to that polynomial with the new point. This effort requires $\mathcal{O}(p^2)$ operations. The pivotal algorithm will simply start by choosing the new optimization iterate to generate the first pivot and then proceed by choosing points which produce the best pivot value until the factorization is complete. The remaining unused point will be the one which is replaced. This approach requires $\mathcal{O}(p^3)$ operations, but it maintains guaranteed Λ-poisedness, which the Lagrange polynomial approach does only if the absolute value of the Lagrange polynomial at the new point happens to be at least 1.

Finally, if a model is suspected to be insufficiently well poised, then Algorithm 6.3 or a repeated application of Algorithm 6.6 can be used. In this case, although not necessary, the global optimization of the corresponding polynomials is desirable, in which case both strategies have comparable complexity.[7]

6.5 Ensuring well poisedness for regression and minimum Frobenius norm models

Let us now consider the cases when the number of points p_1 in Y is not equal to the dimension q_1 of the polynomial space.

Regression models

First, we recall the regression case, $p_1 > q_1$, and the regression Lagrange polynomials discussed in Chapter 4. Algorithm 6.1 for computing and updating interpolation Lagrange polynomials does not apply to the regression case. Regression Lagrange polynomials can be computed from the expression (4.6) for the matrix of the coefficients

$$A_\phi = [M(\phi, Y)^\top M(\phi, Y)]^{-1} M(\phi, Y)^\top.$$

The work involved in this computation is $\mathcal{O}(q^2 p)$. If only one point is replaced in the set Y, then only one row of the matrix $M(\phi, Y)$ is replaced. To update the set of regression Lagrange polynomials one can apply a rank-two update to $[M(\phi, Y)^\top M(\phi, Y)]^{-1}$ via the

---

[7]The optimization of a quadratic polynomial over an ellipsoid is a particular trust-region subproblem. Even in the nonconvex case it is known that the first-order necessary conditions for such trust-region subproblems are sufficient (see Section 10.8). These problems have some form of *hidden convexity* in the sense that the necessary conditions are sufficient (or the duality gap is zero) or there exist convex reformulations. For more details on the complexity bounds and convex reformulations for ellipsoidal trust-region subproblems see [32, 98, 232].

Sherman–Morrison–Woodbury formula (see, e.g., [109, 178]). The overall work required for the update of the regression Lagrange polynomials is $\mathcal{O}(pq)$.

It is, however, not clear if computing the set of regression Lagrange polynomials is useful. Recall that for strong Λ-poisedness we require the maximum value of the Lagrange polynomials to be smaller than 1, when p is sufficiently large. There is no guarantee that an analogue of Algorithm 6.3, based on the regression Lagrange polynomials, can be designed to terminate after a finite number of iterations, since there is no analogue of Theorem 6.3. Moreover, it is not even clear how to fix nonpoised sets with the help of regression Lagrange polynomials. If the set Y is not poised, then it is not possible to invert the matrix $M(\phi, Y)^\top M(\phi, Y)$. Unfortunately, unlike Algorithm 6.1, the proposed method of computing Lagrange polynomials for regression does not have a natural way of detecting the points which cause the set to be nonpoised and of replacing these points. It is possible to do so by considering linear independence of the rows of $M(\phi, Y)$, which leads us to the pivotal algorithms.

We can apply Algorithm 6.4 to the set Y and select or generate a poised subset of q_1 points. The remaining unused points now can be simply added, since their presence will not ruin the poisedness of the set. The same argument applies to generating a Λ-poised regression set. Algorithm 6.5 can be used to select or generate a subset of q_1 points which is $(1/\sqrt{p})\Lambda$-poised in the interpolation sense. Then the remaining points can be added to the set, which will be Λ-poised in the regression sense by the properties of Definition 4.7 (point 2) of Λ-poisedness for regression.

It is also possible to apply Algorithms 6.2 and 6.3 to select poised and $(1/\sqrt{p})\Lambda$-poised subsets of Y, respectively, in the interpolation sense. The resulting interpolation Lagrange polynomials are the interpolation polynomials for the subsets and are not immediately related to the regression polynomials for Y.

For the purpose of providing fully linear and fully quadratic regression models with arbitrarily large sample sets we need to be able to establish strong Λ-poisedness following Definition 4.10. Due to Theorem 4.11 we know that we can guarantee strong Λ-poisedness if we can partition set Y into Λ-poised subsets of q_1 points each. Clearly, this can be done by applying either Algorithm 6.3 or 6.5 repeatedly to select or generate Λ-poised subsets of Y.

Algorithm 6.7 (Computing a strongly Λ-poised set Y).

Initialization: Let $\bar{Y} = Y$ and $i = 1$. Pick $\Lambda > 1$ or threshold $\xi > 0$.

While $|\bar{Y}| \geq q_1$

    1. **Select a Λ-poised subset:** Apply Algorithm 6.3 or 6.5 to $\bar{Y}$ to select or generate a Λ-poised subset Y^i.

    2. **Remove subset:**
$$\bar{Y} \leftarrow \bar{Y} \setminus Y^i, \quad i \leftarrow i + 1.$$

Reset $Y = \bigcup_i Y^i$.

When the algorithm terminates we obtain a possibly smaller set Y, because we do not include the points that fall into the last incomplete subset. Alternatively it is possible to include these points in Y or to generate another Λ-poised subset by adding more points to Y. All of these options are acceptable and there is no significant theoretical difference when large values of p are considered. Hence, only practical considerations should dictate how to treat the remaining points.

It is easy to see that this algorithm generates a strongly Λ-poised set Y (see the discussion before Definition 4.10). Since by the discussion in previous sections we know that Algorithm 6.3 or 6.5 will terminate after a finite number of steps, then so will Algorithm 6.7.

Minimum Frobenius norm models

In the case of the minimum Frobenius norm models discussed in Chapter 5, the set of Lagrange polynomials is also computed by solving a system of linear equations. When the matrix of this system,

$$\begin{bmatrix} M(\bar{\phi}_Q, Y)M(\bar{\phi}_Q, Y)^\top & M(\bar{\phi}_L, Y) \\ M(\bar{\phi}_L, Y)^\top & 0 \end{bmatrix},$$

is singular, then again updating the interpolation set to generate a poised one is not as straightforward as in the full interpolation case. By applying Algorithm 6.2 or 6.4 one cannot generate the minimum Frobenius norm Lagrange polynomials, but one could check and enforce poisedness of Y in the minimum Frobenius norm sense (the details are omitted).

It is more natural, however, to maintain well poisedness of Y using directly the set of minimum Frobenius norm Lagrange polynomials. An efficient update procedure for these polynomials is suggested in [191]. We can use an analogue of Algorithm 6.3 to update the sample set and to maintain Λ-poisedness in the minimum Frobenius norm sense. Notice that the steps that update the Lagrange polynomials have to be changed in Algorithm 6.3 to accommodate the minimum Frobenius norm polynomials. To show that this procedure is also finite one can use an analogue of Theorem 6.3.

Theorem 6.6. *For any given $\Lambda > 1$, a closed ball B, and a fixed polynomial basis ϕ, Algorithm 6.3 for minimum Frobenius norm Lagrange polynomials terminates with a Λ-poised set Y after at most $N = N(\Lambda, \phi)$ iterations, where N is a constant which depends on Λ and ϕ.*

The proof of this theorem is practically identical to the proof of Theorem 6.3 and makes use of the determinant (or volume) ratio interpretation of the minimum Frobenius norm Lagrange polynomials and of Corollary 5.9.

We do not elaborate any further on this issue and just conclude that minimum Frobenius norm models form a fully linear class because of the corresponding error bounds and the existence of finite model-improvement algorithms.

6.6   Other notes and references

Here we present the statement and a proof of the lemma, which we used on several occasions in this chapter and in Chapters 3–5.

Lemma 6.7. *Let $v^\top \bar\phi(x)$ be a quadratic polynomial of degree at most d, where $\|v\|_\infty = 1$ and $\bar\phi$ is the natural basis (defined by (3.1)). Then there exists a constant $\sigma_\infty > 0$ independent of v such that*

$$\max_{x \in B(0;1)} |v^\top \bar\phi(x)| \geq \sigma_\infty.$$

For $d = 1$, $\sigma_\infty \geq 1$, and for $d = 2$, $\sigma_\infty \geq \frac{1}{4}$.

Proof. The first part (general case) has been proved in Lemma 3.10.

When $d = 1$ we have $\bar\phi(x) = [1, x_1, \ldots, x_n]^\top$. Let $w = [v_2, \ldots, v_{n+1}]^\top$. It is easy to see that the optimal solution of problem $\max_{x \in B(0;1)} |v^\top \bar\phi(x)|$ is given either by $w/\|w\|$ (with optimal value $v_1 + \|w\|$) or by $-w/\|w\|$ (with optimal value $-v_1 + \|w\|$). Thus, the optimal value is $|v_1| + \|w\| \geq 1$.

Let us consider the case when $d = 2$. Since $\|v\|_\infty = 1$, at least one of the elements of v is 1 or -1, and thus one of the coefficients of the polynomial $q(x) = v^\top \bar\phi(x)$ is equal to 1, -1, $1/2$, or $-1/2$. Let us consider only the cases where one of the coefficients of $q(x)$ is 1 or $1/2$. The cases -1 or $-1/2$ would be analyzed similarly.

The largest coefficient in absolute value in v corresponds to a term which is either a constant term, a linear term x_i, or a quadratic term $x_i^2/2$ or $x_i x_j$. Let us restrict all variables that do not appear in this term to zero. We will show that the maximum absolute value of the polynomial is at least $1/4$ by considering the four cases of different terms that correspond to the largest coefficient. In each case we will evaluate the restricted polynomial at several points in the unit ball and show that at least at one of these points the polynomial achieves an absolute value of at least $1/4$. It will clearly follow that the maximum absolute value of the unrestricted polynomials is also bounded from below by $1/4$.

- $q(x) = 1$. This case is trivial.

- $q(x) = x_i^2/2 + \alpha x_i + \beta$. In this case we have $q(1) = 1/2 + \alpha + \beta$, $q(-1) = 1/2 - \alpha + \beta$, and $q(0) = \beta$. If $|q(1)| \geq 1/4$ or $|q(-1)| \geq 1/4$, we already have the desired result. If $|q(1)| < 1/4$ and $|q(-1)| < 1/4$, then, by adding these inequalities, we derive $-1/2 < 1 + 2\beta < 1/2$. But then we obtain $q(0) = \beta < -1/4$.

- $q(x) = \alpha x_i^2/2 + x_i + \beta$. This time we have $q(1) = 1 + \alpha/2 + \beta$ and $q(-1) = -1 + \alpha/2 + \beta$, and so
$$\max\{|q(-1)|, |q(1)|\} \geq 1.$$

- $q(x) = \alpha x_i^2/2 + \beta x_j^2/2 + x_i x_j + \gamma x_i + \delta x_j + \epsilon$. In this case we are considering the quadratic function over a two-dimensional ball. By considering four points, $p_1 = (1/\sqrt{2}, 1/\sqrt{2})$, $p_2 = (1/\sqrt{2}, -1/\sqrt{2})$, $p_3 = (-1/\sqrt{2}, 1/\sqrt{2})$, and $p_4 = (-1/\sqrt{2}, -1/\sqrt{2})$, on the boundary of the ball, we get

$$q(p_1) = \frac{\alpha + \beta}{4} + \frac{1}{2} + \frac{\gamma + \delta}{\sqrt{2}} + \epsilon,$$

$$q(p_2) = \frac{\alpha + \beta}{4} - \frac{1}{2} + \frac{\gamma - \delta}{\sqrt{2}} + \epsilon,$$

$$q(p_3) = \frac{\alpha + \beta}{4} - \frac{1}{2} - \frac{\gamma - \delta}{\sqrt{2}} + \epsilon,$$

$$q(p_4) = \frac{\alpha + \beta}{4} + \frac{1}{2} - \frac{\gamma + \delta}{\sqrt{2}} + \epsilon.$$

As a result, we obtain $q(p_1) - q(p_2) = 1 + \sqrt{2}\delta$ and $q(p_3) - q(p_4) = -1 + \sqrt{2}\delta$. In the case where $\delta \geq 0$, we have $q(p_1) - q(p_2) \geq 1$. Thus, if $|q(p_1)| < 1/2$, then $q(p_2) \leq -1/2$. The case $\delta < 0$ is proved analogously. $\square$

6.7   Exercises

1. Prove that Algorithm 6.1 computes the basis of Lagrange polynomials for poised interpolation sets.

2. Show that the gradient of a quadratic interpolation (or regression) polynomial model is Lipschitz continuous, i.e., that

$$\|\nabla m(y_1) - \nabla m(y_2)\| \leq \kappa \|y_1 - y_2\| \quad \forall y_1, y_2 \in B(x; \Delta),$$

where κ depends on the Λ-poisedness constant of the sample set, on the number p_1 of points used to build the model, on an upper bound Δ_{max} for Δ, on an upper bound for the values of f, and on an upper bound for the inverse of $M(\phi, Y)$ (or inverse of the diagonal matrix of singular values of $M(\phi, Y)$).

3. Prove the bound (6.12) when ε_{growth} is given by the product of the norms of the inverses of the L and U factors in the LDU factorization of $M(\bar{\phi}, Y)$.

Part II

Frameworks and algorithms

Chapter 7

Directional direct-search methods

Direct-search methods are derivative-free methods that sample the objective function at a finite number of points at each iteration and decide which actions to take next solely based on those function values and without any explicit or implicit derivative approximation or model building. In this book we divide the presentation of direct-search methods into two chapters. In the next chapter we cover direct-search methods based on simplices and operations over simplices, like reflections, expansions, or contractions. A classical example of a simplicial direct-search algorithm is the Nelder–Mead method.

In this chapter we address direct-search methods where sampling is guided by sets of directions with appropriate features. Of key importance in this chapter are the concepts of positive spanning sets and positive bases (see Section 2.1). The two classes of direct-search methods considered in this book (directional and simplicial) are related to each other. For instance, by recalling what we have seen in Section 2.5, one can easily construct maximal positive bases from any simplex of $n+1$ vertices. Reciprocally, given any positive basis, it is straightforward to identify simplices of $n+1$ vertices. Despite the intimacy of the two concepts (positive spanning and affine independency), the philosophy of the two classes of direct-search methods under consideration differ enough to justify different treatments.

The problem under consideration is the unconstrained optimization of a real-valued function, stated in (1.1). Extensions of directional direct-search methods for various types of derivative-free constrained optimization problems are summarized in Section 13.1.

7.1 The coordinate-search method

One of the simplest directional direct-search methods is called coordinate or compass search. This method makes use of the maximal positive basis $D_\oplus$:

$$D_\oplus = \begin{bmatrix} I & -I \end{bmatrix} = [e_1 \cdots e_n -e_1 \cdots -e_n]. \tag{7.1}$$

Let x_k be a current iterate and α_k a current value for the step size or mesh parameter. Coordinate search evaluates the function f at the points in the set

$$P_k = \{x_k + \alpha_k d : d \in D_\oplus\},$$

following some specified order, trying to find a point in P_k that decreases the objective function value. In the terminology of this chapter, we say that P_k is a set of poll points and $D_\oplus$ is a set of poll vectors or directions. This process of evaluating the objective function is called polling. We illustrate the poll process for coordinate search in Figure 7.1.

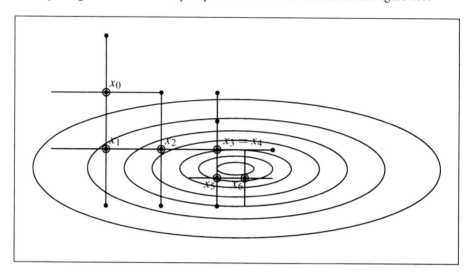

Figure 7.1. *First six iterations of coordinate search with opportunistic polling (following the order North/South/East/West). Function evaluations (a total of 14) occur at circles, but only the bigger circles are iterates. The ellipses depict the level sets of the function.*

Polling is successful when one of the points in P_k is better than the current iterate x_k in terms of the values of f. When that happens, the method defines a new iterate $x_{k+1} = x_k + \alpha_k d_k \in P_k$ such that $f(x_{k+1}) < f(x_k)$ (a *simple decrease* in the objective function). In such a successful case, one either leaves the parameter α_{k+1} unchanged or increases it (say by a factor of 2). If none of the points in P_k leads to a decrease in f, then the parameter α_k is reduced (say by a factor of $1/2$) and the next iteration polls at the same point ($x_{k+1} = x_k$). Polling can be opportunistic, moving to the first encountered better point, or complete, in which case all the poll points are evaluated and the best point is taken (if better than the current iterate). Complete polling is particularly attractive for running on a parallel environment.

Algorithm 7.1 (Coordinate-search method).

Initialization: Choose x_0 and $\alpha_0 > 0$.

For $k = 0, 1, 2, \ldots$

    1. **Poll step:** Order the poll set $P_k = \{x_k + \alpha_k d : d \in D_\oplus\}$. Start evaluating f at the poll points following the order determined. If a poll point $x_k + \alpha_k d_k$ is

found such that $f(x_k + \alpha_k d_k) < f(x_k)$, then stop polling, set $x_{k+1} = x_k + \alpha_k d_k$, and declare the iteration and the poll step successful. Otherwise, declare the iteration (and the poll step) unsuccessful and set $x_{k+1} = x_k$.

2. **Parameter update:** If the iteration was successful, set $\alpha_{k+1} = \alpha_k$ (or $\alpha_{k+1} = 2\alpha_k$). Otherwise, set $\alpha_{k+1} = \alpha_k/2$.

To illustrate what is coming later we also introduce for coordinate search the following set (called a mesh, or a grid):

$$M_k = \left\{ x_k + \alpha_k \left(\sum_{i=1}^{n} u_i e_i + \sum_{i=1}^{n} u_{n+i}(-e_i) \right) : u \in \mathbb{Z}_+^{|2n|} \right\}, \tag{7.2}$$

where $\mathbb{Z}_+$ is the set of nonnegative integers. An example of the mesh is illustrated in Figure 7.2. The mesh is merely conceptual. There is never an attempt in this class of methods to enumerate (computationally or not) points in the mesh.

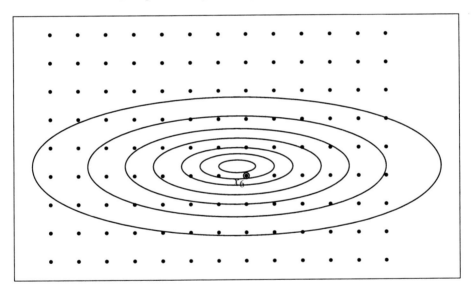

Figure 7.2. *The mesh in coordinate search at x_6.*

For matters of practical efficiency, it is useful to introduce some flexibility in the above coordinate-search framework. Such flexibility can be accommodated by the so-called search step,[8] which is optional and applied just before polling when formulated. Basically, the search step consists of evaluating the objective function at a finite number of points in M_k, trying to find a point $y \in M_k$ such that $f(y) < f(x_k)$. We illustrate this process in Figure 7.3. When the search step is successful so is the iteration (the poll step is skipped and a new iteration starts at $x_{k+1} = y$). The search step is totally optional, not only in the implementation of the method but also when proving its convergence properties. When the

---

[8]We should remark that coordinate search is often described in the literature without a search step.

search step is applied it has no interference in the convergence properties of the method since the points are required to be on the mesh M_k. In the next section, we will describe a class of directional direct-search methods that includes coordinate search as a special case.

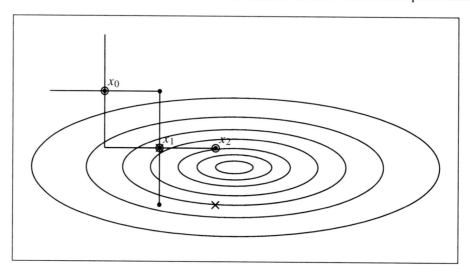

Figure 7.3. *Three iterations of coordinate search with a search step (consisting of trying the South-East point) and opportunistic polling (following the order North/South/East/West). Function evaluations occur at crosses (search step) and circles (a total of 6). Only the bigger circles are iterates. The ellipses depict the level sets of the function.*

7.2   A directional direct-search framework

We now present a class of globally convergent directional direct-search methods. Much of this presentation is based on the generalized pattern-search framework introduced by Audet and Dennis [18] and makes extensive use of the structure of an iteration organized around a search step and a poll step.

To start the presentation let us consider a current iterate x_k and a current value for the step size or mesh parameter α_k. The goal of iteration k of the direct-search methods presented here is to determine a new point x_{k+1} such that $f(x_{k+1}) < f(x_k)$.

The process of finding a new iterate x_{k+1} can be described in two phases (usually designated as the search step and the poll step).

The search step is optional and is not necessary for the convergence properties of the method. It consists of evaluating the objective function at a finite number of points. The choice of points is totally arbitrary as long as their number remains finite (later we will see that the points must be in a mesh M_k if only simple decrease is imposed, but we skip this issue to leave the presentation as conceptual as possible).[9] For example, the points

---

[9]It is obvious that descent in the search step must be controlled in some form. The reader can think of what a sequence of points of the form $x_k = 2 + 1/k$ does to the minimization of $f(x) = x^2$.

might be chosen according to specific application properties or following some heuristic algorithm. The search step can take advantage of the existence of surrogate models for f (see Chapter 12) to improve the efficiency of the direct-search method. The search step and the current iteration are declared successful if a new point x_{k+1} is found such that $f(x_{k+1}) < f(x_k)$.

The poll step is performed only if the search step has been unsuccessful. It consists of a local search around the current iterate, exploring a set of points defined by the step size parameter α_k and by a positive basis D_k:[10]

$$P_k = \{x_k + \alpha_k d : d \in D_k\}. \tag{7.3}$$

The points $x_k + \alpha_k d \in P_k$ are called the poll points and the vectors $d \in D_k$ the poll vectors or directions. Later we will see that the poll points must also lie in the mesh M_k if only simple decrease is imposed, but, again, we skip this issue to concentrate on the geometrical properties of these methods (which are related to those of other derivative-free methods).

The purpose of the poll step is to ensure a decrease of the objective function for a sufficiently small step size parameter α_k. As we saw in Section 2.2, as long as the objective function retains some differentiability properties and unless the current iterate is a stationary point, we know that the poll step must eventually be successful (after a finite number of reductions of the step size parameter). The key ingredient here is the fact that there is at least one descent direction in each positive basis D_k.

The poll step and the current iteration are declared successful if a new point $x_{k+1} \in P_k$ is found such that $f(x_{k+1}) < f(x_k)$. If the poll step fails to produce a point in P_k where the objective function is lower than $f(x_k)$, then both the poll step and the iteration are declared unsuccessful. In these circumstances the step size parameter α_k is typically decreased.

The step size parameter is kept unchanged (or possibly increased) if the iteration is successful (which happens if either in the search step or in the poll step a new iterate is found yielding objective function decrease).

In this class of directional direct-search methods one can consider multiple positive bases and still be able to guarantee global convergence to stationary points. When new iterates are accepted based on *simple decrease* of the objective function (as we have just described), the number of positive bases is required to be finite. As we will point out later, this requirement can be relaxed if one imposes a *sufficient decrease* condition to accept new iterates. Still, in such a case, one can use only an infinite number of positive bases for which the cosine measure is uniformly bounded away from zero.

The class of directional direct-search methods analyzed in this book is described in Algorithm 7.2. Our description follows the one given in [18] for the generalized pattern search, by considering search and poll steps separately. We do not specify for the moment the set $\mathcal{D}$ of positive bases used in the algorithm. Polling is opportunistic, moving to the first encountered better point. The poll vectors (or points) are ordered according to some criterion in the poll step. In many papers and implementations this ordering is the one in which they were originally stored, and it is never changed during the course of the iterations. Consequently, our presentation of directional direct search considers that the poll directions are ordered in some given form before (opportunistic) polling starts. From a theoretical point of view, this ordering does not matter and could change at every

---

[10]The application of this class of direct-search methods and its convergence properties is valid both for positive spanning sets and positive bases (satisfying some properties mentioned later).

iteration. Efficient procedures to order the poll directions include ordering according to the angle proximity to a negative simplex gradient, random ordering, and ordering following the original order but avoiding restarts at new poll iterations (and combinations of these strategies).

Algorithm 7.2 (Directional direct-search method).

Initialization: Choose x_0, $\alpha_0 > 0$, $0 < \beta_1 \leq \beta_2 < 1$, and $\gamma \geq 1$. Let $\mathcal{D}$ be a set of positive bases.

For $k = 0, 1, 2, \ldots$

1. **Search step:** Try to compute a point with $f(x) < f(x_k)$ by evaluating the function f at a finite number of points. If such a point is found, then set $x_{k+1} = x$, declare the iteration and the search step successful, and skip the poll step.

2. **Poll step:** Choose a positive basis D_k from the set $\mathcal{D}$. Order the poll set $P_k = \{x_k + \alpha_k d : d \in D_k\}$. Start evaluating f at the poll points following the chosen order. If a poll point $x_k + \alpha_k d_k$ is found such that $f(x_k + \alpha_k d_k) < f(x_k)$, then stop polling, set $x_{k+1} = x_k + \alpha_k d_k$, and declare the iteration and the poll step successful. Otherwise, declare the iteration (and the poll step) unsuccessful and set $x_{k+1} = x_k$.

3. **Mesh parameter update:** If the iteration was successful, then maintain or increase the step size parameter: $\alpha_{k+1} \in [\alpha_k, \gamma \alpha_k]$. Otherwise, decrease the step size parameter: $\alpha_{k+1} \in [\beta_1 \alpha_k, \beta_2 \alpha_k]$.

The poll step makes at most $|D_k|$ (where $|D_k| \geq n + 1$) function evaluations and exactly that many at all unsuccessful iterations.

The natural stopping criterion in directional direct search is to terminate the run when $\alpha_k < \alpha_{tol}$, for a chosen tolerance $\alpha_{tol} > 0$ (for instance, $\alpha_{tol} = 10^{-5}$).

7.3   Global convergence in the continuously differentiable case

First, we point out that this class of directional direct-search methods is traditionally analyzed under the assumption that all iterates lie in a compact set. Given that the sequence of iterates $\{x_k\}$ is such that $\{f(x_k)\}$ is monotonically decreasing, a convenient way of imposing this assumption is to assume that the level set $L(x_0) = \{x \in \mathbb{R}^n : f(x) \leq f(x_0)\}$ is compact.

Assumption 7.1. *The level set $L(x_0) = \{x \in \mathbb{R}^n : f(x) \leq f(x_0)\}$ is compact.*

We are interested in analyzing the global convergence properties of these methods, meaning convergence to stationary points from arbitrary starting points. In direct search we will deal only with first-order stationary points.

Behavior of the step size parameter

The global convergence analysis for these direct-search methods relies on proving first that there exists a subsequence of step size parameters converging to zero. For this purpose we must be able to ensure the following assumption.

Assumption 7.2. *If there exists an $\alpha > 0$ such that $\alpha_k > \alpha$, for all k, then the algorithm visits only a finite number of points.*

In Sections 7.5 and 7.7, we will discuss how this assumption can be ensured in practical implementations of direct search. We will see in Section 7.5 that when the number of positive bases is finite, some integral/rational structure on the construction of these bases and on the update of the step size parameter will suffice for this purpose. When any number of bases is allowed, then something else is required to achieve Assumption 7.2 (namely a sufficient decrease condition to accept new iterates, as we will prove in Section 7.7).

Based on Assumption 7.2 one can prove that the step size parameter tends to zero.

Theorem 7.1. *Let Assumption 7.2 hold. Then the sequence of step size parameters satisfies*

$$\liminf_{k \longrightarrow +\infty} \alpha_k = 0.$$

Proof. Let us assume, by contradiction, that there exists an $\alpha > 0$ such that $\alpha_k > \alpha$ for all k. Then one knows from Assumption 7.2 that the number of points visited by the algorithm is finite. On the other hand, the algorithm moves to a different point only when a decrease in the objective function is detected. By putting these last two arguments together, we arrive at the conclusion that there must exist an iteration $\bar{k}$ such that $x_k = x_{\bar{k}}$ for all $k \geq \bar{k}$. From the way α_k is updated in unsuccessful iterations, it follows that $\lim_{k \longrightarrow +\infty} \alpha_k = 0$, which contradicts what we have assumed at the beginning of the proof. □

The following corollary follows from Assumption 7.1.

Corollary 7.2. *Let Assumptions 7.1 and 7.2 hold. There exist a point x_* and a subsequence $\{k_i\}$ of unsuccessful iterates for which*

$$\lim_{i \longrightarrow +\infty} \alpha_{k_i} = 0 \qquad and \qquad \lim_{i \longrightarrow +\infty} x_{k_i} = x_*. \tag{7.4}$$

Proof. Theorem 7.1 states the existence of an infinite subsequence of the iterates driving the step size parameter to zero. As a result, there must exist an infinite subsequence of iterations corresponding to unsuccessful poll steps, since the step size parameter is reduced only at such iterations. Let K_u^1 denote the index subsequence of all unsuccessful poll steps.

It follows also from the scheme that updates the step size parameter and from the above observations that there must exist a subsequence $K_u^2 \subset K_u^1$ such that $\alpha_{k+1} \to 0$ for $k \in K_u^2$. Since, $\alpha_k \leq (1/\beta_1)\alpha_{k+1}$ for $k \in K_u^2$, we obtain $\alpha_k \to 0$ for $k \in K_u^2$.

Since $\{x_k\}_{K_u^2}$ is bounded, it contains a convergent subsequence $\{x_k\}_{K_u^3}$. Let $x_* = \lim_{k \in K_u^3} x_k$. Since $K_u^3 \subset K_u^2$, it also holds that $\lim_{k \in K_u^3} \alpha_k = 0$, and the proof is completed by setting $\{k_i\} = K_u^3$. □

The rest of the global convergence analysis of directional direct search relies on the geometrical properties of positive bases and on differentiability properties of f either on the entire level set $L(x_0)$ or at the limit point x_* identified in Corollary 7.2. We will consider the most relevant cases next.

Arbitrary set of positive bases

Directional direct-search methods can make use of an infinite number of positive bases as long as they do not become degenerate, namely, as long as their cosine measures are uniformly bounded away from zero. We frame this assumption next. It is also required to bound the size of all vectors in all positive bases used.

Assumption 7.3. *Let $\xi_1, \xi_2 > 0$ be some fixed positive constants. The positive bases D_k used in the algorithm are chosen from the set*

$$\mathcal{D} = \left\{ \bar{D} \text{ positive basis} : \text{cm}(\bar{D}) > \xi_1, \ \|\bar{d}\| \leq \xi_2, \bar{d} \in \bar{D} \right\}.$$

In addition, when using an infinite number of bases we require that the gradient of f is Lipschitz continuous on the level set $L(x_0)$.

Assumption 7.4. *The gradient ∇f is Lipschitz continuous in an open set containing $L(x_0)$ (with Lipschitz constant $\nu > 0$).*

We will see later that this assumption is not necessary if one uses a finite number of positive bases. By assuming the Lipschitz continuity we can use the result of Theorem 2.8 and, thus, relate the global convergence of this type of direct search to the global convergence of the derivative-free methods based on line searches or trust regions.

Under this assumption we arrive at our first global convergence result for the class of direct-search methods considered.

Theorem 7.3. *Let Assumptions 7.1, 7.2, 7.3, and 7.4 hold. Then*

$$\liminf_{k \longrightarrow +\infty} \|\nabla f(x_k)\| = 0,$$

and the sequence of iterates $\{x_k\}$ has a limit point x_* (given in Corollary 7.2) for which

$$\nabla f(x_*) = 0.$$

Proof. Corollary 7.2 showed the existence of a subsequence $\{k_i\}$ of unsuccessful iterations (or unsuccessful poll steps) for which (7.4) is true.

From Theorem 2.8 (which can be applied at unsuccessful poll steps and for k_i sufficiently large), we get that

$$\|\nabla f(x_{k_i})\| \leq \frac{\nu}{2} \text{cm}(D_{k_i})^{-1} \max_{d \in D_{k_i}} \|d\| \alpha_{k_i}. \tag{7.5}$$

As a result of Assumption 7.3, we obtain

$$\|\nabla f(x_{k_i})\| \leq \frac{\nu \xi_2}{2 \xi_1} \alpha_{k_i}.$$

Thus, we conclude that

$$\lim_{i \longrightarrow +\infty} \|\nabla f(x_{k_i})\| = 0,$$

which shows the first part of the theorem. Since ∇f is continuous, $x_* = \lim_{i \longrightarrow +\infty} x_{k_i}$ is such that $\nabla f(x_*) = 0$. $\quad \square$

One could also obtain the result of Theorem 7.3 by assuming the Lipschitz continuity of ∇f near x_* (meaning in a ball containing x_*), where x_* is the point identified in Corollary 7.2. Note that to obtain this result it is not enough to assume that ∇f is Lipschitz continuous near all the stationary points of f since, as we will see in Section 7.4, the point x_* in Corollary 7.2 might not be a stationary point if smoothness is lacking.

The step size parameter provides a natural stopping criterion for directional direct-search methods, since not only is there a subsequence of step size parameters converging to zero (Theorem 7.1), but one also has (in the continuously differentiable case) that $\nabla f(x_k) = \mathcal{O}(\alpha_k)$ after an unsuccessful iteration (see, again, Theorem 2.8 or (7.5) above). In general, α_k seems to be a reasonable measure of stationarity (even in the nonsmooth case). Dolan, Lewis, and Torczon [80] studied this issue in detail.[11] They reported results indicating that in practice one also observes that $\alpha_k = \mathcal{O}(\nabla f(x_k))$, a hypothesis confirmed by our numerical experience.

Finite set of positive bases

We will now prove global convergence under the assumption that the number of positive bases is finite, using an argument different from the proof of Theorem 7.3.

The proof of Theorem 7.3 goes through when the set of all positive bases is infinite, provided $cm(D_k)^{-1}$ is uniformly bounded. The argument used in Theorem 7.4 below, however, is heavily dependent on a finite number of different D_k's.

Both proofs have their own advantages. The proof of Theorem 7.4 is easily generalized to the nonsmooth case as we will see next. The proof of Theorem 7.3 not only is more of the style of the ones applied to analyze other methods in this book but also allows the use of an infinite number of positive bases (provided their cosine measure is uniformly bounded away from zero).

Assumption 7.5. *The set $\mathcal{D}$ of positive bases used by the algorithm is finite.*

In this case, it is enough to require the continuity of the gradient of f. The following assumption is the counterpart, for continuous differentiability of the objective function, of Assumption 7.4.

Assumption 7.6. *The function f is continuously differentiable in an open set containing $L(x_0)$.*

---

[11] In their paper it is also shown that *pattern-search methods* (directional direct-search methods based on integer lattices, as explained in Section 7.5) produce sequences of iterates for which the subsequence of unsuccessful iterates converges r-linearly to x_* (in the case where α_k is not increased at successful iterations after some finite iteration).

Now we prove, for continuously differentiable functions f, the same result as in Theorem 7.3 but following a different proof.

Theorem 7.4.   *Let Assumptions 7.1, 7.2, 7.5, and 7.6 hold. Then the sequence of iterates* $\{x_k\}$ *has a limit point* x_* *(given in Corollary 7.2) for which*

$$\nabla f(x_*) = 0.$$

Proof. Recall the definitions of the poll set P_k and of an unsuccessful iteration (which includes an unsuccessful poll step). The following is true for any unsuccessful iteration k (such that f is continuously differentiable at x_k):

$$
\begin{aligned}
f(x_k) \leq \min_{x \in P_k} f(x) &= \min_{x \in \{x_k + \alpha_k d : d \in D_k\}} f(x) \\
&= \min_{d \in D_k} f(x_k + \alpha_k d) \\
&= \min_{d \in D_k} f(x_k) + \nabla f(x_k + t_{k,d} \alpha_k d)^\top (\alpha_k d) \\
&= f(x_k) + \alpha_k \min_{d \in D_k} \nabla f(x_k + t_{k,d} \alpha_k d)^\top d,
\end{aligned}
$$

where $t_{k,d} \in (0,1)$ depends on k and d, and consequently

$$0 \leq \min_{d \in D_k} \nabla f(x_k + t_{k,d} \alpha_k d)^\top d.$$

Corollary 7.2 showed the existence of a subsequence of unsuccessful iterations $\{k_i\}$ for which (7.4) is true. The above inequality is true for this subsequence $\{k_i\}$. Since the number of positive bases is finite, there exists at least one $D_* \subset \mathcal{D}$ that is used an infinite number of times in $\{k_i\}$. Thus,

$$0 \leq \min_{d \in D_*} \nabla f(x_*)^\top d. \tag{7.6}$$

Inequality (7.6) and the property of the spanning sets given in Theorem 2.3(iv) necessarily imply $\nabla f(x_*) = 0$.   □

One could also obtain the result of Theorem 7.4 by assuming the continuity of ∇f near x_* (meaning in a ball containing x_*), where x_* is the point identified in Corollary 7.2.

7.4   Global convergence in the nonsmooth case

In the nonsmooth case one cannot expect directional direct search to globally converge to stationarity. In Figure 7.4 we depict the contours of the two-dimensional real function:

$$f(x) = \frac{1}{2} \max \left\{ \|x - c_1\|^2, \|x - c_2\|^2 \right\}, \tag{7.7}$$

where $c_1 = (1, -1)$ and $c_2 = -c_1$. This function, introduced in [145], is a variant of the Dennis–Woods function [78]. The function is continuous and strictly convex everywhere, but its gradient is discontinuous along the line $x_1 = x_2$. The function has a strict minimizer at $(0,0)$.

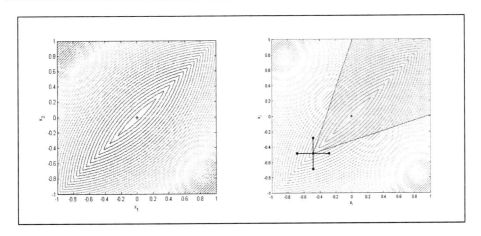

Figure 7.4. *Contours of the Dennis–Woods-type function* (7.7) *for* $c_1 = (1, -1)$ *and* $c_2 = -c_1$. *The cone of descent directions at the poll center is shaded.*

It has also been pointed out in [145] that coordinate search can fail to converge on this function. The reader can immediately see that at any point of the form (a, a), with $a \neq 0$, coordinate search generates an infinite number of unsuccessful iterations without any progress. In fact, none of the elements of $D_\oplus = [e_1 \ e_2 \ -e_1 \ -e_2]$ is a descent direction (see Figure 7.4). The descent directions of f at (a, a), with $a \neq 0$, are marked in the shaded region of the picture. Our numerical experience has not led to the observation (reported in [145]) that coordinate search frequently tends to converge to points of this form where, then, stagnation easily occurs. In fact, we found that stagnation occurs only when the starting points are too close to points on this line, as illustrated in Figure 7.5.

It is possible to prove, though, that directional direct search can generate a sequence of iterates under Assumptions 7.1, 7.2, and 7.5 which has a limit point where directional derivatives are nonnegative for all directions in a positive basis. Such a statement may not be a certificate of any type of stationarity (necessary conditions for optimality), as the example above would immediately show.

Let us consider the point x_* identified in Corollary 7.2. We will assume that f is Lipschitz continuous near x_* (meaning in a neighborhood of x_*), so that the generalized directional derivative (in the Clarke sense [54]) can assume the form

$$f^\circ(x; d) = \limsup_{y \to x, t \downarrow 0} \frac{f(y + td) - f(y)}{t}$$

for all directions $d \in \mathbb{R}^n$. Since f is Lipschitz continuous near x_*, this limit is well defined, and so is the generalized subdifferential (or subgradient)

$$\partial f(x_*) = \{s \in \mathbb{R}^n : f^\circ(x_*; v) \geq v^\top s \quad \forall v \in \mathbb{R}^n\}.$$

Moreover,

$$f^\circ(x_*; d) = \max\{d^\top s : s \in \partial f(x_*)\}.$$

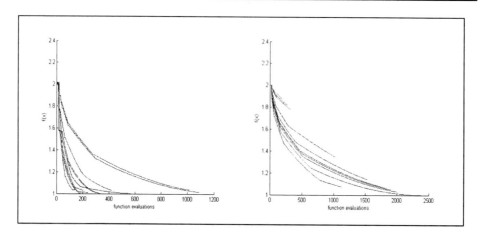

Figure 7.5. *Application of the coordinate-search method to the Dennis–Woods function (7.7) starting around the point $x_0 = (1,1)$. The plots on the left (resp., right) correspond to 10 starting points randomly generated in a box of ℓ_∞ radius 10^{-2} (resp., 10^{-3}) around the point $(1,1)$.*

Assumption 7.7. *Let x_* be the point identified in Corollary 7.2, and let the function f be Lipschitz continuous near x_*.*

Theorem 7.5. *Let Assumptions 7.1, 7.2, 7.5, and 7.7 hold. Then the sequence of iterates $\{x_k\}$ has a limit point x_* (given in Corollary 7.2) for which*

$$f^\circ(x_*;d) \geq 0 \quad \forall d \in D_*,$$

where D_* is one of the positive bases in $\mathcal{D}$.

Proof. Corollary 7.2 showed the existence of a subsequence of unsuccessful iterations $\{k_i\}$ for which (7.4) is true. Since the number of positive bases used is finite, there exists one positive basis $D_* \subset \mathcal{D}$ for which

$$f(x_{k_i} + \alpha_{k_i}d) \geq f(x_{k_i})$$

for all $d \in D_*$ (and all i sufficiently large).

From the definition of the generalized directional derivative, we get, for all $d \in D_*$, that

$$f^\circ(x_*;d) = \limsup_{y \to x_*, t \downarrow 0} \frac{f(y+td) - f(y)}{t} \geq \limsup_{k \in \{k_i\}} \frac{f(x_k + \alpha_k d) - f(x_k)}{\alpha_k} \geq 0.$$

The proof is completed. $\square$

If, in addition to Assumption 7.7, the function f is regular at x_* (meaning that the directional derivative $f'(x_*;v)$ exists and coincides with the generalized directional derivative $f^\circ(x_*;v)$ for all $v \in \mathbb{R}^n$; see [54]), then the result of Theorem 7.5 becomes

$$f'(x_*;d) \geq 0 \quad \forall d \in D_*,$$

where D_* is one of the positive bases in $\mathcal{D}$. Neither the result of Theorem 7.5 nor this particularization for regular functions implies stationarity at x_*, as expected, since $D_* \neq \mathbb{R}^n$ and as the example above demonstrates.

Further, if the function f is, so called, strictly differentiable at x_* (which is equivalent to saying that f is Lipschitz continuous near x_* and there exists a vector $w = \nabla f(x_*)$—"the gradient"—such that

$$f^\circ(x_*; v) = w^\top v$$

for all $v \in \mathbb{R}^n$; see [54]), then the result of Theorem 7.5 becomes $\nabla f(x_*)^\top d \geq 0$ for all $d \in D_*$. Then the property about positive spanning sets given in Theorem 2.3(iv) implies that $\nabla f(x_*) = 0$, which is similar to what we obtained in the continuously differentiable case (Theorem 7.4).

Note that coordinate search can still fail to converge if strict differentiability is not assumed at the point x_* of Corollary 7.2. In [145], the authors provided the example

$$\hat{f}(x) = \left(1 - e^{-10^3 \|x\|^2}\right) f(x),$$

where $f(x)$ is the modified Dennis–Woods function defined in (7.7). The function $\hat{f}(x)$ is just slightly different from $f(x)$ but is now strictly differentiable at the minimizer $(0,0)$ (but still not strictly differentiable at (a,a) with $a \neq 0$). However, the same problem can occur as before: coordinate search might find a point of the form (a,a), with $a \neq 0$, and stop since none of the directions in $D_\oplus$ provides descent at the poll steps, no matter how small the step size parameter is.

7.5 Simple decrease with integer lattices

We start by characterizing the directions in $\mathcal{D}$ used for polling. We assume $\mathcal{D}$ is finite and $\mathcal{D} = D$. As pointed out before in this book, it is convenient to regard D_k as an $n \times |D_k|$ matrix whose columns are the vectors in D_k, and, similarly, we regard the finite set D as an $n \times |D|$ matrix whose columns are the vectors in D.

Assumption 7.8. *The set $\mathcal{D} = D$ of positive bases used by the algorithm is finite. In addition, the columns of D are of the form $G \bar{z}_j$, $j = 1, \ldots, |D|$, where $G \in \mathbb{R}^{n \times n}$ is a nonsingular matrix and each $\bar{z}_j$ is a vector in $\mathbb{Z}^n$.*

Let $\bar{Z}$ denote the matrix whose columns are $\bar{z}_j$, $j = 1, \ldots, |D|$. We can therefore write $D = G\bar{Z}$. The matrix G is called a mesh or pattern generator.

In this section we will impose that all points generated by the algorithm lie on a mesh M_k defined by all possible nonnegative integer combinations of vectors in D:

$$M_k = \left\{ x_k + \alpha_k D u : u \in \mathbb{Z}_+^{|D|} \right\}, \qquad (7.8)$$

where $\mathbb{Z}_+$ is the set of nonnegative integers. The mesh M_k is centered at the current iterate x_k, and its discretization size is defined by the step size or mesh size parameter α_k. It is easy to see that the mesh (7.2) introduced for coordinate search is a particular case of M_k when $D = D_\oplus$.

Note that the range of u in the definition of M_k allows the choice of the vectors in the canonical basis of $\mathbb{R}^{|D|}$. Thus, all points of the form $x_k + \alpha_k d$, $d \in D_k$, are in M_k for any

$D_k \subset \mathcal{D}$. It is clear that $P_k \subset M_k$, and thus we need only impose the following condition on the search step.

Assumption 7.9. *The search step in Algorithm 7.2 evaluates only points in M_k defined by (7.8) for all iterations k.*

A standard way to globalize directional direct-search-type methods is to force the iterates to lie on integer lattices. This intention is accomplished by imposing Assumptions 7.8 and 7.9 and the following additional assumption.

Assumption 7.10. *The step size parameter is updated as follows: Choose a rational number $\tau > 1$, a nonnegative integer $m^+ \geq 0$, and a negative integer $m^- \leq -1$. If the iteration is successful, the step size parameter is maintained or increased by taking $\alpha_{k+1} = \tau^{m_k^+} \alpha_k$, with $m_k^+ \in \{0, \ldots, m^+\}$. Otherwise, the step size parameter is decreased by setting $\alpha_{k+1} = \tau^{m_k^-} \alpha_k$, with $m_k^- \in \{m^-, \ldots, -1\}$.*

Note that these rules respect those of Algorithm 7.2 by setting $\beta_1 = \tau^{m^-}$, $\beta_2 = \tau^{-1}$, and $\gamma = \tau^{m^+}$.

First, we prove an auxiliary result from [18] which is interesting in its own right. This result states that the minimum distance between any two distinct points in the mesh M_k is bounded from below by a multiple of the mesh parameter α_k.

Lemma 7.6. *Let Assumption 7.8 hold. For any integer $k \geq 0$, one has that*

$$\min_{\substack{y,w \in M_k \\ y \neq w}} \|y - w\| \geq \frac{\alpha_k}{\|G^{-1}\|}.$$

Proof. Let $y = x_k + \alpha_k D u_y$ and $w = x_k + \alpha_k D u_w$ be two distinct points in M_k, where $u_y, u_w \in \mathbb{Z}_+^{|D|}$ (with $u_y \neq u_w$). Then

$$\begin{aligned}
0 \neq \|y - w\| &= \alpha_k \|D(u_y - u_w)\| \\
&= \alpha_k \|G\bar{Z}(u_y - u_w)\| \\
&\geq \alpha_k \frac{\|\bar{Z}(u_y - u_w)\|}{\|G^{-1}\|} \\
&\geq \frac{\alpha_k}{\|G^{-1}\|}.
\end{aligned}$$

The last inequality is due to the fact that the norm of a vector of integers not identically zero, like $\bar{Z}(u_y - u_w)$, is never smaller than one. $\square$

It is important to remark that this result is obtained under Assumption 7.8, where the integrality requirement on the generation of the meshes plays a key role. For instance, all the positive integer combinations of directions in $\{-1, +\pi\}$ are dense in the real line, which does not happen with $\{-1, +1\}$. What is important is to guarantee a separation bounded away from zero for a fixed value of the step size parameter, and integrality is a convenient way of guaranteeing that separation.

Now we show that the sequence of step size or mesh parameters is bounded.

Lemma 7.7. *Let Assumptions* 7.1, 7.8, 7.9, *and* 7.10 *hold. There exists a positive integer* r^+ *such that* $\alpha_k \leq \alpha_0 \tau^{r^+}$ *for any* $k \in \mathbb{N}_0$.

Proof. Since $L(x_0)$ is compact, one can consider

$$\theta = \max_{y, w \in L(x_0)} \|y - w\|.$$

Now suppose that $\alpha_k > \theta \|G^{-1}\|$ for some $k \in \mathbb{N}_0$. Then Lemma 7.6, with $w = x_k$, would show us that any $y \in M_k$, different from x_k, would not belong to $L(x_0)$. Thus, if $\alpha_k > \theta \|G^{-1}\|$, then iteration k would not be successful and $x_{k+1} = x_k$.

The step size parameter could pass the bound $\theta \|G^{-1}\|$ when it is lower than it. When it does, it must be at a successful iteration, and it cannot go above $\tau^{m^+} \theta \|G^{-1}\|$, where m^+ is the upper bound on m_k^+. The sequence $\{\alpha_k\}$ must, therefore, be bounded by $\tau^{m^+} \theta \|G^{-1}\|$. Letting r^+ be an integer such that $\tau^{m^+} \theta \|G^{-1}\| \leq \alpha_0 \tau^{r^+}$ completes the proof. $\quad\square$

Since α_{k+1} is obtained by multiplying α_k by an integer power of τ, we can write, for any $k \in \mathbb{N}_0$, that

$$\alpha_k = \alpha_0 \tau^{r_k} \tag{7.9}$$

for some r_k in $\mathbb{Z}$. We now show that under the assumptions imposed in this section, Algorithm 7.2 meets the assumption used before for global convergence.

Theorem 7.8. *Let Assumptions* 7.1, 7.8, 7.9, *and* 7.10 *hold. If there exists an* $\alpha > 0$ *such that* $\alpha_k > \alpha$, *for all* k, *then the algorithm visits only a finite number of points. (In other words, Assumption 7.2 is satisfied.)*

Proof. The step size parameter is of the form (7.9), and hence to show the result we define a negative integer r^- such that $0 < \alpha_0 \tau^{r^-} \leq \alpha_k$ for all $k \in \mathbb{N}_0$. Thus, from Lemma 7.7, we conclude that r_k must take integer values in the set $\{r^-, r^- + 1, \ldots, r^+\}$ for all $k \in \mathbb{N}_0$.

One knows that x_{k+1} can be written, for successful iterations k, as $x_k + \alpha_k D u_k$ for some $u_k \in \mathbb{Z}_+^{|D|}$. In unsuccessful iterations, $x_{k+1} = x_k$ and $u_k = 0$. Replacing α_k by $\alpha_0 \tau^{r_k}$, we get, for any integer $\ell \geq 1$,

$$x_\ell = x_0 + \sum_{k=0}^{\ell-1} \alpha_k D u_k$$

$$= x_0 + \alpha_0 G \sum_{k=0}^{\ell-1} \tau^{r_k} \bar{Z} u_k$$

$$= x_0 + \frac{p^{r^-}}{q^{r^+}} \alpha_0 G \sum_{k=0}^{\ell-1} p^{r_k - r^-} q^{r^+ - r_k} \bar{Z} u_k,$$

where p and q are positive integer numbers satisfying $\tau = p/q$. Since

$$\sum_{k=0}^{\ell-1} p^{r_k - r^-} q^{r^+ - r_k} \bar{Z} u_k$$

is a vector of integers for all $\ell \in \mathbb{N}$, we have just proved that the sequence of iterates $\{x_k\}$ lies in a set of the form (an integer lattice)

$$\mathcal{L} = \{x_0 + G_0 z : z \in \mathbb{Z}^n\},$$

where

$$G_0 = \frac{p^{r^-}}{q^{r^+}} \alpha_0 G$$

is a nonsingular $n \times n$ matrix. Now note that the intersection of $\mathcal{L}$ with the compact $L(x_0)$ is necessarily a finite set, which shows that the algorithm must visit only a finite number of points. $\square$

It is important to stress that no properties of f are specifically required in Theorem 7.8.

The rest of this section focuses on particular cases of the direct-search framework presented and on some extensions which preserve the asymptotic behavior of the step size parameter. A reader not acquainted with the convergence properties of these directional direct-search methods might postpone the rest of this section to a future reading.

Tightness of the integrality and rationality requirements

Assumptions 7.8 and 7.10 are necessary for Theorems 7.1 and 7.8 to hold, when a finite number of positive bases is used and only simple decrease imposed.

It is possible to show that the requirement of integrality stated in Assumption 7.8 for the positive bases cannot be lifted. An example constructed by Audet [12] shows an instance of Algorithm 7.2 which does not meet the integrality requirement of Assumption 7.8 for the positive bases and for which the step size parameter α_k is uniformly bounded away from zero when applied to a particular smooth function f.

Audet [12] also proved that the requirement of rationality on τ is tight. He provided an instance of Algorithm 7.2 for an irrational choice of τ, which for a given function f generates step size parameters α_k uniformly bounded away from zero. We point out that the function f used in this counterexample is discontinuous, which is acceptable under the assumptions of this section.

When $\tau > 1$ is an integer (and not just rational) the analysis above can be further simplified. In fact, one can easily see that $q = 1$ in the proof of Theorem 7.8 and the upper bound r^+ on r_k is no longer necessary.

Other definitions for the mesh

Instead of as in (7.8), the mesh M_k could be defined more generally as

$$M_k = \{x_k + \alpha_k D u : u \in \mathcal{Z}\}$$

as long as the set $\mathcal{Z} \subset \mathbb{Z}^{|D|}$ contains all the vectors of the canonical basis of $\mathbb{R}^{|D|}$ (so that $P_k \subset M_k$). Another possible generalization is sketched in the exercises.

For instance one could set

$$M_k = \{x_k + \alpha_k(jd) : d \in D,\ j \in \mathbb{Z}_+\},$$

which would amount to considering only mesh points along the vectors $d \in D$. Figure 7.6 displays two meshes M_k when $n = 2$, for the cases where D contains one and three maximal positive bases.

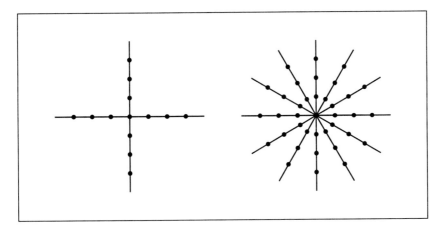

Figure 7.6. *Two pointed meshes, when D has one maximal positive basis (left) and three maximal positive bases (right).*

Complete polling and asymptotic results

Another example is provided by Audet [12], which shows that the application of an instance of Algorithm 7.2 under Assumption 7.8 to a continuously differentiable function can generate an infinite number of limit points, one of them not being stationary. Thus, the result

$$\liminf_{k \longrightarrow +\infty} \|\nabla f(x_k)\| = 0$$

cannot be extended to

$$\lim_{k \longrightarrow +\infty} \|\nabla f(x_k)\| = 0, \tag{7.10}$$

without further assumptions or modifications to the direct-search schemes in Algorithm 7.2. Such an observation is consistent with a similar one in trust-region algorithms for unconstrained nonlinear optimization, pointed out by Yuan [236]. This author constructed an example where a trust-region method based on simple decrease to accept new points (rather than a sufficient decrease condition—see Chapter 10) generates a sequence of iterates that does not satisfy (7.10) either.

Conditions under which it is possible to obtain (7.10) have been analyzed by Torczon [217] (see also [145]). The modifications in the directional direct-search framework are essentially two.

First, it is required that

$$\lim_{k \longrightarrow +\infty} \alpha_k = 0.$$

From Theorem 7.1, one way of guaranteeing this condition is by never increasing the step size parameter at successful iterations.

Second, there is the imposition of the so-called complete polling at all successful iterations. Complete polling requires the new iterate generated in the poll step to minimize the function in the poll set:

$$f(x_{k+1}) = f(x_k + \alpha_k d_k) \leq f(x_k + \alpha_k d) \quad \forall d \in D_k.$$

Complete polling necessarily costs $|D_k|$ (with $|D_k| \geq n+1$) function evaluations at every poll step, and not only at unsuccessful poll steps like in regular polling. The new iterate x_{k+1} could also be computed in a search step as long as $f(x_{k+1}) \leq f(x_k + \alpha_k d)$, for all $d \in D_k$, which means that the search step would have then to follow the (complete) poll step.

The proof of (7.10) under these two modifications is omitted. It can be accomplished in two phases. In a first phase, it is proved that for any $\epsilon > 0$ there exist $\alpha^-, \eta > 0$ such that $f(x_k + \alpha_k d_k) \leq f(x_k) - \eta \alpha_k \|\nabla f(x_k)\|$ if $\|\nabla f(x_k)\| > \epsilon$ and $\alpha_k < \alpha^-$. This inequality can be interpreted as a sufficient-decrease-type condition (see Chapters 9 and 10). A second phase consists of applying the Thomas argument [214] known for deriving lim-type results for trust-region methods (i.e., convergence results for the whole sequence of iterates; see also Chapter 10). The details are in [145].

7.6   The mesh adaptive direct-search method

Audet and Dennis introduced in [19] a class of direct-search algorithms capable of achieving global convergence in the nonsmooth case. This class of methods is called mesh adaptive direct search (MADS) and can be seen as an instance of Algorithm 7.2.

The globalization is achieved by simple decrease with integer lattices. So, let the mesh M_k (given, for instance, as in (7.8) but always by means of a finite D) be defined by Assumptions 7.8 and 7.9. Also let α_k be updated following Assumption 7.10. The key point in MADS is that $\mathcal{D}$ is allowed to be infinite, and thus different from the finite set D—which is important to allow some form of stationarity in the limit for the nonsmooth case—while the poll set P_k (defined in (7.3)) is still defined as a subset of the mesh M_k. (An earlier approach also developed to capture a rich set of directions can be found in [10].)

Thus, MADS first performs a search step by evaluating the objective function at a finite number of points in the mesh M_k. If the search step fails or is skipped, a poll set is tried by evaluating the objective function at the poll set P_k defined by the positive basis D_k chosen from a set of positive bases $\mathcal{D}$ (which is not necessarily explicitly given). However, this set $\mathcal{D}$ is now defined so that the elements $d_k \in D_k$ satisfy the following conditions:

- d_k is a nonnegative integer combination of the columns of D.

- The distance between x_k and the point $x_k + \alpha_k d_k$ tends to zero if and only if α_k does:

$$\lim_{k \in K} \alpha_k \|d_k\| = 0 \iff \lim_{k \in K} \alpha_k = 0 \qquad (7.11)$$

for any infinite subsequence K.

- The limits of all convergent subsequences of $\bar{D}_k = \{d_k/\|d_k\| : d_k \in D_k\}$ are positive bases.

In the spirit of the presentation in [19] we now define the concepts of refining subsequence and refining direction.

Definition 7.9. *A subsequence $\{x_k\}_{k \in K}$ of iterates corresponding to unsuccessful poll steps is said to be a refining subsequence if $\{\alpha_k\}_{k \in K}$ converges to zero.*
   Let x be the limit point of a convergent refining subsequence. If the limit $\lim_{k \in L} d_k/\|d_k\|$ exists, where $L \subseteq K$ and $d_k \in D_k$, then this limit is said to be a refining direction for x.

The existence of a convergent refining subsequence is nothing else than a restatement of Corollary 7.2. It is left as an exercise to confirm that this result is still true for MADS. The next theorem states that the Clarke generalized derivative is nonnegative along any refining direction for x_* (the limit point of Corollary 7.2).

Theorem 7.10. *Let Assumptions 7.1, 7.7, 7.8, 7.9, and 7.10 hold. Then the sequence of iterates $\{x_k\}$ generated by MADS has a limit point x_* (given in Corollary 7.2) for which*

$$f^\circ(x_*; v) \geq 0$$

for all refining directions v for x_*.

Proof. Let $\{x_k\}_{k \in K}$ be the refining subsequence converging to x_* guaranteed by Corollary 7.2, and let $v = \lim_{k \in L} d_k/\|d_k\|$ be a refining direction for x_*, with $d_k \in D_k$ for all $k \in L$. Since f is Lipschitz continuous near x_* and $d_k/\|d_k\| \to v$ and $\alpha_k \|d_k\| \to 0$, for all $k \in L$,

$$f^\circ(x_*; v) \geq \limsup_{k \in L} \frac{f(x_k + \alpha_k \|d_k\| \frac{d_k}{\|d_k\|}) - f(x_k)}{\alpha_k \|d_k\|}. \qquad (7.12)$$

Since x_k is an unsuccessful poll step,

$$\limsup_{k \in L} \frac{f(x_k + \alpha_k d_k) - f(x_k)}{\alpha_k \|d_k\|} \geq 0,$$

and the proof is completed.   □

Audet and Dennis [19] (see also [15]) proposed a scheme to compute the positive bases D_k, called lower triangular matrix based mesh adaptive direct search (LTMADS), which produces a set of refining directions for x_* with union asymptotically dense in $\mathbb{R}^n$ with probability one. From this and Theorem 7.10, MADS is thus able to converge to a point where the Clarke generalized directional derivative is nonnegative for a set of directions dense a.e. in $\mathbb{R}^n$, and not just for a finite set of directions as in Theorem 7.5. And more recently, Abramson et al. [8] proposed an alternative scheme to generate the positive bases D_k (also related to [10]), called OrthoMADS. This strategy also generates an asymptotically dense set of directions, but in a deterministic way, and each positive basis D_k is constructed from an orthogonal basis, thus determining relatively efficiently a reduction of the unexplored regions.

7.7   Imposing sufficient decrease

One alternative to the integrality requirements of Assumption 7.8, which would still provide global convergence for directional direct search, is to accept new iterates only if they satisfy a sufficient decrease condition. We will assume—in this section—that a new point $x_{k+1} \neq x_k$ is accepted (both in search and poll steps) only if

$$f(x_{k+1}) < f(x_k) - \rho(\alpha_k), \tag{7.13}$$

where the *forcing function* $\rho : \mathbb{R}_+ \to \mathbb{R}_+$ is continuous, positive, and satisfies

$$\lim_{t \to 0^+} \frac{\rho(t)}{t} = 0 \qquad \text{and} \qquad \rho(t_1) \leq \rho(t_2) \quad \text{if} \quad t_1 < t_2.$$

A simple example of a forcing function is $\rho(t) = t^2$. Functions of the form $\rho(t) = t^{1+a}$, for $a > 0$, are also in this category.

Theorem 7.11. *Suppose Algorithm 7.2 is modified in order to accept new iterates only if (7.13) holds.*

Let Assumption 7.1 hold. If there exists an $\alpha > 0$ such that $\alpha_k > \alpha$, for all k, then the algorithm visits only a finite number of points. (In other words, Assumption 7.2 is satisfied.)

Proof. Since ρ is monotonically increasing, we know that $0 < \rho(\alpha) \leq \rho(\alpha_k)$ for all $k \in \mathbb{N}_0$.

Suppose that there exists an infinite subsequence of successful iterates. From inequality (7.13) we get, for all successful iterations, that

$$f(x_{k+1}) < f(x_k) - \rho(\alpha_k) \leq f(x_k) - \rho(\alpha).$$

Recall that at unsuccessful iterations $f(x_{k+1}) = f(x_k)$. As a result, the sequence $\{f(x_k)\}$ must converge to $-\infty$, which clearly contradicts Assumption 7.1.   □

To prove a result of the type of Theorem 7.3 for Algorithm 7.2, under the modification given in (7.13), we need to show first that for unsuccessful poll steps k_i one has

$$\|\nabla f(x_{k_i})\| \leq \left(\frac{v}{2} \mathrm{cm}(D_{k_i})^{-1} \max_{d \in D_{k_i}} \|d\| \right) \alpha_{k_i} + \frac{\mathrm{cm}(D_{k_i})^{-1}}{\min_{d \in D_{k_i}} \|d\|} \frac{\rho(\alpha_{k_i})}{\alpha_{k_i}}, \tag{7.14}$$

which is left as an exercise. Now, given the properties of the forcing function ρ, it is clear that $\nabla f(x_{k_i}) \to 0$ when $\alpha_{k_i} \to 0$, provided the minimum size of the vectors in D_k does not approach zero.

When imposing sufficient decrease, one can actually prove that the whole sequence of step size parameters converges to zero.

Theorem 7.12. *Suppose Algorithm 7.2 is modified in order to accept new iterates only if* (7.13) *holds.*
Let Assumption 7.1 hold. Then the sequence of step size parameters satisfies

$$\lim_{k \longrightarrow +\infty} \alpha_k = 0.$$

The proof is also left as an exercise. Note that Assumption 7.1 is more than what is necessary to prove Theorems 7.11 and 7.12. In fact, it would had been sufficient to assume that f is bounded from below on $L(x_0)$.

7.8 Other notes and references

The first reference in the literature to direct search has been attributed to a 1952 report of Fermi and Metropolis [90], in a form that resembles coordinate search (see the preface in Davidon [73]). In the 1950s, Box [42] and Box and Wilson [44] introduced direct-search algorithms related to coordinate search, using positive integer combinations of $D_\oplus$. Their algorithms evaluated points in M_k but not necessarily in P_k. Some of the vocabulary used at this time (like *two-level factorial and composite designs* [44]) was inspired from statistics where much of the early work on direct search was developed.[12]

Hooke and Jeeves [130] seemed to have been the first to use the terminology *direct-search methods*. Hooke and Jeeves [130] are also acknowledged to have been the first to recognize the underlying notion of pattern or integer lattice in direct search, which was then explored by other authors, in particular by Berman [35]. Hooke and Jeeves' *exploratory moves* scheme is a predecessor of the search step. Later, in the 1990s, Torczon [216, 217] showed how to rigorously use integer lattices in the globalization of *pattern-search methods* (which can be defined as directional direct-search methods based on such lattices, as explained in Section 7.5). Audet and Dennis [18] contributed to the field by essentially focusing the analysis of these methods on the subsequence of unsuccessful iterates. The paper of Booker et al. [40] should get most of the credit for the formal statement of the search-poll framework.

But the pioneer work on direct search was not confined to directional methods based on patterns generated by fixed sets of directions. In fact, some of the earliest directional direct-search methods modified the search directions at the end of each iteration by combining, in some form, a previously computed set of directions. Among such methods are the ones by Powell [183] which used conjugate directions (see also the modifications introduced by Zangwill [237] and the analysis in Toint and Callier [215]) and by Rosenbrock [201]. A recent approach has been pursued by Frimannslund and Steihaug [100] by explicitly rotating the direction set based on curvature information extracted from function values.

The introduction of a sufficient decrease condition (involving the step size) in direct search was first made by Yu [235] in 1979. Other authors have explored the use of such a condition in directional direct-search methods, like Grippo, Lampariello, and Lucidi [115], Lucidi and Sciandrone [160], and García-Palomares and Rodríguez [103]. The work of Lucidi and Sciandrone [160], in particular, concerns the development of an algorithmic framework, exploring the use of line-search techniques in directional direct-search

---

[12]J. A. Nelder and R. Mead were also statisticians.

methods. Their convergence theory includes first-order lim-type results derived under reasonable assumptions. A particularity of the approaches in [103, 160] is the consideration of different step sizes along different directions. Diniz-Ehrhardt, Martínez, and Raydan [79] used a sufficient decrease condition in the design of a nonmonotone algorithm.

In the context of globalization of directional direct-search methods by integer lattices (see Section 7.5), it is possible in suitable cases to relax the assumption that the directions are extracted from a finite set D. This has been explored by Coope and Price [66] in their grid-based methods. They have observed that after an unsuccessful iteration one can change D (provided it still satisfies Assumption 7.8) and, thus, gain further flexibility in attempting to capture the curvature of the function. However, there is a price to pay, namely, that $\alpha_k \to 0$ should be imposed, which, for example, can be guaranteed in the context of Theorems 7.1 and 7.8 by never allowing α_k to increase.

There has been some effort in trying to develop efficient serial implementations of pattern-search methods by considering particular instances where the problem structure can be exploited efficiently. Price and Toint [195] examined how to take advantage of partial separability. Alberto et al. [10] have shown ways of incorporating user-provided function evaluations. Abramson, Audet, and Dennis [6] looked at the case where some incomplete form of gradient information is available. Custódio and Vicente [70] suggested several procedures, for general objective functions, to improve the efficiency of pattern-search methods using simplex derivatives. In particular, they showed that ordering the poll directions in opportunistic polling according to a negative simplex gradient can lead to a significant reduction in the overall number of function evaluations (see [68, 70]).

One attractive feature of directional direct-search methods is that it is easy to parallelize the process of evaluating the function during poll steps. Many authors have experimented with different parallel versions of these methods; see [10, 21, 77, 103, 132]. Asynchronous parallel forms of these methods have been proposed and analyzed by García-Palomares and Rodríguez [103] and Hough, Kolda, and Torczon [132] (see also the software produced by Gray and Kolda [110]).

Another attractive feature is the exploration of the directionality aspect to design algorithms for nonsmooth functions with desirable properties. We have mentioned in Section 7.6 that the MADS methods can converge with probability one to a first-order stationary, nonsmooth point. It is shown in [3] how to generalize this result to second-order stationary points with continuous first-order derivatives but nonsmooth second-order derivatives. Other direct-search approaches to deal with nonsmooth functions have recently been proposed [24, 37] but for specific types of nondifferentiability.

The generating search set (GSS) framework

Kolda, Lewis, and Torczon [145] introduced another framework for globally convergent directional direct-search methods. These authors do not make an explicit separation in their algorithmic description between the search step and the poll step. A successful iterate in their GSS framework is of the form $x_k + \alpha_k d_k$, where d_k belongs to a set of directions $G_k \cup H_k$. In GSS, G_k plays the role of our D_k (used in the poll step). The search step is accommodated by the additional set of directions H_k (which might, as in the framework presented in Section 7.2, be empty). When the iterates are accepted solely based on simple decrease of the objective function, integrality requirements similar to those of

Assumption 7.8 (or something equivalent) must be imposed on the finite set of directions that contains all choices of $G_k \cup H_k$ for all $k \in \mathbb{Z}_+$.

The multidirectional search method

Another avenue followed in the development of direct search is simplicial methods (like the Nelder and Mead simplex method [177]). Simplicial direct-search methods, despite sharing features with directional direct search, have their own motivation and, thus, will be treated separately in Chapter 8.

The multidirectional search (MDS) method of Dennis and Torczon [77], described next, can be regarded as both a directional and a simplicial direct-search method. We choose to include MDS in the book essentially for historical reasons and because it will help us frame the modifications necessary to make the Nelder–Mead method globally convergent.

As in the Nelder–Mead method (whose details are not needed now), MDS starts with a simplex of $n + 1$ vertices $Y = \{y^0, y^1, \ldots, y^n\}$. Each iteration is centered at the simplex vertex y^0 with the lowest function value (in contrast with Nelder–Mead which focuses particular attention at the vertex with the highest function value). Then a rotated simplex is formed by rotating the vertices $y^i, i = 1, \ldots, n$, $180°$ around y^0 (see Figure 7.7). (The reader might have already identified a maximal positive basis...) If the best objective value of the rotated vertices is lower than $f(y^0)$, then an expanded simplex is formed in the direction of the rotated one (see Figure 7.7). The next iteration is started from either the rotated or expanded simplex, depending on which is better. If the best objective value of the rotated vertices is no better than $f(y^0)$, then a shrink step is taken just as in Nelder–Mead (see Figure 7.7), and the next iteration is started from the shrunken simplex. We now give more details on the MDS algorithm.

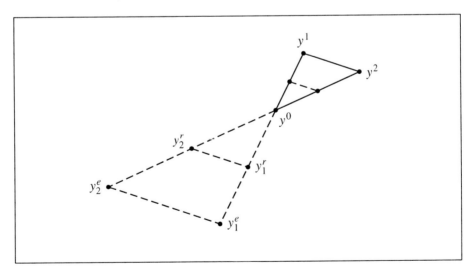

Figure 7.7. *Original simplex, rotated vertices, expanded vertices, shrunken vertices, corresponding to an MDS iteration.*

Algorithm 7.3 (The MDS method).

Initialization: Choose an initial simplex of vertices $Y_0 = \{y_0^0, y_0^1, \dots, y_0^n\}$. Evaluate f at the points in Y_0. Choose constants:

$$0 < \gamma^s < 1 < \gamma^e.$$

For $k = 0, 1, 2, \dots$

    0. Set $Y = Y_k$.

    1. **Find best vertex:** Order the $n + 1$ vertices of $Y = \{y^0, y^1, \dots, y^n\}$ so that $f^0 = f(y^0) \le f(y^i)$, $i = 1, \dots, n$.

    2. **Rotate:** Rotate the simplex around the best vertex y^0:

$$y_i^r = y^0 - (y^i - y^0), \quad i = 1, \dots, n.$$

    Evaluate $f(y_i^r)$, $i = 1, \dots, n$, and set $f^r = \min\{f(y_i^r): i = 1, \dots, n\}$. If $f^r < f^0$, then attempt an expansion (and then take the best of the rotated or expanded simplices). Otherwise, contract the simplex.

    3. **Expand:** Expand the rotated simplex:

$$y_i^e = y^0 - \gamma^e(y^i - y^0), \quad i = 1, \dots, n.$$

    Evaluate $f(y_i^e)$, $i = 1, \dots, n$, and set $f^e = \min\{f(y_i^e): i = 1, \dots, n\}$. If $f^e < f^r$, then accept the expanded simplex and terminate the iteration: $Y_{k+1} = \{y^0, y_1^e, \dots, y_n^e\}$. Otherwise, accept the rotated simplex and terminate the iteration: $Y_{k+1} = \{y^0, y_1^r, \dots, y_n^r\}$.

    4. **Shrink:** Evaluate f at the n points $y^0 + \gamma^s(y^i - y^0)$, $i = 1, \dots, n$, and replace $y^1, \dots, y^n$ by these points, terminating the iteration: $Y_{k+1} = \{y^0 + \gamma^s(y^i - y^0), i = 0, \dots, n\}$.

    Typical values for γ^s and γ^e are $1/2$ and 2, respectively. A stopping criterion could consist of terminating the run when the diameter of the simplex becomes smaller than a chosen tolerance $\Delta_{tol} > 0$ (for instance, $\Delta_{tol} = 10^{-5}$).

    Torczon [216] noted that, provided γ^s and γ^e are rational numbers, all possible vertices visited by the algorithm lie in an integer lattice. This property is independent of the position in each simplex taken by its best vertex. In addition, note that the MDS algorithm enforces a simple decrease to accept new iterates (otherwise, the simplex is shrunk and the best vertex is kept the same). Thus, once having proved the integer lattice statement, the proof of the following theorem follows trivially from the material of Section 7.3.

Theorem 7.13. *Suppose that $\gamma^s, \gamma^e \in \mathbb{Q}$, and let the initial simplex be of the form $Y_0 = G\bar{Z}$, where $G \in \mathbb{R}^{n \times n}$ is nonsingular and the components of $\bar{Z} \in \mathbb{R}^{n \times (n+1)}$ are integers. Assume that $L(y_0^0) = \{x \in \mathbb{R}^n : f(x) \le f(y_0^0)\}$ is compact and that f is continuously differentiable in $L(y_0^0)$. Then the sequence of iterates $\{y_k^0\}$ generated by the MDS method (Algorithm 7.3) has one stationary limit point x_*.*

Proof. We need to convince the reader that we can frame MDS in the format of directional direct search (Algorithm 7.2). Notice, first, that the expanded step can be seen as a search step. Polling is complete and involves a maximal positive basis D_k related to the initial simplex and chosen from the set D formed by

$$\left\{ y_0^j - y_0^i, \ j = 0, \ldots, n, \ j \neq i \right\} \cup \left\{ -(y_0^j - y_0^i), \ j = 0, \ldots, n, \ j \neq i \right\},$$

$i = 0, \ldots, n$. It is then a simple matter to see that the integer lattice requirements (see Section 7.5), i.e., Assumptions 7.8, 7.9, and 7.10, are satisfied. □

Given the pointed nature of the meshes generated by Algorithm 7.3, it is not necessary that Y_0 takes the form given in Theorem 7.13. In fact, the original proof in [216] does not impose this assumption. We could also have lifted it here, but that would require a modification of the mesh/grid framework of Section 7.5.

Another avenue to make MDS globally convergent to stationary points is by imposing sufficient decrease in the acceptance conditions, as is done in Chapter 8 for the modified Nelder–Mead method.

7.9   Exercises

1. In the context of the globalization of the directional direct-search method (Algorithm 7.2) with simple decrease with integer lattices (Section 7.5), prove that the mesh M_k defined by (7.8) can be generalized to

$$M_k = \bigcup_{x \in \mathcal{E}_k} \{ x + \alpha_k D u : u \in \mathbb{Z}_+^{|D|} \}, \tag{7.15}$$

   where $\mathcal{E}_k$ is the set of points where the objective function f has been evaluated by the start of iteration k (and $\mathbb{Z}_+$ is the set of nonnegative integers).

2. Let the mesh be defined by Assumptions 7.8 and 7.9 for the MADS methods. Let α_k be updated following Assumption 7.10. Prove under Assumption 7.1 that the result of Corollary 7.2 is true (in other words that there exists a convergent refining subsequence).

3. Show (7.12). You will need to add and subtract a term and use the Lipschitz continuity of f near x_*.

4. Generalize Theorem 2.8 for unsuccessful poll steps when a sufficient decrease condition of the form (7.13) is imposed. Show that what you get is precisely the bound (7.14).

5. Prove Theorem 7.12.

Chapter 8

Simplicial direct-search methods

The Nelder–Mead algorithm [177] is one of the most popular derivative-free methods. It has been extensively used in the engineering community and is probably the most widely cited of the direct-search methods (the 1965 paper by Nelder and Mead [177] is officially a *Science Citation Classic*). Among the reasons for its success are its simplicity and its ability to adapt to the curvature of the function being minimized. In this chapter we will describe the original Nelder–Mead method for solving (1.1) and some of its features. We will show why it can fail and how it can be fixed to globally converge to stationary points.

8.1 The Nelder–Mead simplex method

The Nelder–Mead algorithm [177] is a direct-search method in the sense that it evaluates the objective function at a finite number of points per iteration and decides which action to take next solely based on those function values and without any explicit or implicit derivative approximation or model building. Every iteration in $\mathbb{R}^n$ is based on a simplex of $n + 1$ vertices $Y = \{y^0, y^1, \ldots, y^n\}$ ordered by increasing values of f. See Section 2.5 for the definition and basic properties of simplices.

The most common Nelder–Mead iterations perform a reflection, an expansion, or a contraction (the latter can be inside or outside the simplex). In such iterations the worst vertex y^n is replaced by a point in the line that connects y^n and y^c,

$$y = y^c + \delta(y^c - y^n), \quad \delta \in \mathbb{R},$$

where $y^c = \sum_{i=0}^{n-1} y^i / n$ is the centroid of the best n vertices. The value of δ indicates the type of iteration. For instance, when $\delta = 1$ we have a (genuine or isometric) reflection, when $\delta = 2$ an expansion, when $\delta = 1/2$ an outside contraction, and when $\delta = -1/2$ an inside contraction. In Figure 8.1, we plot these four situations.

A Nelder–Mead iteration can also perform a simplex shrink, which rarely occurs in practice. When a shrink is performed all the vertices in Y are thrown away except the best one y^0. Then n new vertices are computed by shrinking the simplex at y^0, i.e., by computing, for instance, $y^0 + 1/2(y^i - y^0)$, $i = 1, \ldots, n$. See Figure 8.2. We note that the "shape" of the resulting simplices can change by being stretched or contracted, unless a shrink occurs—as we will study later in detail.

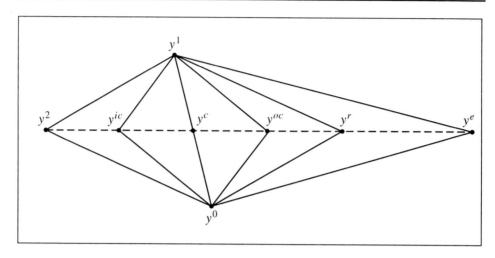

Figure 8.1. *Reflection, expansion, outside contraction, and inside contraction of a simplex, used by the Nelder–Mead method.*

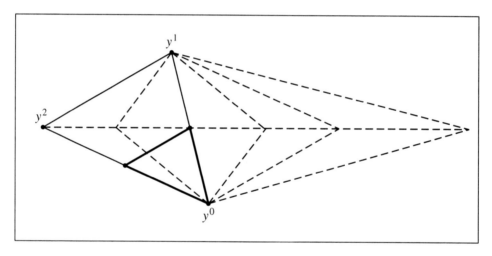

Figure 8.2. *Shrink of a simplex, used by the Nelder–Mead method.*

The Nelder–Mead method is described in Algorithm 8.1. The standard choices for the coefficients used are

$$\gamma^s = \frac{1}{2}, \quad \delta^{ic} = -\frac{1}{2}, \quad \delta^{oc} = \frac{1}{2}, \quad \delta^r = 1, \quad \text{and} \quad \delta^e = 2. \tag{8.1}$$

Note that, except for shrinks, the emphasis is on replacing the worse vertex rather than improving the best. It is also worth mentioning that the Nelder–Mead method does not parallelize well since the sampling procedure is necessarily sequential (except at a shrink).

Algorithm 8.1 (The Nelder–Mead method).

Initialization: Choose an initial simplex of vertices $Y_0 = \{y_0^0, y_0^1, \ldots, y_0^n\}$. Evaluate f at the points in Y_0. Choose constants:

$$0 < \gamma^s < 1, \qquad -1 < \delta^{ic} < 0 < \delta^{oc} < \delta^r < \delta^e.$$

For $k = 0, 1, 2, \ldots$

    0. Set $Y = Y_k$.

    1. **Order:** Order the $n+1$ vertices of $Y = \{y^0, y^1, \ldots, y^n\}$ so that

$$f^0 = f(y^0) \ \leq \ f^1 = f(y^1) \ \leq \ \cdots \ \leq \ f^n = f(y^n).$$

    2. **Reflect:** Reflect the worst vertex y^n over the centroid $y^c = \sum_{i=0}^{n-1} y^i / n$ of the remaining n vertices:

$$y^r \ = \ y^c + \delta^r (y^c - y^n).$$

    Evaluate $f^r = f(y^r)$. If $f^0 \leq f^r < f^{n-1}$, then replace y^n by the reflected point y^r and terminate the iteration: $Y_{k+1} = \{y^0, y^1, \ldots, y^{n-1}, y^r\}$.

    3. **Expand:** If $f^r < f^0$, then calculate the expansion point

$$y^e \ = \ y^c + \delta^e (y^c - y^n)$$

    and evaluate $f^e = f(y^e)$. If $f^e \leq f^r$, replace y^n by the expansion point y^e and terminate the iteration: $Y_{k+1} = \{y^0, y^1, \ldots, y^{n-1}, y^e\}$. Otherwise, replace y^n by the reflected point y^r and terminate the iteration: $Y_{k+1} = \{y^0, y^1, \ldots, y^{n-1}, y^r\}$.

    4. **Contract:** If $f^r \geq f^{n-1}$, then a contraction is performed between the best of y^r and y^n.

      (a) **Outside contraction:** If $f^r < f^n$, perform an outside contraction

$$y^{oc} \ = \ y^c + \delta^{oc} (y^c - y^n)$$

      and evaluate $f^{oc} = f(y^{oc})$. If $f^{oc} \leq f^r$, then replace y^n by the outside contraction point y_k^{oc} and terminate the iteration: $Y_{k+1} = \{y^0, y^1, \ldots, y^{n-1}, y^{oc}\}$. Otherwise, perform a shrink.

      (b) **Inside contraction:** If $f^r \geq f^n$, perform an inside contraction

$$y^{ic} \ = \ y^c + \delta^{ic} (y^c - y^n)$$

      and evaluate $f^{ic} = f(y^{ic})$. If $f^{ic} < f^n$, then replace y^n by the inside contraction point y^{ic} and terminate the iteration: $Y_{k+1} = \{y^0, y^1, \ldots, y^{n-1}, y^{ic}\}$. Otherwise, perform a shrink.

    5. **Shrink:** Evaluate f at the n points $y^0 + \gamma^s(y^i - y^0)$, $i = 1, \ldots, n$, and replace $y^1, \ldots, y^n$ by these points, terminating the iteration: $Y_{k+1} = \{y^0 + \gamma^s(y^i - y^0), i = 0, \ldots, n\}$.

A stopping criterion could consist of terminating the run when the diameter of the simplex becomes smaller than a chosen tolerance $\Delta_{tol} > 0$ (for instance, $\Delta_{tol} = 10^{-5}$).

This algorithmic description is what we can refer to as the "modern interpretation" of the original Nelder–Mead algorithm [177], which had several ambiguities about strictness of inequalities and tie breaking. The only significant difference between Algorithm 8.1 and the original Nelder–Mead method [177] is that in the original version the expansion point y^e is accepted if $f^e < f^0$ (otherwise, the reflection point y^r is accepted). The standard practice nowadays [149, 169] is to accept the best of y^r and y^e if both improve over y^0, as is done in Algorithm 8.1.

The Nelder–Mead algorithm performs the following number of function evaluations per iteration:

1	if the iteration is a reflection,
2	if the iteration is an expansion or contraction,
$n+2$	if the iteration is a shrink.

Lexicographic decrease at nonshrink iterations

We focus our attention now on how ties are broken in Algorithm 8.1 when equal function values occur. The way in which the initial points are originally ordered when ties occur is not relevant to what comes next. It also makes no difference how these ties are broken among the n new points calculated in the shrink step.

However, we need tie-breaking rules if we want to well define the smallest index k^* of a vertex that differs between iterations k and $k + 1$,

$$k^* = \min\left\{i \in \{0, 1, \ldots, n\} : y_k^i \neq y_{k+1}^i\right\}.$$

It is a simple matter to see that such tie-breaking rules involve only the situations reported in the next two paragraphs.

When a new point is brought to the simplex in the reflection, expansion, or contraction steps, there might be a point in the simplex which already has the same objective function value. We need to define tie-breaking rules to avoid unnecessary modifications to the change index k^*. We adopt here the natural rule suggested in [149]. If a new accepted point (y_k^r, y_k^e, y_k^{oc}, or y_k^{ic}) produces an objective function value equal to the value of one (or more than one) of the points $y_k^0, \ldots, y_k^{n-1}$, then it is inserted into Y^{k+1} with an index larger than that of such a point (or points). In this way the change index k^* remains the same whenever points with identical function values are generated in consecutive iterations.

Another situation where tie breaking is necessary to avoid modifications by chance on the change index k^* occurs at a shrink step when the lowest of the values $f(y_k^0 + \gamma^s(y_k^i - y_k^0))$, $i = 1, \ldots, n$, is equal to $f(y_k^0)$. In such a case, we set y_{k+1}^0 to y_k^0.

Thus, k^* takes the following values:

$1 \leq k^* \leq n - 1$	if the iteration ends at a reflection step,
$k^* = 0$	if the iteration ends at an expansion step,
$0 \leq k^* \leq n$	if the iteration ends at a contraction step,
$k^* = 0$ or 1	if the iteration ends at a shrink step.

In addition, the definition of the change index k^*, under the two above-mentioned tie-breaking rules, implies that at a nonshrink iteration

$$f_{k+1}^j = f_k^j \quad \text{and} \quad y_{k+1}^j = y_k^j \quad \text{if } j < k^*,$$

$$f_{k+1}^j < f_k^j \quad \text{and} \quad y_{k+1}^j \neq y_k^j \quad \text{if } j = k^*,$$

$$f_{k+1}^j = f_k^{j-1} \quad \text{and} \quad y_{k+1}^j = y_k^{j-1} \quad \text{if } j > k^*.$$

We observe that the vector $(f_k^0, \ldots, f_k^n)$ decreases lexicographically at nonshrink iterations. It follows from these statements that, at nonshrink iterations,

$$\sum_{j=0}^n f_{k+1}^j < \sum_{j=0}^n f_k^j.$$

This property of the Nelder–Mead algorithm has been explored by several authors. Kelley [140] used it to detect and remedy stagnation in the context of the Nelder–Mead method. Tseng [220] suggested a class of simplex-type methods that includes a modified Nelder–Mead method where this inequality plays a relevant role (see Section 8.3).

Note that the worst vertex function value might not necessarily decrease after a non-shrink iteration. For instance, suppose that $n = 4$ and that the vertex function values are $(f_k^0, f_k^1, f_k^2, f_k^3, f_k^4) = (1, 2, 2, 3, 3)$ at the nonshrink iteration k. Suppose also that the new vertex has function value 2. Then the vertex function values at iteration $k + 1$ are $(f_{k+1}^0, f_{k+1}^1, f_{k+1}^2, f_{k+1}^3, f_{k+1}^4) = (1, 2, 2, 2, 3)$. It is clear from this example that the worst vertex function has not improved. However, one can easily see that the worst function value will necessarily decrease after at most $n + 1$ consecutive nonshrink iterations, unless an optimal value has already been attained.

Nelder–Mead simplices

The Nelder–Mead algorithm was designed with the idea that the simplices would adapt themselves to "the local landscape" [177]. In fact, we can see that the Nelder–Mead moves allow any simplex shape to be approximated. The good practical performance of the Nelder–Mead algorithm, when it works, is directly related to this capability of fitting well the curvature of the function.

However, the simplices can become arbitrarily flat or needle shaped, which is the reason why it is not possible to establish global convergence to stationary points for the Nelder–Mead algorithm (as the example by McKinnon given in Section 8.2 demonstrates). A common procedure used by today's practitioners is to restart Nelder–Mead whenever the geometry or well poisedness of the simplex vertices deteriorates.

One way to monitor the geometry of $Y = \{y^0, y^1, \ldots, y^n\}$ is to check if it is Λ-poised (for some prefixed constant $\Lambda > 0$), i.e., to check if

$$\|\hat{L}(Y)^{-1}\| \leq \Lambda, \tag{8.2}$$

where

$$\hat{L}(Y) = \frac{1}{\Delta(Y)} L(Y) = \frac{1}{\Delta(Y)} \left[y^1 - y^0 \cdots y^n - y^0 \right] \tag{8.3}$$

and $\Delta(Y) = \max_{1 \le i \le n} \|y^i - y^0\|$. It is easy to see that such a simplex measure is consistent with the definition of linear Λ-poisedness (see Sections 2.5, 3.3, and 4.3).

Now recall from Section 2.5 the definition of the diameter of a simplex: $\mathrm{diam}(Y) = \max_{0 \le i < j \le n} \|y^i - y^j\|$. Since $\Delta(Y) \le \mathrm{diam}(Y) \le 2\Delta(Y)$ (see also Section 2.5), it is irrelevant both in practice and in a convergence analysis whether the measure of the scaling of Y is given by $\Delta(Y)$ or by $\mathrm{diam}(Y)$. We choose to work with $\mathrm{diam}(Y)$ in simplex-type methods like Nelder–Mead because it does not depend on a centering point like $\Delta(Y)$ does. Instead of Λ-poisedness, we will work with the normalized volume (see Section 2.5)

$$\mathrm{von}(Y) = \mathrm{vol}\left(\frac{1}{\mathrm{diam}(Y)}Y\right) = \frac{|\det(L(Y))|}{n!\,\mathrm{diam}(Y)^n}.$$

The choices of $\mathrm{diam}(Y)$ and $\mathrm{von}(Y)$ will be mathematically convenient when manipulating a simplex by reflection or shrinkage. Tseng [220] ignores the factor $n!$ in the denominator, which we could also do here.

We end this section with a few basic facts about the volume and normalized volume of Nelder–Mead simplices [149]. Recall from Section 2.5 that the volume of the simplex of vertices $Y_k = \{y_k^0, y_k^1, \ldots, y_k^n\}$ is defined by the (always positive) quantity

$$\mathrm{vol}(Y_k) = \frac{|\det(L_k)|}{n!},$$

where

$$L_k = \left[y_k^0 - y_k^n \cdots y_k^{n-1} - y_k^n\right].$$

Theorem 8.1.

- *If iteration k performs a nonshrink step (reflection, expansion, or contraction), then*

$$\mathrm{vol}(Y_{k+1}) = |\delta|\,\mathrm{vol}(Y_k).$$

- *If iteration k performs a shrink step, then*

$$\mathrm{vol}(Y_{k+1}) = (\gamma^s)^n\,\mathrm{vol}(Y_k).$$

Proof. Let us prove the first statement only. The second statement can be proved trivially.

Let us assume without loss of generality that $y_k^n = 0$. In this case, the vertex computed at a nonshrink step can be written in the form

$$L_k\,t_k(\delta), \quad \text{where} \quad t_k(\delta) = \left[\frac{1+\delta}{n}, \ldots, \frac{1+\delta}{n}\right]^\top.$$

Since the volume of the new simplex Y_{k+1} is independent of the ordering of the vertices, let us assume that the new vertex $L_k t_k(\delta)$ is the last in Y_{k+1}. Thus, recalling that $y_k^n = 0$,

$$|\det(L_{k+1})| = \left|\det\left(L_k - L_k t_k(\delta)e^\top\right)\right| = |\det(L_k)|\left|\det\left(I - t_k(\delta)e^\top\right)\right|,$$

Table 8.1. *Number of times where the diameter of a simplex increased and the normalized volume decreased by isometric reflection. Experiments made on 10^5 simplices in $\mathbb{R}^3$ with $y^0 = 0$ and remaining vertex components randomly generated in $[-1,1]$, using MATLAB® [1] software. The notation used is such that $Y = \{y^0, y^1, y^2, y^3\}$ and $Y^r = \{y^0, y^1, y^2, y^r\}$.*

Difference (power k)	0	2	4	6	8
$\mathrm{diam}(Y^r) > \mathrm{diam}(Y) + 10^{-k}$	0%	24%	26%	26%	26%
$\mathrm{von}(Y^r) < \mathrm{von}(Y) - 10^{-k}$	0%	1%	23%	26%	26%

where e is a vector of ones of dimension n. The eigenvalues of $I - t_k(\delta)e^\top$ are 1 (with multiplicity $n - 1$) and $-\delta$. Thus, $|\det(I - t_k(\delta)e^\top)| = |\delta|$, and the proof is completed.   □

A simple consequence of this result is that all iterations of the Nelder–Mead algorithm generate simplices, i.e., $\mathrm{vol}(Y_k) > 0$, for all k (provided that the vertices of Y_0 form a simplex). Theorem 8.1 also allows us to say, algebraically, that isometric reflections ($\delta = 1$) preserve the volume of the simplices, that contractions and shrinks are volume decreasing, and that expansions are volume increasing.

It is also important to understand how these operations affect the normalized volume of the simplices. When a shrink step occurs one has

$$\mathrm{von}(Y_{k+1}) = \mathrm{von}(Y_k). \tag{8.4}$$

This is also true for isometric reflections ($\delta = 1$) when $n = 2$ or when n is arbitrary but the simplices are equilateral. We leave these simple facts as exercises. Although counterintuitive, isometric reflections do not preserve the normalized volume in general when $n > 2$, and in particular they can lead to a decrease of this measure.[13] We know from above that the volume is kept constant in isometric reflections. However, the diameter can increase and therefore the normalized volume can decrease. The reader can be convinced, for instance, by taking the simplex of vertices $y^0 = (0,0,0)$, $y^1 = (1,1,0)$, $y^2 = (0,1,0)$, and $y^3 = (0,0,1)$. The diameter increases from 1.7321 to 1.7951, and the normalized volume decreases from 0.0321 to 0.0288. We conducted a simple experiment using MATLAB [1] software, reported in Table 8.1, to see how often the normalized volume can change.

One can prove that the decrease in the normalized volume caused by isometric reflections is no worse than

$$\mathrm{von}(Y_{k+1}) \geq \frac{\mathrm{von}(Y_k)}{2^n}. \tag{8.5}$$

In practice, the decrease in the normalized volume after isometric reflections is not significant throughout an optimization run and rarely affects the performance of the Nelder–Mead method.

---

[13]It is unclear whether one could perform an isometric reflection using a centroid point of the form $y^c = \sum_{i=0}^{n-1} \alpha^i y^i$, with $\sum_{i=0}^{n-1} \alpha^i = 1$ and $\alpha^i > 0$, $i = 0, \ldots, n-1$, that would preserve the normalized volume for values of α^i, $i = 0, \ldots, n-1$, bounded away from zero.

8.2   Properties of the Nelder–Mead simplex method

The most general properties of the Nelder–Mead algorithm are stated in the next theorem.

Theorem 8.2. *Consider the application of the Nelder–Mead method (Algorithm 8.1) to a function* f *which is bounded from below on* $\mathbb{R}^n$.

1. *The sequence* $\{f_k^0\}$ *is convergent.*

2. *If only a finite number of shrinks occur, then all the* $n+1$ *sequences* $\{f_k^i\}$, $i = 0,\ldots,n$, *converge and their limits satisfy* $f_*^0 \leq f_*^1 \leq \cdots \leq f_*^n$.

   Moreover, if there is an integer $j \in \{0,\ldots,n-1\}$ *for which* $f_*^j < f_*^{j+1}$ *(a property called broken convergence), then for sufficiently large* k *the change index is such that* $k^* > j$.

3. *If only a finite number of nonshrinks occur, then all the simplex vertices converge to a single point.*

Proof. The proof of the first and second assertions is essentially based on the fact that monotonically decreasing sequences bounded from below are convergent. The proof of the third assertion is also straightforward and left as an exercise. $\square$

Note that the fact that $\{f_k^0\}$ converges does not mean that it converges to the value of f at a stationary point. A consequence of *broken convergence* is that if the change index is equal to zero an infinite number of times, then $f_*^0 = f_*^1 = \cdots = f_*^n$ (assuming that f is bounded from below and no shrinks steps are taken).

If the function is strictly convex, one can show that no shrink steps occur.

Theorem 8.3. *No shrink steps are performed when the Nelder–Mead method (Algorithm 8.1) is applied to a strictly convex function* f.

Proof. Shrink steps are taken only when outside or inside contractions are tried and fail. Let us focus on an outside contraction, which is tried only when $f_k^{n-1} \leq f_k^r < f_k^n$. Now, from the strict convexity of f and the fact that y_k^{oc} is a convex combination of y_k^c and y_k^r for some parameter $\lambda \in (0,1)$,

$$f(y_k^{oc}) = f(\lambda y_k^c + (1-\lambda)y_k^r) < \lambda f(y_k^c) + (1-\lambda)f(y_k^r) \leq \max\{f_k^c, f_k^r\}.$$

But $\max\{f_k^c, f_k^r\} = f_k^r$ since $f_k^{n-1} \leq f_k^r$ and $f_k^c \leq f_k^{n-1}$ (the latter is, again, a consequence of the strict convexity of f). Thus, $f_k^{oc} < f_k^r$ and the outside contraction is applied (and the shrink step is not taken).

If, instead, an inside contraction is to be considered, then a similar argument would be applied, based on the fact that y_k^{ic} is a convex combination of y_k^n and y_k^c. Note that strict convexity is required for this argument. $\square$

Lagarias et al. [149] proved that the Nelder–Mead method (Algorithm 8.1) is globally convergent when $n = 1$. An alternative and much shorter proof, mentioned in [145] for the standard choices (8.1), is sketched in the exercises.

Convergence of the Nelder–Mead method to nonstationary points

Woods [230] constructed a nonconvex differentiable function in two variables for which the Nelder–Mead method is claimed to fail. The reason for this failure is that the method applies consecutive shrinks towards a point that is not a minimizer.

McKinnon [169] has derived a family of strictly convex examples for which the Nelder–Mead method (Algorithm 8.1) converges to a nonstationary point. From Theorem 8.3, shrink steps are immediately ruled out. In these examples the inside contraction step is applied repeatedly with the best vertex remaining fixed. McKinnon referred to this behavior as *repeated focused inside contraction* (RFIC). It is shown in [169] that no other type of step is taken in these examples. The simplices generated by the Nelder–Mead method collapse along a direction orthogonal to the steepest descent direction. The functions are defined in $\mathbb{R}^2$ as follows:

$$f(x_1, x_2) = \begin{cases} \theta \phi |x_1|^\tau + x_2 + x_2^2 & \text{if } x_1 \leq 0, \\ \theta x_1^\tau + x_2 + x_2^2 & \text{if } x_1 > 0. \end{cases} \tag{8.6}$$

The function is strictly convex if $\tau > 1$. It has continuous first derivatives if $\tau > 1$, continuous second derivatives if $\tau > 2$, and continuous third derivatives if $\tau > 3$. Note that $(0, -1)$ is a descent direction from the origin. The Nelder–Mead algorithm is started with the simplex of vertices

$$y_0^0 = \begin{bmatrix} 0 \\ 0 \end{bmatrix}, \quad y_0^1 = \begin{bmatrix} \lambda_1^0 \\ \lambda_2^0 \end{bmatrix} = \begin{bmatrix} 1 \\ 1 \end{bmatrix}, \quad \text{and} \quad y_0^2 = \begin{bmatrix} \lambda_1^1 \\ \lambda_2^1 \end{bmatrix}, \tag{8.7}$$

where $\lambda_1^1 = (1 + \sqrt{33})/8 \simeq 0.84$ and $\lambda_2^1 = (1 - \sqrt{33})/8 \simeq -0.59$. For values of τ, θ, and ϕ satisfying certain conditions, the method can be shown to converge to the origin which is not a stationary point. An example of values of τ, θ, and ϕ that satisfy these conditions is $\tau = 2, \theta = 6$, and $\phi = 60$. The contours of the function (8.6) are shown in Figure 8.3 for these values of τ, θ, and ϕ. Another set of parameter values for which this type of counterexample works is $\tau = 3, \theta = 6$, and $\phi = 400$. The RFIC behavior generates a sequence of simplices whose vertices are not uniformly Λ-poised (for any fixed $\Lambda > 0$).

We ran the MATLAB [1] implementation of the Nelder–Mead method to minimize the McKinnon function (8.6) for the choices $\tau = 2, \theta = 6$, and $\phi = 60$. First, we selected the initial simplex as in (8.7). As expected, we can see from Figure 8.3 that the method never moved the best vertex from the origin. Then we changed the initial simplex to

$$y_0^0 = \begin{bmatrix} 0 \\ 0 \end{bmatrix}, \quad y_0^1 = \begin{bmatrix} 1 \\ 0 \end{bmatrix}, \quad \text{and} \quad y_0^2 = \begin{bmatrix} 0 \\ 1 \end{bmatrix}, \tag{8.8}$$

and it can be observed that the Nelder–Mead method was able to move away from the origin and to converge to the minimizer $(x_* = (0, -0.5), f(x_*) = -0.25)$.

8.3   A globally convergent variant of the Nelder–Mead method

There are a number of issues that must be taken care of in the Nelder–Mead method (Algorithm 8.1) to make it globally convergent to stationary points.

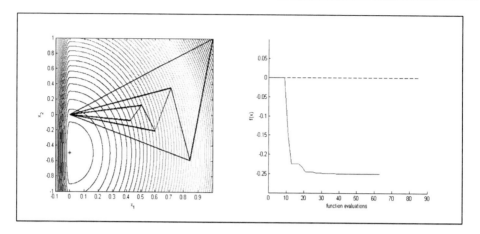

Figure 8.3. *The left plot depicts the contours of the McKinnon function* (8.6) *for* $\tau = 2$, $\theta = 6$, *and* $\phi = 60$ *and illustrates the RFIC when starting from the initial simplex* (8.7). *The right plot describes the application of the Nelder–Mead method to this function. The dashed line corresponds to the initial simplex* (8.7) *and the solid line to* (8.8).

First, the quality of the geometry of the simplices must be monitored for all operations, with the exception of shrinks for which we know that the normalized volume is preserved (see (8.4)). Thus, when a shrink occurs, if the normalized volume of Y_k satisfies $\text{von}(Y_k) \geq \xi$ for some constant $\xi > 0$ independent of k, so does the normalized volume of Y_{k+1}. However, there is no guarantee that this will happen for reflections (even isometric ones), expansions, and contractions. A threshold condition like $\text{von}(Y_{k+1}) \geq \xi$ must therefore be imposed in these steps.

Expansions or contractions might then be skipped because of failure in determining expansion or contraction simplices that pass the geometry threshold. However, special provision must be taken for reflections since these are essential for ensuring global convergence, due to their positive spanning effect. One must guarantee that some form of reflection is always feasible in the sense that it does not deteriorate the geometry of the simplices (i.e., does not lead to a decrease in their normalized volumes). Unfortunately, isometric reflections are not enough for this purpose because they might decrease the normalized volume. Several strategies are then possible. To simplify matters, we will assume that an isometric reflection is always tried first. If the isometric reflected point satisfies $\text{diam}\left(\{y^0, y^1, \ldots, y^{n-1}\} \cup \{y^r\}\right) \leq \gamma^e \Delta$ and $\text{von}\left(\{y^0, y^1, \ldots, y^{n-1}\} \cup \{y^r\}\right) \geq \xi$, then no special provision is taken and the method proceeds by evaluating the function at y^r. Otherwise, we attempt a safeguard step, by rotating the vertices y^i, $i = 1, \ldots, n$, 180° around y^0. This rotation is the same as the one applied by the MDS method (see the end of Chapter 7). As in MDS, we could also consider an expansion step by enlarging this rotated simplex, but we will skip it for the sake of brevity.

On the other hand, we know from Chapter 7 that avoiding degeneracy in the geometry is not sufficient for direct-search methods which accept new iterates solely based on simple decrease. In Chapter 7 we described two possibilities to fortify the decrease in the objective function: (i) to ask the iterates to lie on a sequence of meshes defined as integer lattices, where the minimal separation of the mesh points is proportional to the step size α_k; (ii) to ask the iterates to satisfy a sufficient decrease condition of the type $f(x_{k+1}) < f(x_k) - \rho(\alpha_k)$. (Recall that $\rho : (0, +\infty) \to \mathbb{R}_+$ was called a *forcing function* and was asked to be continuous and positive and to satisfy

$$\lim_{t \longrightarrow 0^+} \frac{\rho(t)}{t} = 0 \quad \text{and} \quad \rho(t_1) \leq \rho(t_2) \quad \text{if} \quad t_1 < t_2.$$

A simple example of a forcing function presented was $\rho(t) = t^2$.)

Forcing the iterates to lie on integer lattices seems an intractable task in the Nelder–Mead context for $n > 1$, given the diversity of steps that operate on the simplices. Thus, the approach we follow in this book for a modified Nelder–Mead method is based on the imposition of sufficient decrease. However, in the Nelder–Mead context we do not have a situation like the one we have in the directional direct-search methods of Chapter 7, where the current iterate is the best point found so far. In the Nelder–Mead algorithm, one makes comparisons among several objective function values, and sufficient decrease must be applied to the different situations. Also, the step size parameter α_k used in the sufficient decrease condition of Chapter 7 is now replaced by the diameter of the current simplex $\Delta_k = \text{diam}(Y_k)$—but, as we have mentioned before, we could had chosen $\Delta_k = \Delta(Y_k)$.

The modified Nelder–Mead method described in Algorithm 8.2 is essentially one of the instances suggested and analyzed by Tseng [220]. To simplify matters, the two contraction steps (inside and outside) have been restated as a single contraction step. There is also a relevant difference in the shrink step compared to the original Nelder–Mead method. We have seen that the shrink step in Algorithm 8.1 is guaranteed not to increase the minimal simplex value ($f^0_{k+1} \leq f^0_k$), which follows trivially from the fact that the best vertex of Y_k is kept in Y_{k+1}. This is not enough now because we need sufficient decrease; in other words, we need something like $f^0_{k+1} \leq f^0_k - \rho(\Delta_k)$, where $\Delta_k = \text{diam}(Y_k)$. When this sufficient decrease condition is not satisfied, the iteration is repeated but using the shrunken simplex. Thus, we must take into account the possibility of having an infinite number of cycles within an iteration by repeatedly applying shrink steps. When that happens we will show that the algorithm returns a stationary limit point.

Algorithm 8.2 (A modified Nelder–Mead method).

Initialization: Choose $\xi > 0$. Choose an initial simplex of vertices $Y_0 = \{y_0^0, y_0^1, \ldots, y_0^n\}$ such that $\text{von}(Y_0) \geq \xi$. Evaluate f at the points in Y_0. Choose constants:

$$0 < \gamma^s < 1 < \gamma^e, \qquad -1 < \delta^{ic} < 0 < \delta^{oc} < \delta^r < \delta^e.$$

For $k = 0, 1, 2, \ldots$

   0. Set $Y = Y_k$.

   1. **Order:** Order the $n + 1$ vertices of $Y = \{y^0, y^1, \ldots, y^n\}$ so that

$$f^0 = f(y^0) \leq f^1 = f(y^1) \leq \cdots \leq f^n = f(y^n).$$

   Set $\Delta = \text{diam}(Y)$.

2. **Reflect:** Calculate an isometric reflected point y^r (as in Algorithm 8.1 with $\delta^r = 1$). If

$$\begin{aligned}
\text{diam}\left(\{y^0, y^1, \dots, y^{n-1}\} \cup \{y^r\}\right) &\leq \gamma^e \Delta, \\
\text{von}\left(\{y^0, y^1, \dots, y^{n-1}\} \cup \{y^r\}\right) &\geq \xi,
\end{aligned} \qquad (8.9)$$

then evaluate $f^r = f(y^r)$. If $f^r \leq f^{n-1} - \rho(\Delta)$, then attempt an expansion (and then accept either the reflected or the expanded point). Otherwise, attempt a contraction.

Safeguard rotation: If the isometric reflection failed to satisfy (8.9), then rotate the simplex around the best vertex y^0:

$$y^{rot,i} = y^0 - (y^i - y^0), \quad i = 1, \dots, n.$$

Evaluate $f(y^{rot,i})$, $i = 1, \dots, n$, and set $f^{rot} = \min\{f(y^{rot,i}) : i = 1, \dots, n\}$. If $f^{rot} \leq f^0 - \rho(\Delta)$, then terminate the iteration and take the rotated simplex: $Y_{k+1} = \{y^0, y^{rot,1}, \dots, y^{rot,n}\}$. Otherwise, attempt a contraction.

3. **Expand:** Calculate an expansion point y^e (for instance, as in Algorithm 8.1). If

$$\begin{aligned}
\text{diam}\left(\{y^0, y^1, \dots, y^{n-1}\} \cup \{y^e\}\right) &\leq \gamma^e \Delta, \\
\text{von}\left(\{y^0, y^1, \dots, y^{n-1}\} \cup \{y^e\}\right) &\geq \xi,
\end{aligned}$$

then evaluate $f^e = f(y^e)$, and if $f^e \leq f^r$, replace y^n by the expansion point y^e, and terminate the iteration: $Y_{k+1} = \{y^0, y^1, \dots, y^{n-1}, y^e\}$. Otherwise, replace y^n by the reflected point y^r, and terminate the iteration: $Y_{k+1} = \{y^0, y^1, \dots, y_k^{n-1}, y^r\}$.

4. **Contract:** Calculate a contraction point y^{cc} (such as an outside or inside contraction in Algorithm 8.1). If

$$\begin{aligned}
\text{diam}\left(\{y^0, y^1, \dots, y^{n-1}\} \cup \{y^{cc}\}\right) &\leq \Delta, \\
\text{von}\left(\{y^0, y^1, \dots, y^{n-1}\} \cup \{y^{cc}\}\right) &\geq \xi,
\end{aligned}$$

then evaluate $f^{cc} = f(y^{cc})$, and if $f^{cc} \leq f^n - \rho(\Delta)$, then replace y^n by the contraction point y^{cc} and terminate the iteration: $Y_{k+1} = \{y^0, y^1, \dots, y^{n-1}, y^{cc}\}$. Otherwise, perform a shrink.

5. **Shrink:** Evaluate f at the n points $y^0 + \gamma^s(y^i - y^0)$, $i = 1, \dots, n$, and let f^s be the lowest of these values. If $f^s \leq f^0 - \rho(\Delta)$, then accept the shrunken simplex and terminate the iteration: $Y_{k+1} = \{y^0 + \gamma^s(y^i - y^0), i = 0, \dots, n\}$. Otherwise, go back to Step 0 with $Y = \{y^0 + \gamma^s(y^i - y^0), i = 0, \dots, n\}$.

In practice we could choose γ^e close to 1 for reflections and around 2 for expansions, similarly as in the original Nelder–Mead method. Note also that the normalized volume of the simplices does not change after safeguard rotations and shrinks. In safeguard rotations the diameter of the simplex is unaltered, whereas for shrinks it is reduced by a factor of γ^s. Once again, a stopping criterion could consist of terminating the run when the diameter Δ_k of the simplex becomes smaller than a chosen tolerance $\Delta_{tol} > 0$ (for instance, $\Delta_{tol} = 10^{-5}$).

We define an index n_k depending on the operation in which the iteration has terminated:

$$n_k = n \quad \text{for (isometric) reflections, expansions, and contractions,}$$

$$n_k = 0 \quad \text{for shrinks and safeguard rotations.}$$

Then the sequence of simplices generated by the modified Nelder–Mead method (Algorithm 8.2) satisfies

$$f_{k+1}^i \leq f_k^i, \quad i = 0, \ldots, n_k, \tag{8.10}$$

and

$$\sum_{i=0}^{n_k} f_{k+1}^i \leq \sum_{i=0}^{n_k} f_k^i - \rho(\Delta_k). \tag{8.11}$$

Theorem 8.4 below, which plays a central role in the analysis of the modified Nelder–Mead method, is essentially based on conditions (8.10)–(8.11) and thus is valid for other (possibly more elaborated) simplex-based direct-search methods as long as they satisfy these conditions for any $n_k \in \{0, \ldots, n\}$.

What is typically done in the convergence analysis of algorithms for nonlinear optimization is to impose smoothness and boundedness requirements for f on a level set of the form

$$L(x_0) = \{x \in \mathbb{R}^n : f(x) \leq f(x_0)\}.$$

A natural candidate for x_0 in the context of the modified Nelder–Mead method would be y_0^n. However, although (isometric) reflection, rotation, expansion, and contraction steps generate simplex vertices for which the objective function values are below f_0^n, a shrink might not necessarily do so. It is possible to define a value f^{max} such that all the simplex vertices lie in $\{x \in \mathbb{R}^n : f(x) \leq f^{max}\}$, but such a definition would unnecessarily complicate the presentation. We will impose our assumptions on f in $\mathbb{R}^n$; in particular, in what comes next, we assume that f is bounded from below and uniformly continuous in $\mathbb{R}^n$.

The next theorem states under these assumptions on f that the diameter of the simplices generated by the Nelder–Mead algorithm converges to zero. Its proof is due to Tseng [220]—and it is surprisingly complicated. The difficulty comes from the fact that, for steps like isometric reflection, expansion, or contraction, the sufficient decrease condition is imposed at simplex vertices different from the one with the best objective function value. Note that such a result would be proved in a relatively straightforward way for an algorithm that generates a sequence of points $\{x_k\}$ for which $f(x_{k+1}) < f(x_k) - \rho(\Delta_k)$. In the context of simplex-type methods that would correspond, for instance, to having $x_k = y_k^0$ and $f(y_{k+1}^0) < f(y_k^0) - \rho(\Delta_k)$ (a condition we impose only for shrinks and safeguard rotations in the modified Nelder–Mead method).

Theorem 8.4. *If f is bounded from below and uniformly continuous in $\mathbb{R}^n$, then the modified Nelder–Mead method (Algorithm 8.2) generates a sequence $\{Y_k\}$ of simplices whose diameters converge to zero:*

$$\lim_{k \to +\infty} \text{diam}(Y_k) = 0.$$

Proof. The proof is done by contradiction, assuming that Δ_k does not converge to zero. For each $i \in \{0, 1, \ldots, n\}$, we define

$$K_i = \left\{ k \in \{0, 1, \ldots\} : f_{k+1}^i \leq f_k^i - \rho(\Delta_k)/(n_k + 1) \right\}.$$

The fact that both $n_k \le n$ and (8.11) hold at every iteration guarantees, for all k, that there exists at least one $i \in \{0, 1, \ldots, n\}$ such that $k \in K_i$. Thus, $\cup_{i=0}^{n} K_i = \{0, 1, \ldots\}$ and the following index is well defined:

$$i_{min} = \min \left\{ i \in \{0, 1, \ldots, n\} : |K_i| = +\infty, \lim_{k \in K_i} \Delta_k \ne 0 \right\}.$$

Now, since $\Delta_k \not\to 0$ in $K_{i_{min}}$, there exists a subsequence $K \subset K_{i_{min}}$ and a positive constant κ such that $\rho(\Delta_k) \ge \kappa$ for all $k \in K$. As a result of this,

$$f_{k+1}^0 \le f_{k+1}^{i_{min}} \le f_k^{i_{min}} - \kappa/(n+1) \quad \forall k \in K. \tag{8.12}$$

For each k now let ℓ_k be the largest index $\ell \in \{1, \ldots, k\}$ for which $f_\ell^{i_{min}} > f_{\ell-1}^{i_{min}}$ (with $\ell_k = 0$ if no such ℓ exists). Note that ℓ_k must tend to infinity; otherwise, $\{f_k^{i_{min}}\}$ would have a nonincreasing tail; i.e., there would be an index k_{tail} such that $\{f_k^{i_{min}}\}_{k \ge k_{tail}}$ is non-increasing. Then $\{f_k^{i_{min}}\}_{k \ge k_{tail}, k \in K}$ would also be nonincreasing and thus convergent (since f is bounded from below). By taking limits in (8.12) a contradiction would be reached.

The relation (8.12) and the definition ℓ_k trivially imply

$$f_{k+1}^0 \le f_{\ell_k}^{i_{min}} - \kappa/(n+1) \quad \forall k \in K. \tag{8.13}$$

The definition of ℓ_k also implies that $f_{\ell_k}^{i_{min}} > f_{\ell_k-1}^{i_{min}}$ (for k sufficiently large such that $\ell_k > 0$). Thus $\ell_k - 1 \notin K_{i_{min}}$. On the other hand, we have seen before that $\ell_k - 1$ must be in K_j for some j, which must satisfy $j < i_{min}$. Since $\ell_k \to +\infty$ when $k \to +\infty$, by passing at a subsequence if necessary, we can assume that this index j is the same for all indices $\ell_k - 1$. We also have, for the same reason, that $|K_j| = +\infty$. From the choice of i_{min} and the fact that $i_{min} \ne j$, we necessarily have that $\Delta_{\ell_k-1} \to 0$ for $k \in K$. Since $\Delta_{k+1} \le \gamma^e \Delta_k$ for all k, it turns out that $\Delta_{\ell_k} = \text{diam}(Y_{\ell_k}) \to 0$ for $k \in K$.

One can now conclude the proof by arriving at a statement that contradicts (8.13). First, we write

$$f_{k+1}^0 - f_{\ell_k}^{i_{min}} = \left(f_{k+1}^0 - f_{\ell_k}^0 \right) + \left(f_{\ell_k}^0 - f_{\ell_k}^{i_{min}} \right).$$

Note that the first term converges to zero since both f_{k+1}^0 and $f_{\ell_k}^0$ converge to the same value (here we use the fact that $\{f_k^0\}$ is decreasing and f is bounded from below but also that ℓ_k tends to infinity). The second term also converges to zero since f is uniformly continuous and $\text{diam}(Y_k) \to 0$ for $k \in K$. Thus, $f_{k+1}^0 - f_{\ell_k}^{i_{min}}$ converges to zero in K, which contradicts (8.13). $\quad \Box$

In the next theorem we prove that if the sequence of simplex vertices is bounded, then it has at least one limit point which is stationary. The proof of this result relies on the fact that the set of vectors $y^n - y^i$, $i = 0, \ldots, n-1$, and $y^r - y^i$, $i = 0, \ldots, n-1$, taken together form a positive spanning set (in fact, a maximal positive basis; see Figure 8.4.) It is simple to see that this set (linearly) spans $\mathbb{R}^n$. It can be also trivially verified that

$$\sum_{i=0}^{n-1} (y^n - y^i) + \sum_{i=0}^{n-1} (y^r - y^i) = 0, \tag{8.14}$$

and, hence, from Theorem 2.3(iii), we conclude that this set spans $\mathbb{R}^n$ positively.

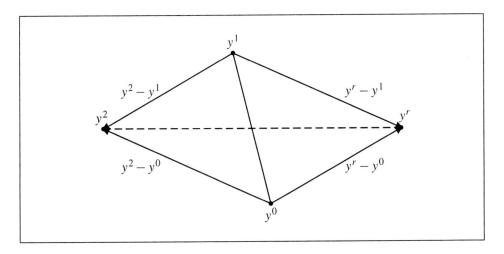

Figure 8.4. *The vectors $y^r - y^0, y^r - y^1$ (right) and $y^2 - y^0, y^2 - y^1$ (left). (This picture is misleading in the sense that isometric reflections are guaranteed only to preserve the diameter and the normalized volume of simplices when $n = 2$ or the simplices are equilateral.)*

When we say that the sequence of simplex vertices $\{Y_k\}$ has a limit point x_* we mean that there is a sequence of vertices of the form $\{x_k\}$, with $x_k \in Y_k$, which has a subsequence converging to x_*.

Theorem 8.5. *Let f be bounded from below, uniformly continuous, and continuously differentiable in $\mathbb{R}^n$. Assume that the sequence of simplex vertices $\{Y_k\}$ generated by the modified Nelder–Mead method (Algorithm 8.2) lies in a compact set. Then $\{Y_k\}$ has at least one stationary limit point x_*.*

Proof. From the hypotheses of the theorem and the fact that $\Delta_k \to 0$ (see Theorem 8.4), there exists a subsequence K_1 of iterations consisting of contraction or shrink steps for which all the vertices of Y_k converge to a point x_*.

From the ordering of the vertices in the simplices, we know that

$$f(y_k^n) \geq f(y_k^i), \quad i = 0, \ldots, n-1,$$

which in turn implies

$$\nabla f(a_k^i)^\top (y_k^n - y_k^i) \geq 0, \quad i = 0, \ldots, n-1, \tag{8.15}$$

for some a_k^i in the line segment connecting y_k^n and y_k^i. Since the sequences $\{(y_k^n - y_k^i)/\Delta_k\}$ are bounded, by passing to nested subsequences if necessary we can assure the existence of a subsequence $K_2 \subseteq K_1$ such that $(y_k^n - y_k^i)/\Delta_k \to z^i, i = 0, \ldots, n-1$. Thus, dividing (8.15) by Δ_k and taking limits in K_2 leads to

$$\nabla f(x_*)^\top z^i \geq 0, \quad i = 0, \ldots, n-1. \tag{8.16}$$

On the other hand, contraction or shrink steps are attempted only when either

$$f(y_k^r) > f(y_k^{n-1}) - \rho(\Delta_k) \tag{8.17}$$

or

$$f_k^{rot} > f(y_k^0) - \rho(\Delta_k). \tag{8.18}$$

Thus, there exists a subsequence K_3 of K_2 such that either (8.17) or (8.18) holds.

In the (8.17) case, we get for $k \in K_3$

$$f(y_k^r) > f(y_k^i) - \rho(\Delta_k), \quad i = 0, \ldots, n-1. \tag{8.19}$$

Thus, for $k \in K_3$,

$$\nabla f(b_k^i)^\top (y_k^r - y_k^i) > -\rho(\Delta_k), \quad i = 0, \ldots, n-1,$$

for some b_k^i in the line segment connecting y_k^r and y_k^i. Once again, since the sequences $\{(y_k^r - y_k^i)/\Delta_k\}$ are bounded, by passing to nested subsequences if necessary we can assure the existence of a subsequence $K_4 \subseteq K_3$ such that $(y_k^r - y_k^i)/\Delta_k \to w^i$, $i = 0, \ldots, n-1$. Also, from the properties of the forcing function ρ we know that $\rho(\Delta_k)/\Delta_k$ tends to zero in K_4. If we now divide (8.19) by Δ_k and take limits in K_4, we get

$$\nabla f(x_*)^\top w^i \geq 0, \quad i = 0, \ldots, n-1. \tag{8.20}$$

We now remark that $[z^0 \cdots z^{n-1} w^0 \cdots w^{n-1}]$ is a positive spanning set. One possible argument is the following. First, we point out that both $[z^0 \cdots z^{n-1}]$ and $[w^0 \cdots w^{n-1}]$ (linearly) span $\mathbb{R}^n$, since they are limits of uniform linearly independent sets. Then we divide (8.14) by Δ_k and take limits, resulting in

$$z^0 + \cdots + z^{n-1} + w^0 + \cdots + w^{n-1} = 0.$$

Thus, from Theorem 2.3(iii), $[z^0 \cdots z^{n-1} w^0 \cdots w^{n-1}]$ is a positive spanning set. The property about spanning sets given in Theorem 2.3(iv) and inequalities (8.16) and (8.20) then imply that $\nabla f(x_*) = 0$.

In the (8.18) case, we have that, for all $k \in K_3$,

$$f(y_k^{rot,i}) > f(y_k^0) - \rho(\Delta_k), \quad i = 1, \ldots, n.$$

From the ordering of the vertices in the simplices, we also known that, for all $k \in K_3$,

$$f(y_k^i) > f(y_k^0), \quad i = 1, \ldots, n.$$

Thus, for $k \in K_3$,

$$\nabla f(c_k^i)^\top (y_k^{rot,i} - y_k^0) > -\rho(\Delta_k), \quad i = 1, \ldots, n, \tag{8.21}$$

and

$$\nabla f(c_k^i)^\top (y_k^i - y_k^0) \geq 0, \quad i = 1, \ldots, n, \tag{8.22}$$

for some c_k^i in the line segment connecting $y_k^{rot,i}$ and y_k^0, and for some d_k^i in the line segment connecting y_k^i and y_k^0. Since the sequences $\{(y_k^{rot,i} - y_k^0)/\Delta_k\}$ and $\{(y_k^i - y_k^0)/\Delta_k\}$

are bounded, by passing to nested subsequences if necessary we can assure the existence of a subsequence $K_4 \subseteq K_3$ such that $(y_k^{rot,i} - y_k^0)/\Delta_k \rightarrow -u^i$, $i = 1,\ldots,n$, and $(y_k^i - y_k^0)/\Delta_k \rightarrow u^i$, $i = 1,\ldots,n$. Thus, dividing (8.21) and (8.22) by Δ_k and taking limits in K_3 yields

$$\nabla f(x_*)^\top(-u^i) \geq 0 \quad \text{and} \quad \nabla f(x_*)^\top u^i \geq 0, \quad i = 1,\ldots,n. \tag{8.23}$$

Now note that $[u^1 \cdots u^n]$ (linearly) spans $\mathbb{R}^n$, since it is a limit of uniform linearly independent sets. Thus, from what has been said after Theorem 2.4, we know that $[u^1 \cdots u^n -u^1 \cdots - u^n]$ is a positive spanning set. Theorem 2.3(iv) and inequalities (8.23) together imply that $\nabla f(x_*) = 0$.

It remains to analyze what happens when an infinite number of cycles occur within an iteration (by consecutive application of shrink steps). Using the same arguments as those above, it is possible to prove that the vertices of the shrunken simplices converge to a stationary point. $\square$

It is actually possible to prove that all limit points of the sequence of vertices are stationary. In [220] this result is proved for a broader class of simplicial direct-search methods. For this purpose, one needs to impose one additional condition to accept isometric reflections or expansions.[14] As in [220], one performs isometric reflections or expansions if both conditions are satisfied:

$$f^r \leq f^{n-1} - \rho(\Delta) \quad \text{and} \quad f^r \leq f^{n-1} - \left(f^n - \frac{1}{n}\sum_{i=0}^{n-1} f^i\right) + \rho(\Delta). \tag{8.24}$$

Thus, in the isometric reflection or expansion cases we have both (8.11) for $n_k = n$ and

$$\sum_{i=0}^n f_{k+1}^i \leq \sum_{i=0}^n f_k^i - \left(f_k^n - \frac{1}{n}\sum_{i=0}^{n-1} f_k^i\right) + \rho(\Delta_k). \tag{8.25}$$

Theorem 8.6. *Let f be bounded from below, uniformly continuous, and continuously differentiable in $\mathbb{R}^n$. Assume that the sequence of simplex vertices $\{Y_k\}$ generated by the modified Nelder–Mead method (Algorithm 8.2, further modified to accept only isometric reflections or expansions if (8.24) occurs) lies in a compact set. Assume that isometric reflections always satisfy (8.9). Then all the limit points of $\{Y_k\}$ are stationary.*

Proof. The first part of the proof consists of showing that Theorem 8.5 is still valid under the modification introduced by (8.24). Now contraction or shrink steps can be attempted because of either (8.17) or

$$f_k^r > f_k^{n-1} - \left(f_k^n - \frac{1}{n}\sum_{i=0}^{n-1} f_k^i\right) + \rho(\Delta_k). \tag{8.26}$$

If we have an infinite subsequence of K_2 for which the condition (8.17) is true, then the proof of Theorem 8.5 remains valid by passing first to a subsequence of K_2 if necessary.

---

[14]The need for additional conditions arises in other direct-search methods too (see, for instance, Section 7.5, where it is pointed out that it is possible to prove (7.10) for complete polling).

Thus, we just need to consider the case where K_2, or a subsequence of K_2, satisfies (8.26). However, this case is treated similarly. First, we write

$$f(y_k^r) > f(y_k^i) - \left(f_k^n - \frac{1}{n} \sum_{i=0}^{n-1} f_k^i \right) + \rho(\Delta_k), \quad i = 0, \dots, n-1.$$

Thus, for $k \in K_2$,

$$\nabla f(b_k^i)^\top (y_k^r - y_k^i) > \frac{1}{n} \left(\sum_{i=0}^{n-1} \nabla f(a_k^i)^\top (y_k^i - y_k^n) \right) + \rho(\Delta_k), \quad i = 0, \dots, n-1, \quad (8.27)$$

where a_k^i is in the line segment connecting y_k^n and y_k^i, and b_k^i is in the line segment connecting y_k^r and y_k^i. We already know that $\{(y_k^n - y_k^i)/\Delta_k\}$ converges to z_i in K_2 for $i = 0, \dots, n-1$. Once again, since the sequences $\{(y_k^r - y_k^i)/\Delta_k\}$ are bounded, by passing to nested subsequences if necessary we can assure the existence of a subsequence $K_3 \subseteq K_2$ such that $(y_k^r - y_k^i)/\Delta_k \to w^i$, $i = 0, \dots, n-1$. So, by taking limits in (8.27) for $k \in K_3$, we obtain

$$\nabla f(x_*)^\top w^i \geq \nabla f(x_*)^\top \left(\frac{1}{n} \sum_{i=0}^{n-1} (-z_i) \right), \quad i = 0, \dots, n-1,$$

or, equivalently,

$$\nabla f(x_*)^\top \left(w_i + \frac{1}{n} \sum_{i=0}^{n-1} z_i \right) \geq 0, \quad i = 0, \dots, n-1. \quad (8.28)$$

From (8.16) and (8.28), we conclude that $\nabla f(x_*) = 0$ (the argument used here is similar to the one presented before).

Now suppose that there is a limit point x_∞ which is not stationary. Then, from the continuous differentiability of f, there exists a ball $B(x_\infty; \Delta_\infty)$ of radius $\Delta_\infty > 0$ centered at x_∞ where there are no stationary points.

We focus our attention on one (necessarily infinite) subsequence $\{x_k\}_{k \in K_\infty}$, with $x_k \in Y_k$, that lies in this ball. Note that we can guarantee that for sufficiently large k there are no contraction or shrink iterations in K_∞, since otherwise we would apply a line of thought similar to that of the proof of Theorem 8.5 and conclude that there would be a stationary point in $B(x_\infty; \Delta_\infty)$. So, we can assume without loss of generality that K_∞ is formed by iterations where an isometric reflection or an expansion necessarily occurs. Thus, we can assume, for all $k \in K_\infty$, that inequality (8.25) holds.

We point out that there must exist a constant $\kappa > 0$ such that, for $k \in K_\infty$ sufficiently large,

$$\frac{f_k^n - \frac{1}{n} \sum_{i=0}^{n-1} f_k^i}{\Delta_k} \geq 2\kappa. \quad (8.29)$$

(Otherwise, we would apply an argument similar to the one of Theorem 8.5 and conclude that $\nabla f(x_\infty)^\top (-z_i) \geq 0$, $i = 0, \dots, n-1$, which, together with $\nabla f(x_\infty)^\top z_i \geq 0$, $i = 0, \dots, n-1$, would imply $\nabla f(x_\infty) = 0$.) Thus, by applying inequality (8.29) and the properties of the forcing function ρ to (8.25), we can assure, for k sufficiently large, that

$$\sum_{i=0}^{n} f_{k+1}^i - \sum_{i=0}^{n} f_k^i \leq -\kappa \Delta_k.$$

Now we divide the ball $B(x_\infty; \Delta_\infty)$ into three mutually exclusive sets:

$$
\begin{aligned}
R_1 &= \{x \in \mathbb{R}^n : 2\Delta_\infty/3 < \|x - x_\infty\| \leq \Delta_\infty\}, \\
R_2 &= \{x \in \mathbb{R}^n : \Delta_\infty/3 < \|x - x_\infty\| \leq 2\Delta_\infty/3\}, \\
R_3 &= \{x \in \mathbb{R}^n : \|x - x_\infty\| \leq \Delta_\infty/3\}.
\end{aligned}
$$

Since, from Theorem 8.4, $\Delta_k \to 0$, we know that the number of contractions or shrinks must be infinite. Thus, the sequence of vertices $\{x_k\}$ enters and leaves the ball $B(x_\infty; \Delta_\infty)$ an infinite number of times. Because $\Delta_k \to 0$ this implies that the sequence of vertices $\{x_k\}$ must cross between R_1 and R_3 through R_2 also an infinite number of times. So, there must be subsequences $\{k_j\}$ and $\{k_\ell\}$ of K_∞ such that

$$x_{k_j} \in R_1, \quad x_{k_j+1} \in R_2, \dots, x_{k_\ell-1} \in R_2, \quad \text{and} \quad x_{k_\ell} \in R_3.$$

Then, from the fact that $\|x_{k+1} - x_k\| \leq \Delta_{k+1} + \Delta_k \leq (1 + \gamma^e)\Delta_k$ and that the distance between points in R_1 and R_3 is at least $\Delta_\infty/3$, we obtain

$$
\begin{aligned}
\sum_{i=0}^{n} f_{k_\ell}^i - \sum_{i=0}^{n} f_{k_j}^i &= \left\{\sum_{i=0}^{n} f_{k_\ell}^i - \sum_{i=0}^{n} f_{k_\ell-1}^i\right\} + \cdots + \left\{\sum_{i=0}^{n} f_{k_j+1}^i - \sum_{i=0}^{n} f_{k_j}^i\right\} \\
&\leq -\kappa \left(\Delta_{k_\ell-1} + \cdots + \Delta_{k_j}\right) \\
&\leq -\frac{\kappa}{1+\gamma^e} \left(\|x_{k_\ell} - x_{k_\ell-1}\| + \cdots + \|x_{k_j+1} - x_{k_j}\|\right) \\
&\leq -\frac{\kappa}{1+\gamma^e} \|x_{k_\ell} - x_{k_j}\| \\
&\leq -\frac{\kappa \Delta_\infty}{3(1+\gamma^e)}.
\end{aligned}
$$

One can now arrive at a contradiction. From the above inequality, the monotone decreasing subsequence $\{\sum_{i=0}^{n} f_k^i\}_{k \in K_\infty}$ cannot converge, which is a contradiction. In fact, we know from (8.10)–(8.11) that $\{f_k^0\}_{k \in K_\infty}$, under the boundedness of f, must be convergent. Since $\Delta_k \to 0$ and f is uniformly continuous, then the subsequences $\{f_k^i\}_{k \in K_\infty}$ are also convergent for $i = 1, \dots, n$, and $\{\sum_{i=0}^{n} f_k^i\}_{k \in K_\infty}$ is convergent. $\quad\square$

A similar result can be proved when safeguard rotations are always attempted (see the exercises).

Other modifications to the Nelder–Mead method

We know that the Nelder–Mead method can stagnate and fail to converge to a stationary point due to the deterioration of the simplex geometry or lack of sufficient decrease. One approach followed by some authors is to let Nelder–Mead run relatively freely, as long as it provides some form of sufficient decrease, and to take action only when failure to satisfy such a condition is identified.

For instance, the modified Nelder–Mead method of Price, Coope, and Byatt [194] is in this category. Basically, they let Nelder–Mead run (without shrinks) as long as the worst vertex results in sufficient decrease relatively to the size of Y_k, e.g., $f_{k+1}^n < f_k^n - \rho(\text{diam}(Y_k))$, where ρ is a forcing function. These internal Nelder–Mead iterations are not counted as regular iterations. After a finite number of these internal Nelder–Mead steps either a new simplex of vertices Y_{k+1} is found yielding sufficient decrease $f_{k+1}^0 < f_k^0 - \rho(\text{diam}(Y_k))$, in which case a new iteration is started from Y_{k+1}, or else the algorithm attempts to form a *quasi-minimal frame* (see below) around the vertex y_{k+1}^0, for which we know that $f_{k+1}^0 \le f_k^0$.

It is the process of identifying this quasi-minimal frame that deviates from the Nelder–Mead course of action. A quasi-minimal frame in the Coope and Price terminology is a polling set of the form

$$P_k = \{x_k + \alpha_k d : d \in D_k\},$$

where D_k is a positive basis or positive spanning set, and $f(x_k + \alpha_k d) + \rho(\alpha_k) \ge f(x_k)$, for all $d \in D_k$. In the above context we have that $x_k = y_{k+1}^0$ and α_k is of the order of $\text{diam}(Y_k)$. One choice for D_k would be a maximal positive basis formed by positive multiples of $y_{k+1}^i - y_{k+1}^0$, $i = 1, \ldots, n$, and their negative counterparts, if the cosine measure $\text{cm}(D_k)$ is above a uniform positive threshold, or a suitable replacement otherwise. Other choices are possible in the Nelder–Mead simplex geometry. Note that the process of attempting to identify a *quasi-minimal frame* either succeeds or generates a new point y_{k+1}^0 for which $f_{k+1}^0 < f_k^0 - \rho(\text{diam}(Y_k))$. It is shown in [194], based on previous work by Coope and Price [65], that the resulting algorithm generates a sequence of iterates for which all the limit points are stationary, provided the iterates are contained in an appropriate level set in which f is continuously differentiable and has a Lipschitz continuous gradient. However, their analysis requires one to algorithmically enforce $\alpha_k \to 0$, which in the Nelder–Mead environment is equivalent to enforcing $\text{diam}(Y_k) \to 0$.

In the context of the Nelder–Mead method, Kelley [140] used the simplex gradient in a sufficient decrease-type condition to detect stagnation as well as in determining the orientation of the new simplices to restart the process. More precisely, he suggested restarting Nelder–Mead when

$$\sum_{j=0}^n f_{k+1}^j < \sum_{j=0}^n f_k^j$$

holds but

$$\sum_{j=0}^n f_{k+1}^j < \sum_{j=0}^n f_k^j - \eta \|\nabla_s f(y_k^0)\|^2$$

fails, where η is a small positive number and $\nabla_s f(y_k^0)$ is the simplex gradient (see Section 2.6) calculated using Y_k. For the restarts, the vertices $y_k^1, \ldots, y_k^n$ are replaced by $y_k^0 \pm (0.5 \min_{1 \le i \le n} \|y_k^i - y_k^0\|) e_i$, $i = 1, \ldots, n$, where e_i is the ith column of the identity matrix of order n. The signs $\pm$ are chosen depending of the sign of the ith component of $\nabla_s f(y_k^0)$. (In the same spirit but in a different context, Mifflin [171] had suggested using the signs of the centered simplex gradients as descent indicators.)

8.4   Other notes and references

The work by Nelder and Mead [177] profited by the earlier contribution of Spendley, Hext, and Himsworth [210] in 1962, where simplex-based operations were first introduced for the purpose of optimization. In their approach, Spendley, Hext, and Himsworth tried to improve the worst vertex of a simplex (in terms of the values of the objective function) by isometrically reflecting it with respect to the centroid of the other n vertices or else by repeating such operations but now reflecting the second worst vertex. The Nelder–Mead algorithm [177] incorporates similar types of reflections but "improves" over the Spendley–Hext–Himsworth, by allowing nonisometric reflections, which can be regarded as expansions and contractions, and thus permitting arbitrary simplex shapes. Of course, it is this additional flexibility that makes convergence more difficult to consider. In the year of the publication of the Nelder–Mead paper, Box [45] published a "simplicial" method based on reflecting the worst vertex over the centroid of the remaining vertices. The method allowed a number of points between $n + 1$ and $2n$ and took simple bounds on the variables into consideration.

Several other variants of the original methods by Nelder and Mead [177] and Spendley, Hext, and Himsworth [210] have been proposed and analyzed, in particular in the Russian literature. Dambrauskas [71] suggested an extension of the Spendley–Hext–Himsworth method in which the simplex may also contract toward its centroid. Yu [234] proved global convergence to a stationary point of a modified version of the Spendley–Hext–Himsworth method (where the condition to accept reflections was already based on a sufficient decrease condition). Rykov (see [202] and the references therein) proposed direct-search algorithms based on reflections, expansions, and contractions of simplices. Tseng [220] lists in detail the differences between his general simplex-based framework and Rykov's. One fundamental difference is that Rykov's analysis requires the objective function to be convex. Woods [230] and Nazareth and Tseng [176] also proved properties for modified Nelder–Mead algorithms under forms of convexity.

Hvattum and Glover [135] developed a method inspired by several direct-search methods of simplicial and directional types which works with sample sets of various sizes and is enhanced by techniques from scatter search to handle the selection of the sample sets.

The MDS method was tested and applied by a number of authors. Hough and Meza [133], for instance, applied the MDS method to the derivative-free solution of a modified trust-region subproblem within a derivative-based trust-region framework. Buckley and Ma [48] studied practical improvements of MDS by quadratic interpolation over sample sets generated by the algorithm.

8.5   Exercises

1. The Nelder–Mead method is invariant under affine transformations [149]. To prove this property consider $g(x) = Ax + b$, where $b \in \mathbb{R}^n$ and $A \in \mathbb{R}^{n \times n}$ is a nonsingular matrix. Show that the Nelder–Mead method (Algorithm 8.1) applied to $f(x)$ from the starting simplex Y_0 and to $f(g(x))$ from the starting simplex $g^{-1}(Y_0) = A^{-1}Y_0 - b$ generate the same sequence of simplex vertices.

2. Show that shrink steps preserve the normalized volume of simplices.

3. Prove that isometric reflections preserve the normalized volume of simplices when either $n = 2$ or the simplices are equilateral (meaning that the distance between the vertices is constant).

4. Prove that isometric reflections in general yield

$$\text{diam}(Y_{k+1}) \leq \left(\frac{2n-1}{n}\right)\text{diam}(Y_k).$$

Use this bound to show (8.5).

5. Show that the rotations and expansions of the MDS method (see the end of Chapter 7) preserve the normalized volume of simplices.

6. Show that the three assertions of Theorem 8.2 are true.

7. Frame the Nelder–Mead method (Algorithm 8.1), when $n = 1$ and the parameters are given by the standard values (8.1), as a directional direct-search method of the type of Algorithm 7.2. By showing that Assumptions 7.8, 7.9, and 7.10 are satisfied (globalization by simple decrease with integer lattices) the Nelder–Mead method produces a sequence of iterates $\{x_k\}$ for which a subsequence of $\{\|\nabla f(x_k)\|\} = \{|f'(x_k)|\}$ converges to zero.

8. Why do the sequences $\{a_k^i\}$, $\{b_k^i\}$, and $\{c_k^i\}$ converge to x_* (proof of Theorem 8.5)?

9. Using similar arguments as in the proof of Theorem 8.5, prove that if an infinite loop occurs at a given iteration of Algorithm 8.2, then the vertices of the shrunken simplices converge to a stationary point.

10. In the context of the proof of Theorem 8.6, show that (8.16) and (8.28) imply that $\nabla f(x_*) = 0$.

11. Explain why a simplified version of the modified Nelder–Mead method that considers only isometric reflections and shrinks is globally convergent (in the sense of Theorems 8.5 and 8.6) if it starts from an equilateral simplex.

12. Prove the following alternative for Theorem 8.6: Let f be bounded from below, uniformly continuous, and continuously differentiable in $\mathbb{R}^n$. Assume that the sequence of simplex vertices $\{Y_k\}$ generated by the modified Nelder–Mead method (Algorithm 8.2) lies in a compact set. Assume that safeguarded rotations are always attempted (meaning that the reflection step would consist only of safeguard rotations). Then all the limit points of $\{Y_k\}$ are stationary. (The proof follows the lines of the proof of Theorem 8.6 but is simpler since (8.24) is not needed and $f(y_{k+1}^0) > f(y_k^0) - \rho(\Delta_k)$ can be used directly in the contradicting argument.)

Chapter 9

Line-search methods based on simplex derivatives

The implicit-filtering method of Kelley et al. in [229] (see also [141]) can be viewed as a line-search method based on the simplex gradient. In this chapter, we present a modified version of the implicit-filtering method that is guaranteed to be globally convergent to first-order stationary points of (1.1). The main modification relies on the application of the backtracking scheme used to achieve sufficient decrease, which can be shown to eventually be successful. Under such a modification, this derivative-free algorithm can be seen as a line-search counterpart of the trust-region derivative-free method considered in Chapter 10.

9.1 A line-search framework

For simplicity let us consider, at each iteration k of the algorithm, a sample set $Y_k = \{y_k^0, y_k^1, \ldots, y_k^n\}$, formed by $n + 1$ points. We will assume that this set is poised in the sense of linear interpolation. In other words, the points $y_k^0, y_k^1, \ldots, y_k^n$ are assumed to be the vertices of a simplex set.

Now we consider the simplex gradient based at y_k^0 and computed from this sample set. Such a simplex gradient is given by

$$\nabla_s f(x_k) = L_k^{-1} \delta f(Y_k),$$

where

$$L_k = \left[y_k^1 - y_k^0 \cdots y_k^n - y_k^0 \right]^\top$$

and

$$\delta f(Y_k) = \begin{bmatrix} f(y_k^1) - f(y_k^0) \\ \vdots \\ f(y_k^n) - f(y_k^0) \end{bmatrix}.$$

Let us also define

$$\Delta_k = \max_{1 \le i \le n} \| y_k^i - y_k^0 \|.$$

As we will see later in this chapter, other more elaborate simplex gradients could be used, following a regression approach (see also Chapters 2 and 4) or a centered difference scheme.

No matter which simplex gradient calculation is chosen, all we need to ensure is that the following error bound can be satisfied:

$$\|\nabla f(x_k) - \nabla_s f(x_k)\| \leq \kappa_{eg} \Delta_k, \tag{9.1}$$

where, as we have seen in Chapters 2 and 3, κ_{eg} is a positive constant depending on the geometry of the sample points. The algorithmic framework presented in Chapter 6 can be used to improve the geometry of the sample set. In order to make the statement of the algorithm simultaneously rigorous and close to a practical implementation, we introduce the following assumption, where by an improvement step of the simplex geometry we mean recomputation of one of the points in the set $\{y_k^1, \ldots, y_k^n\}$.

Assumption 9.1. *We assume that* (9.1) *can be satisfied for some fixed positive* $\kappa_{eg} > 0$ *and for any value of* $\Delta_k > 0$ *in a finite, uniformly bounded number of improvement steps of the sample set.*

Basically, one geometry improvement step consists of applying the algorithms described in Chapter 6 to replace one vertex of the simplex. Thus, in the interpolation case ($n + 1$ sample points), for any positive value of Δ_k, the error bound (9.1) can be achieved using at most n improvement steps.

At each iteration of the line-search derivative-free method that follows, a sufficient decrease condition is imposed on the computation of the new point. When using the simplex gradient $\nabla_s f(x_k)$, with $x_k = y_k^0$, this sufficient decrease condition is of the form

$$f(x_k - \alpha \nabla_s f(x_k)) - f(x_k) \leq -\eta \alpha \|\nabla_s f(x_k)\|^2, \tag{9.2}$$

where η is a constant in the interval $(0, 1)$ for all k and $\alpha > 0$. The new point x_{k+1} is *in principle* of the form $x_{k+1} = x_k - \alpha_k \nabla_s f(x_k)$, where α_k is chosen by a backtracking procedure to ensure (9.2) with $\alpha = \alpha_k$. However, the line-search version of the method analyzed here considers the possibility of accepting a point different from $x_k - \alpha_k \nabla_s f(x_k)$ as long as it provides a lower objective value.

The line-search derivative-free method is presented below and includes a standard backtracking scheme. However, this line search can fail. When it fails, the size of Δ_k is reduced compared to the size of $\|\nabla_s f(x_k)\|$ (which involves a number of improvement steps and the recomputation of the simplex gradient) and the line search is restarted from the same point (with a likely more accurate simplex gradient).

Algorithm 9.1 (Line-search derivative-free method based on simplex gradients).

Initialization: Choose an initial point x_0 and an initial poised sample set $\{y_0^0(= x_0), y_0^1, \ldots, y_0^n\}$. Choose β, η, and ω in $(0, 1)$. Select $j_{max} \in \mathbb{N}$.

For $k = 0, 1, 2, \ldots$

    1. **Simplex gradient calculation:** Compute a simplex gradient $\nabla_s f(x_k)$ such that $\Delta_k \leq \|\nabla_s f(x_k)\|$ (apply Algorithm 9.2 below). Set $j_{current} = j_{max}$ and $\mu = 1$.

2. **Line search:** For $j = 0, 1, 2, \ldots, j_{current}$

   (a) Set $\alpha = \beta^j$. Evaluate f at $x_k - \alpha \nabla_s f(x_k)$.

   (b) If the sufficient decrease condition (9.2) is satisfied for α, then stop this step with $\alpha_k = \alpha$ (and go to Step 4).

3. **Line-search failure:** If the line search failed, then divide μ by two, recompute a simplex gradient $\nabla_s f(x_k)$ such that $\Delta_k \leq \mu \| \nabla_s f(x_k) \|$ (apply Algorithm 9.2 below), increase $j_{current}$ by one, and repeat the line search (go back to Step 2).

4. **New point:** Set

$$x_{k+1} = \operatorname*{argmin}_{x \in \mathcal{X}_k} \{ f(x_k - \alpha_k \nabla_s f(x_k)), f(x) \},$$

where $\mathcal{X}_k$ is the set of points where f has possibly been evaluated during the course of Steps 1 and 3. Set $y_{k+1}^0 = x_{k+1}$. Update $y_{k+1}^1, \ldots, y_{k+1}^n$ from $y_k^0, y_k^1, \ldots, y_k^n$ by dropping one of these points.

A possible stopping criterion is to terminate the run when Δ_k becomes smaller than a chosen tolerance $\Delta_{tol} > 0$ (for instance $\Delta_{tol} = 10^{-5}$).

The algorithm that recomputes the sample set at x_k and the corresponding simplex gradient such that $\Delta_k \leq \mu \| \nabla_s f(x_k) \|$ is described next.

Algorithm 9.2 (Criticality step). *This algorithm is applied only when $\Delta_k > \mu \| \nabla_s f(x_k) \|$. The constant $\omega \in (0, 1)$ should be chosen in the initialization of Algorithm 9.1.*

Initialization: Set $i = 0$. Set $\nabla_s f(x_k)^{(0)} = \nabla_s f(x_k)$.

Repeat Increment i by one. Compute a new simplex gradient $\nabla_s f(x_k)^{(i)}$ based on a sample set containing x_k and contained in $B(x_k; \omega^i \mu \| \nabla_s f(x_k)^{(0)} \|)$ such that

$$\| \nabla f(x_k) - \nabla_s f(x_k)^{(i)} \| \leq \kappa_{eg} \left(\omega^i \mu \| \nabla_s f(x_k)^{(0)} \| \right)$$

(notice that this can be done in a finite, uniformly bounded number of steps). Set $\Delta_k = \omega^i \mu \| \nabla_s f(x_k)^{(0)} \|$ and $\nabla_s f(x_k) = \nabla_s f(x_k)^{(i)}$.

Until $\Delta_k \leq \mu \| \nabla_s f(x_k)^{(i)} \|$.

9.2 Global convergence for first-order critical points

We need to assume that ∇f is Lipschitz continuous on the level set (where the iterates must necessarily lie):

$$L(x_0) = \{ x \in \mathbb{R}^n : f(x) \leq f(x_0) \}.$$

However, we also need to take into account that the points used in the simplex gradient calculations might lie outside $L(x_0)$, especially at the early iterations of the method. Thus, we need to enlarge $L(x_0)$. If we impose that Δ_k never exceeds a given positive constant Δ_{max}, then all points used by the algorithm will lie in the following set:

$$L_{enl}(x_0) = L(x_0) \cup \bigcup_{x \in L(x_0)} B(x; \Delta_{max}) = \bigcup_{x \in L(x_0)} B(x; \Delta_{max}).$$

We will assume that f is a continuously differentiable function in an open set containing $L_{enl}(x_0)$ and that ∇f is Lipschitz continuous on $L_{enl}(x_0)$ with constant $\nu > 0$.

The purpose of the first lemma is to show that Steps 1 and 3 of Algorithm 9.1 are well defined, in the sense that Algorithm 9.2 will take a finite number of steps.

Lemma 9.1. *If $\nabla f(x_k) \neq 0$, Steps 1 and 3 of Algorithm 9.1 will satisfy $\Delta_k \leq \mu \|\nabla_s f(x_k)\|$ in a finite number of improvement steps (by applying Algorithm 9.2).*

Proof. The proof is postponed to Chapter 10 (see Lemma 10.5), where it is done in the trust-region environment, in a slightly more general context. The simplex gradient $\nabla_s f(x_k)$ plays the same role here as the gradient g_k of the model plays there. $\quad\square$

Now we need to analyze under what conditions the sufficient decrease (9.2) is attained and to make sure that the line search in Step 2 of Algorithm 9.1 can be accomplished after a finite number of reductions of α and μ. In other words, we will prove that one cannot loop infinitely between Steps 2 and 3 unless the point is stationary.

Lemma 9.2. *Let f be a continuously differentiable function in an open set containing $L_{enl}(x_0)$. Assume that ∇f is Lipschitz continuous on $L_{enl}(x_0)$ with constant $\nu > 0$. Let x_k be such that $\nabla f(x_k) \neq 0$. The sufficient decrease condition (9.2) is satisfied for all α and μ such that*

$$0 < \alpha \leq \frac{2(1 - \eta - \kappa_{eg}\mu)}{\nu} \tag{9.3}$$

and

$$\mu < \frac{1 - \eta}{\kappa_{eg}}. \tag{9.4}$$

Proof. First, we know that (see, e.g., the proof of Theorem 2.8)

$$f(x_k + \alpha d) - f(x_k)$$
$$\leq \alpha \nabla f(x_k)^\top d + \frac{\nu \alpha^2}{2} \|d\|^2$$
$$= \alpha (\nabla f(x_k) - \nabla_s f(x_k))^\top d + \alpha \nabla_s f(x_k)^\top d + \frac{\nu \alpha^2}{2} \|d\|^2.$$

By replacing d by $-\nabla_s f(x_k)$ and using (9.1) and $\alpha > 0$, we obtain

$$f(x_k - \alpha \nabla_s f(x_k)) - f(x_k)$$
$$\leq \alpha (\kappa_{eg} \Delta_k / \|\nabla_s f(x_k)\| - 1 + \nu \alpha/2) \|\nabla_s f(x_k)\|^2.$$

Combining this inequality with $\Delta_k \leq \mu \|\nabla_s f(x_k)\|$ and $\alpha > 0$ yields

$$f(x_k - \alpha \nabla_s f(x_k)) - f(x_k) \leq \alpha (\kappa_{eg}\mu - 1 + \nu \alpha/2) \|\nabla_s f(x_k)\|^2.$$

Thus, the sufficient decrease condition (9.2) is satisfied if

$$\kappa_{eg}\mu - 1 + \frac{\nu\alpha}{2} \leq -\eta,$$

and the proof is completed.   □

As a result of Lemma 9.2 and of the scheme in Step 2 to update α, one can guarantee that α_k is bounded from below by

$$\alpha_k \geq \bar{\alpha} = \frac{2(1 - \eta - \kappa_{eg}\bar{\mu})\beta}{\nu}, \tag{9.5}$$

where $\bar{\mu}$ is any number such that

$$0 < \bar{\mu} < \frac{1 - \eta}{\kappa_{eg}}.$$

The global convergence of Algorithm 9.1 to stationary points is stated in the following theorem.

Theorem 9.3.   *Let f be a continuously differentiable function in an open set containing $L_{enl}(x_0)$. Assume that f is bounded from below in $L(x_0)$ and that ∇f is Lipschitz continuous on $L_{enl}(x_0)$ with constant $\nu > 0$. Then the sequence of iterates generated by Algorithm 9.1 satisfies*

$$\lim_{k \longrightarrow +\infty} \|\nabla f(x_k)\| = 0. \tag{9.6}$$

Proof. The proof of

$$\lim_{k \longrightarrow +\infty} \|\nabla_s f(x_k)\| = 0 \tag{9.7}$$

follows the classical arguments for line-search methods. The sequence $\{f(x_k)\}$ is decreasing (by construction) and bounded from below in $L(x_0)$ (by hypothesis). Thus, the left-hand side in (9.2) (with $\alpha = \alpha_k$) converges to zero. The limit (9.7) then follows from

$$f(x_{k+1}) - f(x_k) \leq f(x_k - \alpha_k \nabla_s f(x_k)) - f(x_k) \leq -\eta\bar{\alpha}\|\nabla_s f(x_k)\|^2,$$

where $\bar{\alpha}$ is given in (9.5), and the first inequality comes from Step 4 of the algorithm.

Since $\nabla_s f(x_k)$ converges to zero, we know from (9.1) and Steps 1 and 3 of the algorithm that, for k sufficiently large,

$$\|\nabla f(x_k) - \nabla_s f(x_k)\| \leq \kappa_{eg}\|\nabla_s f(x_k)\|.$$

This inequality together with (9.7) implies (9.6).   □

9.3   Analysis for noise

In many applications the noise level in the objective function is not zero, and what is evaluated in practice can be represented as

$$f(x) = f_{smooth}(x) + \varepsilon(x), \tag{9.8}$$

where f_{smooth} is an underlying smooth function and $\varepsilon(x)$ represents the noise in its evaluation.

We are interested in extending the analysis of the line-search derivative-free method to a noisy function f of the form (9.8). So, let us consider Algorithm 9.1 applied to the minimization of f in (9.8).

In order to retain global convergence to first-order critical points, we require the noise level in the objective function to satisfy

$$\max_{0 \le i \le n} |\varepsilon(y_k^i)| \le c \Delta_k^2 \qquad (9.9)$$

for all iterations, where $c > 0$ is independent of k. In addition, since for sufficient decrease purposes one needs to evaluate f at points which might not be used for simplex gradient calculations, we also ask the noise level to satisfy

$$\varepsilon(x_k - \alpha \nabla_s f(x_k)) \le c \Delta_k^2 \qquad (9.10)$$

for all values of α considered in the backtracking scheme.

When the noise level satisfies the two conditions stated above, it is possible to prove, similarly to Theorem 9.3, that the limit of the gradient of the underlying function f_{smooth} converges to zero.

9.4   The implicit-filtering algorithm

The implicit-filtering method of Kelley et al. in [229] (see also [141]) differs from Algorithm 9.1 in a number of aspects. First, the sample set is not dynamically updated on an iteration base. Instead, a new sample set is chosen at each iteration for the purpose of the simplex gradient calculation. Such a choice may make the algorithm appealing for parallel computation but might compromise its efficiency in a serial environment.

No provision is made to link the accuracy of the simplex gradient and the quality of the geometry of the sample set to the line-search scheme itself, as in Algorithm 9.1. Therefore, it is not possible to guarantee success for the line search (which can terminate unsuccessfully after the predetermined finite number of steps j_{max}).

In addition to the basic line-search scheme, implicit filtering incorporates a quasi-Newton update (see, e.g., [76, 178]) for the Hessian approximation based on the simplex gradients that are being computed. When the line search fails the quasi-Newton matrix is reset to the initial choice. The method has been applied to problems with noisy functions, for which it has been shown to be numerically robust. We present it below especially having in mind the case where f is of the form (9.8) and the noise obeys (9.9) and (9.10).

Algorithm 9.3 (Implicit filtering method).

Initialization: Choose β and η in $(0,1)$. Choose an initial point x_0 and an initial Hessian approximation H_0 (for instance, the identity matrix). Select $j_{max} \in \mathbb{N}$.

For $k = 0, 1, 2, \ldots$

1. **Simplex gradient calculation:** Compute a simplex gradient $\nabla_s f(x_k)$ such that $\Delta_k \le \|\nabla_s f(x_k)\|$. Compute $d_k = -H_k^{-1} \nabla_s f(x_k)$.

2. **Line search:** For $j = 0, 1, 2, \ldots, j_{max}$

   (a) Set $\alpha = \beta^j$. Evaluate f at $x_k + \alpha d_k$.

   (b) If the sufficient decrease condition

   $$f(x_k + \alpha d_k) - f(x_k) \leq \eta \alpha \nabla_s f(x_k)^\top d_k$$

   is satisfied for α, then stop this step with $\alpha_k = \alpha$.

3. **New point:** If the line search succeeded, then $x_{k+1} = x_k + \alpha_k d_k$.

4. **Hessian update:** If the line search failed set $H_{k+1} = H_0$. Otherwise, update H_{k+1} from H_k using a quasi-Newton update based on $x_{k+1} - x_k$ and $\nabla_s f(x_{k+1}) - \nabla_s f(x_k)$.

Other provisions may be necessary to make this algorithm practical when line-search failures occur. For instance, one might have to recompute the sampling points (by scaling them towards x_k so that $\Delta_k \leq \mu \|\nabla_s f(x_k)\|$ for some $\mu > 0$ smaller than one, as in Algorithm 9.1) and to repeat an iteration after a line-search failure.

9.5   Other simplex derivatives

There exist alternatives for the computation of simplex gradients based on $n + 1$ points. For example, if the function is evaluated at more than $n + 1$ points, one can compute simplex gradients in the regression sense, as explained in Chapter 2. In both cases, the order of accuracy is linear in Δ (the size of the ball containing the points).

Centered simplex gradients

When the number of points is $2n + 1$ and the sampling set Y is of the form

$$\left\{ y^0, y^0 + (y^1 - y^0), \ldots, y^0 + (y^n - y^0), y^0 - (y^1 - y^0), \ldots, y^0 - (y^n - y^0) \right\}$$

it is possible to compute a centered simplex gradient with Δ^2 accuracy. First, note that the sampling set given above is obtained by retaining the original points and adding their reflection through y^0. This geometrical structure has been seen before:

$$\left[y^1 - y^0 \cdots y^n - y^0 \ -(y^1 - y^0) \cdots -(y^n - y^0) \right]$$

forms a (maximal) positive basis (see Section 2.1). When $y^0 = 0$, this sampling set reduces to

$$\left\{ 0, y^1, \ldots, y^n, -y^1, \ldots, -y^n \right\}.$$

Consider, again, the matrix $L = \left[y^1 - y^0 \cdots y^n - y^0 \right]^\top$. The centered simplex gradient is defined by

$$\nabla_{cs} f(y^0) = L^{-1} \delta_{cs} f(Y),$$

where

$$\delta_{cs} f(Y) = \frac{1}{2} \begin{bmatrix} f(y^0 + (y^1 - y^0)) - f(y^0 - (y^1 - y^0)) \\ \vdots \\ f(y^0 + (y^n - y^0)) - f(y^0 - (y^n - y^0)) \end{bmatrix}.$$

One can easily show that if $\nabla^2 f$ is Lipschitz continuous with constant $\nu > 0$ in an open set containing the ball $B(y^0; \Delta)$, where

$$\Delta = \max_{1 \leq i \leq n} \|y^i - y^0\|,$$

then

$$\|\nabla f(y^0) - \nabla_{cs} f(y^0)\| \leq \kappa_{eg} \Delta^2,$$

where $\kappa_{eg} = n^{\frac{1}{2}} \nu \|\hat{L}^{-1}\|$ and $\hat{L} = L/\Delta$. It is important to stress that this improvement in the order of accuracy of the gradient approximation does not have consequences for second-order approximations. In fact, it is not possible in general, given only a number of points linear in n, to compute a simplex Hessian (see the coming paragraphs) that approximates the true Hessian within an error of the order of Δ.

Simplex Hessians

Given a sample set $Y = \{y^0, y^1, \ldots, y^p\}$, with $p = (n+1)(n+2)/2 - 1$, poised in the sense of quadratic interpolation, one can compute a simplex gradient $\nabla_s f(y^0)$ and a simplex (symmetric) Hessian $\nabla_s^2 f(y^0)$ from the system of linear equations

$$(y^i - y^0)^\top \nabla_s f(y^0) + \frac{1}{2}(y^i - y^0)^\top \nabla_s^2 f(y^0)(y^i - y^0) = f(y^i) - f(y^0), \qquad (9.11)$$

$i = 1, \ldots, p$.

One can observe that the simplex gradient and simplex Hessian defined above are nothing else than the coefficients of the quadratic interpolation model $m(x) = c + g^\top x + (1/2)x^\top Hx$:

$$\nabla_s f(y^0) = g \quad \text{and} \quad \nabla_s^2 f(y^0) = H.$$

When $p = 2n$, it is possible to neglect all the off-diagonal elements of the Hessian and compute a simplex gradient and a diagonal simplex Hessian.

9.6   Other notes and references

The implicit-filtering algorithm was first described in the already cited paper [229] and later, in more detail, in the journal publications by Stoneking at al. [212] and Gilmore and Kelley [105]. The global convergence properties of the method were analyzed by Bortz and Kelley [41]. Choi and Kelley [52] studied the rate of local convergence.

Mifflin [171] suggested in 1975 a line-search algorithm based on centered simplex gradients and approximated simplex Hessians, for which he proved global convergence to first-order stationary points and studied the rate of local convergence. Mifflin's algorithm is a hybrid approach, sharing features with direct-search methods (use of the coordinate-search directions to compute the simplex derivatives) and with line-search algorithms.

9.7 Exercises

1. Prove that Theorem 9.3 remains true for f_{smooth} when Algorithm 9.1 is applied to f given in (9.8) if conditions (9.9) and (9.10) are satisfied.

Chapter 10

Trust-region methods based on derivative-free models

Trust-region methods are a well-studied class of algorithms for the solution of nonlinear programming problems [57, 178]. These methods have a number of attractive features. The fact that they are intrinsically based on quadratic models makes them particularly attractive to deal with curvature information. Their robustness is partially associated with the regularization effect of minimizing quadratic models over regions of predetermined size. Extensive research on solving trust-region subproblems and related numerical issues has led to efficient implementations and commercial codes. On the other hand, the convergence theory of trust-region methods is both comprehensive and elegant in the sense that it covers many problem classes and particularizes from one problem class to a subclass in a natural way. Many extensions have been developed and analyzed to deal with different algorithmic adaptations or problem features (see [57]).

In this chapter we address trust-region methods for unconstrained derivative-free optimization. These methods maintain quadratic (or linear) models which are based only on the objective function values computed at sample points. The corresponding models can be constructed by means of polynomial interpolation or regression or by any other approximation technique. The approach taken in this chapter abstracts from the specifics of model building. In fact, it is not even required that these models be polynomial functions as long as appropriate decreases (such as Cauchy and eigenstep decreases) can be extracted from the trust-region subproblems. Instead, it is required that the derivative-free models have a uniform local behavior (possibly after a finite number of modifications of the sample set) similar to what is observed by Taylor models in the presence of derivatives. In Chapter 6, we called such models, depending on their accuracy, *fully linear* and *fully quadratic*. It has been rigorously shown in Chapters 3–6 how such *fully linear* and *fully quadratic* models can be constructed in the context of polynomial interpolation or regression.

Again, the problem we are considering is (1.1), where f is a real-valued function, assumed to be once (or twice) continuously differentiable and bounded from below.

10.1 The trust-region framework basics

The fundamentals of trust-region methods are rather simple. As in traditional derivative-based trust-region methods, the main idea is to use a model for the objective function which one, hopefully, is able to trust in a neighborhood of the current point. To be useful it must be significantly easier to optimize the model within the neighborhood than solving the original problem. The neighborhood considered is called the trust region. The model has to be fully linear in order to ensure global convergence to a first-order critical point. One would also like to have something approaching a fully quadratic model, to allow global convergence to a second-order critical point (and to speed up local convergence). Typically, the model is quadratic, written in the form

$$m_k(x_k + s) = m_k(x_k) + s^\top g_k + \frac{1}{2} s^\top H_k s. \tag{10.1}$$

The derivatives of this quadratic model with respect to the s variables are given by $\nabla m_k(x_k + s) = H_k s + g_k$, $\nabla m_k(x_k) = g_k$, and $\nabla^2 m_k(x_k) = H_k$. Clearly, g_k is the gradient of the model at $s = 0$.

If m_k is a first-order Taylor model, then $m_k(x_k) = f(x_k)$ and $g_k = \nabla f(x_k)$, and if it is a second-order Taylor model, one has, in addition, $H_k = \nabla^2 f(x_k)$. In general, even in the derivative case, H_k is a symmetric approximation to $\nabla^2 f(x_k)$. In the derivative-free case, we use models where $H_k \neq \nabla^2 f(x_k)$, $g_k \neq \nabla f(x_k)$, and, in the absence of interpolation, $m_k(x_k) \neq f(x_k)$.

At each iterate k, we consider the model $m_k(x_k + s)$ that is intended to approximate the true objective f within a suitable neighborhood of x_k—the trust region. This region is taken for simplicity as the set of all points

$$B(x_k; \Delta_k) = \{x \in \mathbb{R}^n : \|x - x_k\| \leq \Delta_k\},$$

where Δ_k is called the trust-region radius, and where $\| \cdot \|$ could be an iteration-dependent norm, but usually is fixed and in our case will be taken as the standard Euclidean norm. Figure 10.1 illustrates a linear model and a quadratic model of a nonlinear function in a trust region (a simple ball), both built by interpolation. As expected, the quadratic model captures the curvature of the function.

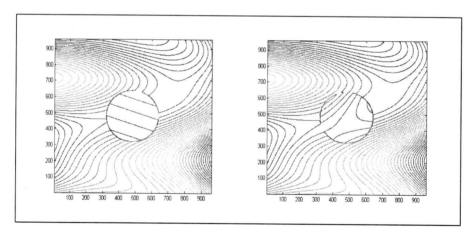

Figure 10.1. *Contours of a linear model (left) and a quadratic model (right) of a nonlinear function in a trust region.*

Thus, in the unconstrained case, the local model problem (called trust-region subproblem) we are considering is stated as

$$\min_{s \in B(0;\Delta_k)} m_k(x_k + s), \tag{10.2}$$

where $m_k(x_k + s)$ is the model for the objective function given at (10.1) and $B(0; \Delta_k)$ is our trust region of radius Δ_k, now centered at 0 and expressed in terms of $s = x - x_k$.

The Cauchy step

In some sense, the driving force for all optimization techniques is steepest descent since it defines the locally best direction of descent. It turns out that it is crucial, from the point of view of global convergence, that one minimizes the model at least as well as something related to steepest descent. On this basis, one defines something called the Cauchy step s_k^C, which is actually the step to the minimum of the model along the steepest descent direction within the trust region. Thus, if we define

$$t_k^C = \operatorname*{argmin}_{t \geq 0: x_k - t g_k \in B(x_k; \Delta_k)} m_k(x_k - t g_k),$$

then the Cauchy step is a step given by

$$s_k^C = -t_k^C g_k. \tag{10.3}$$

A fundamental result that drives trust-region methods to first-order criticality is stated and proved below.

Theorem 10.1. *Consider the model* (10.1) *and the Cauchy step* (10.3). *Then*

$$m_k(x_k) - m_k(x_k + s_k^C) \geq \frac{1}{2} \|g_k\| \min\left\{ \frac{\|g_k\|}{\|H_k\|}, \Delta_k \right\}, \tag{10.4}$$

where we assume that $\|g_k\|/\|H_k\| = +\infty$ *when* $H_k = 0$.

Proof. We first note that

$$m_k(x_k - \alpha g_k) = m_k(x_k) - \alpha \|g_k\|^2 + \frac{1}{2} \alpha^2 g_k^\top H_k g_k. \tag{10.5}$$

In the case where the curvature of the model along the steepest descent direction $-g_k$ is positive, that is, when $g_k^\top H_k g_k > 0$, we know that the model is convex along that direction, and so a stationary point will necessarily be the global minimizer in that direction. Denoting the optimal parameter by α_k^* we have that $-\|g_k\|^2 + \alpha_k^* g_k^\top H_k g_k = 0$ and

$$\alpha_k^* = \frac{\|g_k\|^2}{g_k^\top H_k g_k}.$$

Thus, if $\|-\alpha_k^* g_k\| = \|g_k\|^3 / g_k^\top H_k g_k \leq \Delta_k$, then the unique minimizer lies in the trust region and we can conclude that

$$m_k(x_k + s_k^C) - m_k(x_k) = -\frac{1}{2} \frac{\|g_k\|^4}{g_k^\top H_k g_k},$$

and, consequently,

$$m_k(x_k) - m_k(x_k + s_k^C) \geq \frac{1}{2} \|g_k\|^2 / \|H_k\|. \tag{10.6}$$

If $\|-\alpha_k^* g_k\| > \Delta_k$, then we take $\|-t_k^C g_k\| = \Delta_k$; i.e., we go to the boundary of the trust region. In this case

$$m_k(x_k + s_k^C) - m_k(x_k) = -\frac{\Delta_k \|g_k\|^2}{\|g_k\|} + \frac{1}{2}\frac{\Delta_k^2 \|g_k\|}{\|g_k\|^3} g_k^\top H_k g_k.$$

But $\|g_k\|^3 / g_k^\top H_k g_k > \Delta_k$ (since the optimal step was outside the trust region) then gives

$$m_k(x_k) - m_k(x_k + s_k^C) \geq \frac{1}{2}\Delta_k \|g_k\|. \tag{10.7}$$

It remains to consider the case when the one-dimensional problem in α is not convex. In this case, again we know that we will terminate at the boundary of the trust region (because the second and third terms on the right-hand side of (10.5) are both negative for all positive α). Furthermore, $g_k^\top H_k g_k \leq 0$ implies that

$$m_k(x_k + s_k^C) - m_k(x_k) \leq -t_k^C \|g_k\|^2 = -\Delta_k \|g_k\| < -\frac{1}{2}\Delta_k \|g_k\|. \tag{10.8}$$

The derived bounds (10.6), (10.7), and (10.8) on the change in the model imply that (10.4) holds.   □

In fact, it is not necessary to actually find the Cauchy step to achieve global convergence to first-order stationarity. It is sufficient to relate the step computed to the Cauchy step, and thus what is required is the following assumption.

Assumption 10.1. *For all iterations k,*

$$m_k(x_k) - m_k(x_k + s_k) \geq \kappa_{fcd} \left[m_k(x_k) - m_k(x_k + s_k^C) \right] \tag{10.9}$$

for some constant $\kappa_{fcd} \in (0, 1]$.

The steps computed under Assumption 10.1 will therefore provide a fraction of Cauchy decrease, which from Theorem 10.1 can be bounded from below as

$$m_k(x_k) - m_k(x_k + s_k) \geq \frac{\kappa_{fcd}}{2} \|g_k\| \min \left\{ \frac{\|g_k\|}{\|H_k\|}, \Delta_k \right\}. \tag{10.10}$$

If $m_k(x_k + s)$ is not a linear or a quadratic function, then Theorem 10.1 is not directly applicable. In this case one could, for instance, define a Cauchy step by applying a line search at $s = 0$ along $-g_k$ to the model $m_k(x_k + s)$, stopping when some type of sufficient decrease condition is satisfied (see [57, Section 6.3.3] or Section 12.2). Calculating a step yielding a decrease better than the Cauchy decrease could be achieved by approximately solving the trust-region subproblem, which now involves the minimization of a nonlinear function within a trust region.

Assumption 10.1 is the minimum requirement for how well one has to do at solving (10.2) to achieve global convergence to first-order critical points. If we would like to guarantee more, then we must drive the algorithm with more than just the steepest descent direction. We will consider this case next.

The eigenstep

When considering a quadratic model and global convergence to second-order critical points, the model reduction that is required can be achieved along a direction related to the greatest negative curvature. Let us assume that H_k has at least one negative eigenvalue, and let $\tau_k = \lambda_{min}(H_k)$ be the most negative eigenvalue of H_k. In this case, we can determine a step of negative curvature s_k^{E}, such that

$$(s_k^{\mathrm{E}})^\top (g_k) \leq 0, \quad \|s_k^{\mathrm{E}}\| = \Delta_k, \quad \text{and} \quad (s_k^{\mathrm{E}})^\top H_k(s_k^{\mathrm{E}}) = \tau_k \Delta_k^2. \tag{10.11}$$

We refer to s_k^{E} as the eigenstep.

The eigenstep s_k^{E} is the eigenvector of H_k corresponding to the most negative eigenvalue τ_k, whose sign and scale are chosen to ensure that the first two parts of (10.11) are satisfied. Note that due to the presence of negative curvature, s_k^{E} is the minimizer of the quadratic function along that direction inside the trust region. The eigenstep induces the following decrease in the model.

Lemma 10.2. *Suppose that the model Hessian H_k has negative eigenvalues. Then we have that*

$$m_k(x_k) - m_k(x_k + s_k^E) \geq -\frac{1}{2}\tau_k \Delta_k^2. \tag{10.12}$$

Proof. It suffices to point out that

$$
\begin{aligned}
m_k(x_k) - m_k(x_k + s_k^{\mathrm{E}}) &= -(s_k^{\mathrm{E}})^\top (g_k) - \tfrac{1}{2}(s_k^{\mathrm{E}})^\top H_k(s_k^{\mathrm{E}}) \\
&\geq -\tfrac{1}{2}(s_k^{\mathrm{E}})^\top H_k(s_k^{\mathrm{E}}) \\
&= -\tfrac{1}{2}\tau_k \Delta_k^2. \quad \square
\end{aligned}
$$

The eigenstep plays a role similar to that of the Cauchy step, in that, provided negative curvature is present in the model, we now require the model decrease at $x_k + s_k$ to satisfy

$$m_k(x_k) - m_k(x_k + s_k) \geq \kappa_{fed}[m_k(x_k) - m_k(x_k + s_k^{\mathrm{E}})]$$

for some constant $\kappa_{fed} \in (0, 1]$. Since we also want the step to yield a fraction of Cauchy decrease, we will consider the following assumption.

Assumption 10.2. *For all iterations k,*

$$m_k(x_k) - m_k(x_k + s_k) \geq \kappa_{fod}\left[m_k(x_k) - \min\{m_k(x_k + s_k^C), m_k(x_k + s_k^{\mathrm{E}})\}\right] \tag{10.13}$$

for some constant $\kappa_{fod} \in (0, 1]$.

A step satisfying this assumption is given by computing both the Cauchy step and, in the presence of negative curvature in the model, the eigenstep, and by choosing the one that provides the larger reduction in the model. By combining (10.4), (10.12), and (10.13), we obtain that

$$m_k(x_k) - m_k(x_k + s_k) \geq \frac{\kappa_{fod}}{2} \max\left\{ \|g_k\| \min\left\{ \frac{\|g_k\|}{\|H_k\|}, \Delta_k \right\}, -\tau_k \Delta_k^2 \right\}. \tag{10.14}$$

In some trust-region literature what is required for global convergence to second-order critical points is a fraction of the decrease obtained by the optimal trust-region step (i.e, an optimal solution of (10.2)). Note that a fraction of optimal decrease condition is stronger than (10.14) for the same value of κ_{fod}.

If $m_k(x_k + s)$ is not a quadratic function, then Theorem 10.1 and Lemma 10.2 are not directly applicable. Similarly to the Cauchy step case, one could here define an eigenstep by applying a line search to the model $m_k(x_k + s)$, at $s = 0$ and along a direction of negative (or most negative) curvature of H_k, stopping when some type of sufficient decrease condition is satisfied (see [57, Section 6.6.2] or Section 12.2). Calculating a step yielding a decrease better than the Cauchy and eigenstep decreases could be achieved by approximately solving the trust-region subproblem, which, again, now involves the minimization of a nonlinear function within a trust region.

The update of the trust-region radius

The other essential ingredient of a trust-region method is the so-called trust-region management. The basic idea is to compare the truth, that is, the actual reduction in the objective function, to the predicted reduction in the model. If the comparison is good, we take the new step and (possibly) increase the trust-region radius. If the comparison is bad, we reject the new step and decrease the trust-region radius. Formally, this procedure can be stated as follows. We introduce a distinction between simple and sufficient decreases in the objective function. In the former case, when $\eta_0 = 0$, the step is accepted as long as it provides a simple decrease in the objective function, which might be a natural thing to do in derivative-free optimization when functions are expensive to evaluate.

Suppose that the current iterate is x_k and that the candidate for the next iterate is $x_k + s_k$. Assume, also, that one is given constants η_0, η_1, and γ satisfying $\gamma \in (0, 1)$ and $0 \le \eta_0 \le \eta_1 < 1$ (with $\eta_1 \ne 0$).

Truth versus prediction: Define

$$\rho_k = \frac{f(x_k) - f(x_k + s_k)}{m_k(x_k) - m_k(x_k + s_k)}.$$

Step acceptance (sufficient decrease): If $\rho_k \ge \eta_1$, then accept the new point $x_k + s_k$; otherwise, the new candidate point is rejected (and the trust-region radius is reduced, as below).

Step acceptance (possibly based on simple decrease): If $\rho_k \ge \eta_0$, then accept the new point $x_k + s_k$ (but reduce the trust-region radius if $\rho_k < \eta_1$, as below); otherwise, the new candidate point is rejected (and the trust-region radius is reduced, as below).

Trust-region management: Set

$$\Delta_{k+1} \in \begin{cases} [\Delta_k, +\infty) & \text{if } \rho_k \ge \eta_1, \\ \{\gamma \Delta_k\} & \text{if } \rho_k < \eta_1. \end{cases}$$

An important property which ensures convergence is the following: if the model is based on, for example, (some reasonable approximation to) a truncated Taylor series expansion, then we know that as the trust-region radius becomes smaller the model necessarily becomes better. This guarantees that the trust-region radius is bounded away from zero, away from stationarity. In what follows we are not using Taylor models because we are interested in the case where although derivatives may exist they are not available.

10.2 Conditions on the trust-region models

In the derivative-free case, away from stationary points, it is necessary to ensure that the trust-region radius remains bounded away from zero. In the case of models based on a Taylor series approximation, this is ensured by the fact that the model becomes progressively better as the neighborhood (trust region) becomes smaller. The management of the trust-region radius guarantees that it stays bounded away from zero as long as the current iterate is not a stationary point. Models like polynomial interpolation or regression models do not necessarily become better when the radius of the trust region is reduced. Hence, we have to ensure that we reduce only the trust-region radius when we are certain that the failure of a current step is due to the size of the trust region and not to the poor quality of the model itself. With this safeguard, we can prove, once again, that, as the neighborhood becomes small, the prediction becomes good and thus the trust-region radius remains bounded away from zero and one can obtain, via Assumptions 10.1 and 10.2, similar results to those one is able to obtain in the case with derivatives. As we know from Chapters 3–6, what one requires in these cases is Taylor-like error bounds with a uniformly bounded constant that characterizes the geometry of the sample sets.

In the remainder of this section, we will describe the assumptions on the function and on the models which we use, in this chapter, to prove the global convergence of the derivative-free trust-region algorithms. We will impose only those requirements on the models that are essential for the convergence theory. The models might not necessarily be quadratic functions as mentioned in Section 10.1. (We will cover the use of nonlinear models in trust-region methods in Section 12.2.)

For the purposes of convergence to first-order critical points, we assume that the function f and its gradient are Lipschitz continuous in the domain considered by the algorithms. To better define this region, we suppose that x_0 (the initial iterate) is given and that new iterates correspond to reductions in the value of the objective function. Thus, the iterates must necessarily belong to the level set

$$L(x_0) = \left\{ x \in \mathbb{R}^n : f(x) \le f(x_0) \right\}.$$

However, when considering models based on sampling it is possible (especially at the early iterations) that the function f is evaluated outside $L(x_0)$. Let us assume that sampling is restricted to regions of the form $B(x_k; \Delta_k)$ and that Δ_k never exceeds a given (possibly large) positive constant Δ_{max}. Under this scenario, the region where f is sampled is within the set

$$L_{enl}(x_0) = L(x_0) \cup \bigcup_{x \in L(x_0)} B(x; \Delta_{max}) = \bigcup_{x \in L(x_0)} B(x; \Delta_{max}).$$

Thus, what we need are the requirements already stated in Assumption 6.1, making sure, however, that the open domain mentioned there contains the larger set $L_{enl}(x_0)$.

Assumption 10.3. *Suppose x_0 and Δ_{max} are given. Assume that f is continuously differentiable with Lipschitz continuous gradient in an open domain containing the set $L_{enl}(x_0)$.*

The algorithmic framework which will be described in Section 10.3 for interpolation-based trust-region methods also requires the selection of a fully linear class $\mathcal{M}$—see Definition 6.1. We reproduce this definition below in the notation of this chapter (where y is given by $x + s$ and the set S is $L(x_0)$).

Definition 10.3. *Let a function $f : \mathbb{R}^n \to \mathbb{R}$, that satisfies Assumption 10.3, be given. A set of model functions $\mathcal{M} = \{m : \mathbb{R}^n \to \mathbb{R}, m \in C^1\}$ is called a fully linear class of models if the following hold:*

1. *There exist positive constants κ_{ef}, κ_{eg}, and ν_1^m such that for any $x \in L(x_0)$ and $\Delta \in (0, \Delta_{max}]$ there exists a model function $m(x + s)$ in $\mathcal{M}$, with Lipschitz continuous gradient and corresponding Lipschitz constant bounded by ν_1^m, and such that*

   - *the error between the gradient of the model and the gradient of the function satisfies*

$$\|\nabla f(x + s) - \nabla m(x + s)\| \leq \kappa_{eg} \Delta \quad \forall s \in B(0; \Delta), \tag{10.15}$$

   and

   - *the error between the model and the function satisfies*

$$|f(x + s) - m(x + s)| \leq \kappa_{ef} \Delta^2 \quad \forall s \in B(0; \Delta). \tag{10.16}$$

   Such a model m is called fully linear on $B(x; \Delta)$.

2. *For this class $\mathcal{M}$ there exists an algorithm, which we will call a "model-improvement" algorithm, that in a finite, uniformly bounded (with respect to x and Δ) number of steps can*

   - *either establish that a given model $m \in \mathcal{M}$ is fully linear on $B(x; \Delta)$ (we will say that a certificate has been provided and the model is certifiably fully linear),*

   - *or find a model $\tilde{m} \in \mathcal{M}$ that is fully linear on $B(x; \Delta)$.*

For the remainder of this chapter we will assume, without loss of generality, that the constants κ_{ef}, κ_{eg}, and ν_1^m of any fully linear class $\mathcal{M}$ which we use in our algorithmic framework are such that Lemma 10.25 below holds. In this way, we make sure that if a model is fully linear in a ball, it will be so in any larger concentric one, as happens with Taylor models defined by first-order derivatives.

To analyze the convergence to second-order critical points, we require, in addition, the Lipschitz continuity of the Hessian of f. The overall smoothness requirement has been stated in Assumption 6.2, but we need to consider here, however, that the open domain mentioned there now contains the larger set $L_{enl}(x_0)$.

Assumption 10.4. *Suppose x_0 and Δ_{max} are given. Assume that f is twice continuously differentiable with Lipschitz continuous Hessian in an open domain containing the set $L_{enl}(x_0)$.*

The algorithmic framework which will be described in Section 10.5 for interpolation-based trust-region methods also requires the selection of a fully quadratic class $\mathcal{M}$—see Definition 6.2. We repeat this definition below using the notation of this chapter (where y is given by $x + s$ and the set S is $L(x_0)$).

Definition 10.4. *Let a function f, that satisfies Assumption 10.4, be given. A set of model functions $\mathcal{M} = \{m : \mathbb{R}^n \to \mathbb{R}, m \in C^2\}$ is called a fully quadratic class of models if the following hold:*

1. *There exist positive constants κ_{ef}, κ_{eg}, κ_{eh}, and ν_2^m such that for any $x \in L(x_0)$ and $\Delta \in (0, \Delta_{max}]$ there exists a model function $m(x + s)$ in $\mathcal{M}$, with Lipschitz continuous Hessian and corresponding Lipschitz constant bounded by ν_2^m, and such that*

   - *the error between the Hessian of the model and the Hessian of the function satisfies*

$$\|\nabla^2 f(x+s) - \nabla^2 m(x+s)\| \leq \kappa_{eh} \Delta \quad \forall s \in B(0; \Delta), \tag{10.17}$$

   - *the error between the gradient of the model and the gradient of the function satisfies*

$$\|\nabla f(x+s) - \nabla m(x+s)\| \leq \kappa_{eg} \Delta^2 \quad \forall s \in B(0; \Delta), \tag{10.18}$$

   and

   - *the error between the model and the function satisfies*

$$|f(x+s) - m(x+s)| \leq \kappa_{ef} \Delta^3 \quad \forall s \in B(0; \Delta). \tag{10.19}$$

   Such a model m is called fully quadratic on $B(x; \Delta)$.

2. *For this class $\mathcal{M}$ there exists an algorithm, which we will call a "model-improvement" algorithm, that in a finite, uniformly bounded (with respect to x and Δ) number of steps can*

   - *either establish that a given model $m \in \mathcal{M}$ is fully quadratic on $B(x; \Delta)$ (we will say that a certificate has been provided and the model is certifiably fully quadratic),*
   - *or find a model $\tilde{m} \in \mathcal{M}$ that is fully quadratic on $B(x; \Delta)$.*

For the remainder of this chapter we will assume, without loss of generality, that the constants κ_{ef}, κ_{eg}, κ_{eh}, and ν_2^m of any fully quadratic class $\mathcal{M}$ which we use in our algorithmic framework are such that Lemma 10.26 below holds. By proceeding in this way, we guarantee that if a model is fully quadratic in a ball, it remains so in any larger concentric ball (as in Taylor models defined by first- and second-order derivatives).

10.3 Derivative-free trust-region methods (first order)

Derivative-free trust-region methods can be roughly classified into two categories: the methods which target good practical performance, such as the methods in [163, 190] (see

Chapter 11); and the methods for which global convergence was shown, but at the expense of practicality, such as described in [57, 59]. In this book we try to bridge the gap by describing an algorithmic framework in the spirit of the first category of methods, while retaining all the same global convergence properties of the second category. We list next the features that make this algorithmic framework closer to a practical one when compared to the methods in [57, 59].

The trust-region maintenance that we will use is different from the approaches in derivative-based methods [57]. In derivative-based methods, under appropriate conditions, the trust-region radius becomes bounded away from zero when the iterates converge to a local minimizer [57]; hence, its radius can remain unchanged or increase near optimality. This is not the case in trust-region derivative-free methods. The trust region for these methods serves two purposes: it restricts the step size to the neighborhood where the model is assumed to be good, and it also defines the neighborhood in which the points are sampled for the construction of the model. Powell in [190] suggests using two different trust regions, which makes the method and its implementation more complicated. We choose to maintain only one trust region. However, it is important to keep the radius of the trust region comparable to some measure of stationarity so that when the measure of stationarity is close to zero (that is, the current iterate may be close to a stationary point) the models become more accurate, a procedure that is accomplished by the so-called *criticality step*. The update of the trust-region radius at the criticality step forces it to converge to zero, hence defining a natural stopping criterion for this class of methods.

Another feature of this algorithmic framework is the acceptance of new iterates that provide a simple decrease in the objective function, rather than a sufficient decrease. This feature is of particular relevance in the derivative-free context, especially when function evaluations are expensive. As in the derivative case [184], the standard liminf-type results are obtained for general trust-region radius updating schemes (such as the simple one described in Section 10.1). In particular, it is possible to update the trust-region radius freely at the end of successful iterations (as long as it is not decreased). However, to derive the classical lim-type global convergence result [214] in the derivative case, an additional requirement is imposed on the update of the trust-region radius at successful iterations, to avoid a cycling effect of the type described in [236]. But, as we will see, because of the update of the trust-region radius at the criticality step mentioned in the previous paragraph, such further provisions are not needed to achieve lim-type global convergence to first-order critical points even when iterates are accepted based on simple decrease.[15]

In our framework it is possible to take steps, and for the algorithm to progress, without insisting that the model be made fully linear or fully quadratic on *every* iteration. In contrast with [57, 59], we require only (i) that the models can be made fully linear or fully quadratic during a finite, uniformly bounded number of iterations and (ii) that if a model is not fully linear or fully quadratic (depending on the order of optimality desired) in a given iteration, then the new iterate can be accepted as long as it provides a decrease in the objective function (sufficient decrease for the lim-result). This modification slightly complicates the convergence analysis, but it reflects much better the typical implementation of a trust-region derivative-free algorithm.

---

[15]We point out that a modification to derivative-based trust-region algorithms based on a criticality step would produce a similar lim-type result. However, forcing the trust-region radius to converge to zero may jeopardize the fast rates of local convergence under the presence of derivatives.

We now formally state the first-order version of the algorithm that we consider. The algorithm contemplates acceptance of new iterates based on simple decrease by selecting $\eta_0 = 0$. We have already noted that accepting new iterates when function evaluations are expensive based on simple decrease is particularly appropriate in derivative-free optimization. We also point out that the model m_k and the trust-region radius Δ_k are set only at the end of the criticality step (Step 1). The iteration ends by defining an incumbent model m_{k+1}^{icb} and an incumbent trust-region radius Δ_{k+1}^{icb} for the next iteration, which then might be changed or might not by the criticality step.

Algorithm 10.1 (Derivative-free trust-region method (first order)).

Step 0 (initialization): Choose a fully linear class of models $\mathcal{M}$ and a corresponding model-improvement algorithm (see, e.g., Chapter 6). Choose an initial point x_0 and $\Delta_{max} > 0$. We assume that an initial model $m_0^{icb}(x_0 + s)$ (with gradient and possibly the Hessian at $s = 0$ given by g_0^{icb} and H_0^{icb}, respectively) and a trust-region radius $\Delta_0^{icb} \in (0, \Delta_{max}]$ are given.

The constants η_0, η_1, γ, γ_{inc}, ϵ_c, β, μ, and ω are also given and satisfy the conditions $0 \le \eta_0 \le \eta_1 < 1$ (with $\eta_1 \ne 0$), $0 < \gamma < 1 < \gamma_{inc}$, $\epsilon_c > 0$, $\mu > \beta > 0$, and $\omega \in (0, 1)$. Set $k = 0$.

Step 1 (criticality step): If $\|g_k^{icb}\| > \epsilon_c$, then $m_k = m_k^{icb}$ and $\Delta_k = \Delta_k^{icb}$.

If $\|g_k^{icb}\| \le \epsilon_c$, then proceed as follows. Call the model-improvement algorithm to attempt to certify if the model m_k^{icb} is fully linear on $B(x_k; \Delta_k^{icb})$. If at least one of the following conditions holds,

- the model m_k^{icb} is not certifiably fully linear on $B(x_k; \Delta_k^{icb})$,

- $\Delta_k^{icb} > \mu\|g_k^{icb}\|$,

then apply Algorithm 10.2 (described below) to construct a model $\tilde{m}_k(x_k + s)$ (with gradient and possibly the Hessian at $s = 0$ given by $\tilde{g}_k$ and $\tilde{H}_k$, respectively), which is fully linear (for some constants κ_{ef}, κ_{eg}, and ν_1^m, which remain the same for all iterations of Algorithm 10.1) on the ball $B(x_k; \tilde{\Delta}_k)$, for some $\tilde{\Delta}_k \in (0, \mu\|\tilde{g}_k\|]$ given by Algorithm 10.2. In such a case set[16]

$$m_k = \tilde{m}_k \quad \text{and} \quad \Delta_k = \min\{\max\{\tilde{\Delta}_k, \beta\|\tilde{g}_k\|\}, \Delta_k^{icb}\}.$$

Otherwise, set $m_k = m_k^{icb}$ and $\Delta_k = \Delta_k^{icb}$.

Step 2 (step calculation): Compute a step s_k that sufficiently reduces the model m_k (in the sense of (10.9)) and such that $x_k + s_k \in B(x_k; \Delta_k)$.

Step 3 (acceptance of the trial point): Compute $f(x_k + s_k)$ and define

$$\rho_k = \frac{f(x_k) - f(x_k + s_k)}{m_k(x_k) - m_k(x_k + s_k)}.$$

---

[16]Note that Δ_k is selected to be the number in $[\tilde{\Delta}_k, \Delta_k^{icb}]$ closest to $\beta\|\tilde{g}_k\|$.

If $\rho_k \geq \eta_1$ or if both $\rho_k \geq \eta_0$ and the model is fully linear (for the positive constants κ_{ef}, κ_{eg}, and v_1^m) on $B(x_k; \Delta_k)$, then $x_{k+1} = x_k + s_k$ and the model is updated to include the new iterate into the sample set, resulting in a new model $m_{k+1}^{icb}(x_{k+1} + s)$ (with gradient and possibly the Hessian at $s = 0$ given by g_{k+1}^{icb} and H_{k+1}^{icb}, respectively); otherwise, the model and the iterate remain unchanged ($m_{k+1}^{icb} = m_k$ and $x_{k+1} = x_k$).

Step 4 (model improvement): If $\rho_k < \eta_1$, use the model-improvement algorithm to

- attempt to certify that m_k is fully linear on $B(x_k; \Delta_k)$,
- if such a certificate is not obtained, we say that m_k is not certifiably fully linear and make one or more suitable improvement steps.

Define m_{k+1}^{icb} to be the (possibly improved) model.

Step 5 (trust-region radius update): Set

$$
\Delta_{k+1}^{icb} \in \begin{cases} [\Delta_k, \min\{\gamma_{inc}\Delta_k, \Delta_{max}\}] & \text{if } \rho_k \geq \eta_1, \\ \{\gamma\Delta_k\} & \text{if } \rho_k < \eta_1 \text{ and } m_k \text{ is fully linear,} \\ \{\Delta_k\} & \text{if } \rho_k < \eta_1 \text{ and } m_k \\ & \text{is not certifiably fully linear.} \end{cases}
$$

Increment k by one and go to Step 1.

The procedure invoked in the criticality step (Step 1 of Algorithm 10.1) is described in the following algorithm.

Algorithm 10.2 (Criticality step: first order). *This algorithm is applied only if $\|g_k^{icb}\| \leq \epsilon_c$ and at least one of the following holds: the model m_k^{icb} is not certifiably fully linear on $B(x_k; \Delta_k^{icb})$ or $\Delta_k^{icb} > \mu\|g_k^{icb}\|$. The constant $\omega \in (0,1)$ is chosen at Step 0 of Algorithm 10.1.*

Initialization: Set $i = 0$. Set $m_k^{(0)} = m_k^{icb}$.

Repeat Increment i by one. Use the model-improvement algorithm to improve the previous model $m_k^{(i-1)}$ until it is fully linear on $B(x_k; \omega^{i-1}\Delta_k^{icb})$ (notice that this can be done in a finite, uniformly bounded number of steps given the choice of the model-improvement algorithm in Step 0 of Algorithm 10.1). Denote the new model by $m_k^{(i)}$. Set $\tilde{\Delta}_k = \omega^{i-1}\Delta_k^{icb}$ and $\tilde{m}_k = m_k^{(i)}$.

Until $\tilde{\Delta}_k \leq \mu\|g_k^{(i)}\|$.

We will prove in the next section that Algorithm 10.2 terminates after a finite number of steps if $\|\nabla f(x_k)\| \neq 0$. If $\|\nabla f(x_k)\| = 0$, then we will cycle in the criticality step until some stopping criterion is met.

Note that if $\|g_k^{icb}\| \leq \epsilon_c$ in the criticality step of Algorithm 10.1 and Algorithm 10.2 is invoked, the model m_k is fully linear on $B(x_k; \tilde{\Delta}_k)$ with $\tilde{\Delta}_k \leq \Delta_k$. Then, by Lemma 10.25, m_k is also fully linear on $B(x_k; \Delta_k)$ (as well as on $B(x_k; \mu\|g_k\|)$).

After Step 3 of Algorithm 10.1, we may have the following possible situations at each iteration:

1. $\rho_k \geq \eta_1$; hence, the new iterate is accepted and the trust-region radius is retained or increased. We will call such iterations **successful**. We will denote the set of indices of all successful iterations by $\mathcal{S}$.

2. $\eta_1 > \rho_k \geq \eta_0$ and m_k is fully linear. Hence, the new iterate is accepted and the trust-region radius is decreased. We will call such iterations **acceptable**. (There are no acceptable iterations when $\eta_0 = \eta_1 \in (0,1)$.)

3. $\eta_1 > \rho_k$ and m_k is not certifiably fully linear. Hence, the model is improved. The new point might be included in the sample set but is not accepted as a new iterate. We will call such iterations **model improving**.

4. $\rho_k < \eta_0$ and m_k is fully linear. This is the case when no (acceptable) decrease was obtained and there is no need to improve the model. The trust-region radius is reduced, and nothing else changes. We will call such iterations **unsuccessful**.

10.4 Global convergence for first-order critical points

We will first show that unless the current iterate is a first-order stationary point, then the algorithm will not loop infinitely in the criticality step of Algorithm 10.1 (Algorithm 10.2). The proof is very similar to the one in [59], but we repeat the details here for completeness.

Lemma 10.5. *If $\nabla f(x_k) \neq 0$, Step 1 of Algorithm 10.1 will terminate in a finite number of improvement steps (by applying Algorithm 10.2).*

Proof. Assume that the loop in Algorithm 10.2 is infinite. We will show that $\nabla f(x_k)$ has to be zero in this case. At the start, we know that we do not have a certifiably fully linear model m_k^{icb} or that the radius Δ_k^{icb} exceeds $\mu \| g_k^{icb} \|$. We then define $m_k^{(0)} = m_k^{icb}$, and the model is improved until it is fully linear on the ball $B(x_k; \omega^0 \Delta_k^{icb})$ (in a finite number of improvement steps). If the gradient $g_k^{(1)}$ of the resulting model $m_k^{(1)}$ satisfies $\mu \| g_k^{(1)} \| \geq \omega^0 \Delta_k^{icb}$, the procedure stops with

$$\tilde{\Delta}_k^{icb} = \omega^0 \Delta_k^{icb} \leq \mu \| g_k^{(1)} \|.$$

Otherwise, that is, if $\mu \| g_k^{(1)} \| < \omega^0 \Delta_k^{icb}$, the model is improved until it is fully linear on the ball $B(x_k; \omega \Delta_k^{icb})$. Then, again, either the procedure stops or the radius is again multiplied by ω, and so on.

The only way for this procedure to be infinite (and to require an infinite number of improvement steps) is if

$$\mu \| g_k^{(i)} \| < \omega^{i-1} \Delta_k^{icb},$$

for all $i \geq 1$, where $g_k^{(i)}$ is the gradient of the model $m_k^{(i)}$. This argument shows that $\lim_{i \to +\infty} \| g_k^{(i)} \| = 0$. Since each model $m_k^{(i)}$ was fully linear on $B(x_k; \omega^{i-1} \Delta_k^{icb})$, (10.15) with $s = 0$ and $x = x_k$ implies that

$$\| \nabla f(x_k) - g_k^{(i)} \| \leq \kappa_{eg} \omega^{i-1} \Delta_k^{icb}$$

for each $i \geq 1$. Thus, using the triangle inequality, it holds for all $i \geq 1$ that

$$\|\nabla f(x_k)\| \leq \|\nabla f(x_k) - g_k^{(i)}\| + \|g_k^{(i)}\| \leq \left(\kappa_{eg} + \frac{1}{\mu}\right)\omega^{i-1}\Delta_k^{icb}.$$

Since $\omega \in (0,1)$, this implies that $\nabla f(x_k) = 0$. □

We will now prove the results related to global convergence to first-order critical points. For minimization we need to assume that f is bounded from below.

Assumption 10.5. *Assume that f is bounded from below on $L(x_0)$; that is, there exists a constant κ_* such that, for all $x \in L(x_0)$, $f(x) \geq \kappa_*$.*

We will make use of the assumptions on the boundedness of f from below and on the Lipschitz continuity of the gradient of f (i.e., Assumptions 10.5 and 10.3) and of the existence of fully linear models (Definition 10.3). For simplicity of the presentation, we also require the model Hessian $H_k = \nabla^2 m_k(x_k)$ to be uniformly bounded. In general, fully linear models are required only to have continuous first-order derivatives (κ_{bhm} below can then be regarded as a bound on the Lipschitz constant of the gradient of these models).

Assumption 10.6. *There exists a constant $\kappa_{bhm} > 0$ such that, for all x_k generated by the algorithm,*

$$\|H_k\| \leq \kappa_{bhm}.$$

We start the main part of the analysis with the following key lemma.

Lemma 10.6. *If m_k is fully linear on $B(x_k; \Delta_k)$ and*

$$\Delta_k \leq \min\left\{\frac{\|g_k\|}{\kappa_{bhm}}, \frac{\kappa_{fcd}\|g_k\|(1-\eta_1)}{4\kappa_{ef}}\right\},$$

then the kth iteration is successful.

Proof. Since

$$\Delta_k \leq \frac{\|g_k\|}{\kappa_{bhm}},$$

the fraction of Cauchy decrease condition (10.9)–(10.10) immediately gives that

$$m_k(x_k) - m_k(x_k + s_k) \geq \frac{\kappa_{fcd}}{2}\|g_k\|\min\left\{\frac{\|g_k\|}{\kappa_{bhm}}, \Delta_k\right\} = \frac{\kappa_{fcd}}{2}\|g_k\|\Delta_k. \qquad (10.20)$$

On the other hand, since the current model is fully linear on $B(x_k; \Delta_k)$, then from the bound (10.16) on the error between the function and the model and from (10.20) we have

$$\begin{aligned}
|\rho_k - 1| &\leq \left|\frac{f(x_k + s_k) - m_k(x_k + s_k)}{m_k(x_k) - m_k(x_k + s_k)}\right| + \left|\frac{f(x_k) - m_k(x_k)}{m_k(x_k) - m_k(x_k + s_k)}\right| \\
&\leq \frac{4\kappa_{ef}\Delta_k^2}{\kappa_{fcd}\|g_k\|\Delta_k} \\
&\leq 1 - \eta_1,
\end{aligned}$$

where we have used the assumption $\Delta_k \leq \kappa_{fcd}\|g_k\|(1 - \eta_1)/(4\kappa_{ef})$ to deduce the last inequality. Therefore, $\rho_k \geq \eta_1$, and iteration k is successful. ☐

It now follows that if the gradient of the model is bounded away from zero, then so is the trust-region radius.

Lemma 10.7. *Suppose that there exists a constant $\kappa_1 > 0$ such that $\|g_k\| \geq \kappa_1$ for all k. Then there exists a constant $\kappa_2 > 0$ such that*

$$\Delta_k \geq \kappa_2$$

for all k.

Proof. We know from Step 1 of Algorithm 10.1 (independently of whether Algorithm 10.2 has been invoked) that

$$\Delta_k \geq \min\{\beta\|g_k\|, \Delta_k^{icb}\}.$$

Thus,

$$\Delta_k \geq \min\{\beta\kappa_1, \Delta_k^{icb}\}. \tag{10.21}$$

By Lemma 10.6 and by the assumption that $\|g_k\| \geq \kappa_1$ for all k, whenever Δ_k falls below a certain value given by

$$\bar{\kappa}_2 = \min\left\{\frac{\kappa_1}{\kappa_{bhm}}, \frac{\kappa_{fcd}\kappa_1(1 - \eta_1)}{4\kappa_{ef}}\right\},$$

the kth iteration has to be either successful or model improving (when it is not successful and m_k is not certifiably fully linear) and hence, from Step 5, $\Delta_{k+1}^{icb} \geq \Delta_k$. We conclude from this, (10.21), and the rules of Step 5 that $\Delta_k \geq \min\{\Delta_0^{icb}, \beta\kappa_1, \gamma\bar{\kappa}_2\} = \kappa_2$. ☐

We will now consider what happens when the number of successful iterations is finite.

Lemma 10.8. *If the number of successful iterations is finite, then*

$$\lim_{k \to +\infty} \|\nabla f(x_k)\| = 0.$$

Proof. Let us consider iterations that come after the last successful iteration. We know that we can have only a finite (uniformly bounded, say by N) number of model-improving iterations before the model becomes fully linear, and hence there is an infinite number of iterations that are either acceptable or unsuccessful and in either case the trust region is reduced. Since there are no more successful iterations, Δ_k is never increased for sufficiently large k. Moreover, Δ_k is decreased at least once every N iterations by a factor of γ. Thus, Δ_k converges to zero.

Now, for each j, let i_j be the index of the first iteration after the jth iteration for which the model m_j is fully linear. Then

$$\|x_j - x_{i_j}\| \leq N\Delta_j \to 0$$

as j goes to $+\infty$.

Let us now observe that

$$\|\nabla f(x_j)\| \leq \|\nabla f(x_j) - \nabla f(x_{i_j})\| + \|\nabla f(x_{i_j}) - g_{i_j}\| + \|g_{i_j}\|.$$

What remains to be shown is that all three terms on the right-hand side are converging to zero. The first term converges to zero because of the Lipschitz continuity of ∇f and the fact that $\|x_{i_j} - x_j\| \to 0$. The second term is converging to zero because of the bound (10.15) on the error between the gradients of a fully linear model and the function f and because of the fact that m_{i_j} is fully linear. Finally, the third term can be shown to converge to zero by Lemma 10.6, since if it was bounded away from zero for a subsequence, then for small enough Δ_{i_j} (recall that $\Delta_{i_j} \to 0$), i_j would be a successful iteration, which would then yield a contradiction. □

We now prove another useful lemma, namely, that the trust-region radius converges to zero, which is particularly relevant in the derivative-free context.

Lemma 10.9.

$$\lim_{k \to +\infty} \Delta_k = 0. \tag{10.22}$$

Proof. When S is finite the result is shown in the proof of Lemma 10.8. Let us consider the case when S is infinite. For any $k \in S$ we have

$$f(x_k) - f(x_{k+1}) \geq \eta_1 [m_k(x_k) - m_k(x_k + s_k)].$$

By using the bound on the fraction of Cauchy decrease (10.10), we have that

$$f(x_k) - f(x_{k+1}) \geq \eta_1 \frac{\kappa_{fcd}}{2} \|g_k\| \min \left\{ \frac{\|g_k\|}{\|H_k\|}, \Delta_k \right\}.$$

Due to Step 1 of Algorithm 10.1 we have that $\|g_k\| \geq \min\{\epsilon_c, \mu^{-1}\Delta_k\}$; hence

$$f(x_k) - f(x_{k+1}) \geq \eta_1 \frac{\kappa_{fcd}}{2} \min\{\epsilon_c, \mu^{-1}\Delta_k\} \min \left\{ \frac{\min\{\epsilon_c, \mu^{-1}\Delta_k\}}{\|H_k\|}, \Delta_k \right\}.$$

Since S is infinite and f is bounded from below, and by using Assumption 10.6, the right-hand side of the above expression has to converge to zero. Hence, $\lim_{k \in S} \Delta_k = 0$, and the proof is completed if all iterations are successful. Now recall that the trust-region radius can be increased only during a successful iteration, and it can be increased only by a ratio of at most γ_{inc}. Let $k \notin S$ be the index of an iteration (after the first successful one). Then $\Delta_k \leq \gamma_{inc}\Delta_{s_k}$, where s_k is the index of the last successful iteration before k. Since $\Delta_{s_k} \to 0$, then $\Delta_k \to 0$ for $k \notin S$. □

The following lemma now follows easily.

Lemma 10.10.

$$\liminf_{k \to +\infty} \|g_k\| = 0. \tag{10.23}$$

Proof. Assume, for the purpose of deriving a contradiction, that, for all k,

$$\|g_k\| \geq \kappa_1 \tag{10.24}$$

for some $\kappa_1 > 0$. By Lemma 10.7 we have that $\Delta_k \geq \kappa_2$ for all k. We obtain a contradiction with Lemma 10.9. □

We now show that if the model gradient $\|g_k\|$ converges to zero on a subsequence, then so does the true gradient $\|\nabla f(x_k)\|$.

Lemma 10.11. *For any subsequence $\{k_i\}$ such that*

$$\lim_{i \to +\infty} \|g_{k_i}\| = 0 \tag{10.25}$$

it also holds that

$$\lim_{i \to +\infty} \|\nabla f(x_{k_i})\| = 0. \tag{10.26}$$

Proof. First, we note that, by (10.25), $\|g_{k_i}\| \leq \epsilon_c$ for i sufficiently large. Thus, the mechanism of the criticality step (Step 1) ensures that the model m_{k_i} is fully linear on a ball $B(x_{k_i}; \Delta_{k_i})$ with $\Delta_{k_i} \leq \mu \|g_{k_i}\|$ for all i sufficiently large (if $\nabla f(x_{k_i}) \neq 0$). Then, using the bound (10.15) on the error between the gradients of the function and the model, we have

$$\|\nabla f(x_{k_i}) - g_{k_i}\| \leq \kappa_{eg} \Delta_{k_i} \leq \kappa_{eg} \mu \|g_{k_i}\|.$$

As a consequence, we have

$$\|\nabla f(x_{k_i})\| \leq \|\nabla f(x_{k_i}) - g_{k_i}\| + \|g_{k_i}\| \leq (\kappa_{eg}\mu + 1)\|g_{k_i}\|$$

for all i sufficiently large. But since $\|g_{k_i}\| \to 0$ then this implies (10.26). □

Lemmas 10.10 and 10.11 immediately give the following global convergence result.

Theorem 10.12. *Let Assumptions 10.3, 10.5, and 10.6 hold. Then*

$$\liminf_{k \to +\infty} \nabla f(x_k) = 0.$$

If the sequence of iterates is bounded, then this result implies the existence of one limit point that is first-order critical. In fact we are able to prove that all limit points of the sequence of iterates are first-order critical.

Theorem 10.13. *Let Assumptions 10.3, 10.5, and 10.6 hold. Then*

$$\lim_{k \to +\infty} \nabla f(x_k) = 0.$$

Proof. We have established by Lemma 10.8 that in the case when S is finite the theorem holds. Hence, we will assume that S is infinite. Suppose, for the purpose of establishing a contradiction, that there exists a subsequence $\{k_i\}$ of successful or acceptable iterations such that

$$\|\nabla f(x_{k_i})\| \geq \epsilon_0 > 0 \tag{10.27}$$

for some $\epsilon_0 > 0$ and for all i (we can ignore the other types of iterations, since x_k does not change during such iterations). Then, because of Lemma 10.11, we obtain that

$$\|g_{k_i}\| \geq \epsilon > 0$$

for some $\epsilon > 0$ and for all i sufficiently large. Without loss of generality, we pick ϵ such that

$$\epsilon \leq \min\left\{\frac{\epsilon_0}{2(2+\kappa_{eg}\mu)}, \epsilon_c\right\}. \tag{10.28}$$

Lemma 10.10 then ensures the existence, for each k_i in the subsequence, of a first iteration $\ell_i > k_i$ such that $\|g_{\ell_i}\| < \epsilon$. By removing elements from $\{k_i\}$, without loss of generality and without a change of notation, we thus obtain that there exists another subsequence indexed by $\{\ell_i\}$ such that

$$\|g_k\| \geq \epsilon \text{ for } k_i \leq k < \ell_i \text{ and } \|g_{\ell_i}\| < \epsilon \tag{10.29}$$

for sufficiently large i.

We now restrict our attention to the set $\mathcal{K}$ corresponding to the subsequence of iterations whose indices are in the set

$$\bigcup_{i\in\mathbb{N}_0}\{k \in \mathbb{N}_0 : k_i \leq k < \ell_i\},$$

where k_i and ℓ_i belong to the two subsequences given above in (10.29).

We know that $\|g_k\| \geq \epsilon$ for $k \in \mathcal{K}$. From Lemma 10.9, $\lim_{k\to+\infty} \Delta_k = 0$, and by Lemma 10.6 we conclude that for any large enough $k \in \mathcal{K}$ the iteration k is either successful, if the model is fully linear, or model improving, otherwise.

Moreover, for each $k \in \mathcal{K} \cap \mathcal{S}$ we have

$$f(x_k) - f(x_{k+1}) \geq \eta_1[m_k(x_k) - m_k(x_k + s_k)] \geq \eta_1 \frac{\kappa_{fcd}}{2} \|g_k\| \min\left\{\frac{\|g_k\|}{\kappa_{bhm}}, \Delta_k\right\}, \tag{10.30}$$

and, for any such k large enough, $\Delta_k \leq \frac{\epsilon}{\kappa_{bhm}}$. Hence, we have, for $k \in \mathcal{K} \cap \mathcal{S}$ sufficiently large,

$$\Delta_k \leq \frac{2}{\eta_1 \kappa_{fcd}\epsilon}[f(x_k) - f(x_{k+1})].$$

Since for any $k \in \mathcal{K}$ large enough the iteration is either successful or model improving and since for a model-improving iteration $x_k = x_{k+1}$, we have, for all i sufficiently large,

$$\|x_{k_i} - x_{\ell_i}\| \leq \sum_{\substack{j=k_i \\ j\in\mathcal{K}\cap\mathcal{S}}}^{\ell_i-1} \|x_j - x_{j+1}\| \leq \sum_{\substack{j=k_i \\ j\in\mathcal{K}\cap\mathcal{S}}}^{\ell_i-1} \Delta_j \leq \frac{2}{\eta_1 \kappa_{fcd}\epsilon}[f(x_{k_i}) - f(x_{\ell_i})].$$

Since the sequence $\{f(x_k)\}$ is bounded from below (Assumption 10.5) and monotonic decreasing, we see that the right-hand side of this inequality must converge to zero, and we therefore obtain that

$$\lim_{i \to +\infty} \|x_{k_i} - x_{\ell_i}\| = 0.$$

Now

$$\|\nabla f(x_{k_i})\| \le \|\nabla f(x_{k_i}) - \nabla f(x_{\ell_i})\| + \|\nabla f(x_{\ell_i}) - g_{\ell_i}\| + \|g_{\ell_i}\|.$$

The first term of the right-hand side tends to zero because of the Lipschitz continuity of the gradient of f (Assumption 10.3), and it is thus bounded by ϵ for i sufficiently large. The third term is bounded by ϵ by (10.29). For the second term we use the fact that from (10.28) and the mechanism of the criticality step (Step 1) at iteration ℓ_i the model m_{ℓ_i} is fully linear on $B(x_{\ell_i}; \mu\|g_{\ell_i}\|)$. Thus, using (10.15) and (10.29), we also deduce that the second term is bounded by $\kappa_{eg}\mu\epsilon$ (for i sufficiently large). As a consequence, we obtain from these bounds and (10.28) that

$$\|\nabla f(x_{k_i})\| \le (2 + \kappa_{eg}\mu)\epsilon \le \frac{1}{2}\epsilon_0$$

for i large enough, which contradicts (10.27). Hence, our initial assumption must be false, and the theorem follows.   □

This last theorem is the only result for which we need to use the fact that $x_k = x_{k+1}$ at the model-improving iterations. So, this requirement could be lifted from the algorithm if only a liminf-type result is desired. The advantage of this is that it becomes possible to accept simple decrease in the function value even when the model is not fully linear. The disadvantage, aside from the weaker convergence result, is in the inherent difficulty of producing fully linear models after at most N consecutive model-improvement steps when the region where each such model has to be fully linear can change at each iteration.

10.5 Derivative-free trust-region methods (second order)

In order to achieve global convergence to second-order critical points, the algorithm must attempt to drive to zero a quantity that expresses second-order stationarity. Following [57, Section 9.3], one possibility is to work with

$$\sigma_k^m = \max\{\|g_k\|, -\lambda_{min}(H_k)\},$$

which measures the second-order stationarity of the model.

The algorithm follows mostly the same arguments as those of Algorithm 10.1. One fundamental difference is that σ_k^m now plays the role of $\|g_k\|$. Another is the need to work with fully quadratic models. A third main modification is the need to be able to solve the trust-region subproblem better, so that the step yields both a fraction of Cauchy decrease and a fraction of the eigenstep decrease when negative curvature is present. Finally, to prove the lim-type convergence result in the second-order case, we also need to increase the trust-region radius on some of the successful iterations, whereas in the first-order case that was optional. Unlike the case of traditional trust-region methods that seek second-order convergence results [57], we do not increase the trust-region radius on *every* successful iteration. We insist on such an increase only when the size of the trust-region radius is small when compared to the measure of stationarity.

We state the version of the algorithm we wish to consider.

Algorithm 10.3 (Derivative-free trust-region method (second order)).

Step 0 (initialization): Choose a fully quadratic class of models $\mathcal{M}$ and a corresponding model-improvement algorithm (see, e.g., Chapter 6). Choose an initial point x_0 and $\Delta_{max} > 0$. We assume that an initial model $m_0^{icb}(x_0+s)$ (with gradient and Hessian at $s = 0$ given by g_0^{icb} and H_0^{icb}, respectively), with $\sigma_0^{m,icb} = \max\{\|g_0^{icb}\|, -\lambda_{min}(H_0^{icb})\}$, and a trust-region radius $\Delta_0^{icb} \in (0, \Delta_{max}]$ are given.

The constants $\eta_0, \eta_1, \gamma, \gamma_{inc}, \epsilon_c, \beta, \mu,$ and ω are also given and satisfy the conditions $0 \leq \eta_0 \leq \eta_1 < 1$ (with $\eta_1 \neq 0$), $0 < \gamma < 1 < \gamma_{inc}, \epsilon_c > 0, \mu > \beta > 0,$ and $\omega \in (0,1)$. Set $k = 0$.

Step 1 (criticality step): If $\sigma_k^{m,icb} > \epsilon_c$, then $m_k = m_k^{icb}$ and $\Delta_k = \Delta_k^{icb}$.

If $\sigma_k^{m,icb} \leq \epsilon_c$, then proceed as follows. Call the model-improvement algorithm to attempt to certify if the model m_k^{icb} is fully quadratic on $B(x_k; \Delta_k^{icb})$. If at least one of the following conditions holds,

- the model m_k^{icb} is not certifiably fully quadratic on $B(x_k; \Delta_k^{icb})$,
- $\Delta_k^{icb} > \mu \sigma_k^{m,icb}$,

then apply Algorithm 10.4 (described below) to construct a model $\tilde{m}_k(x_k + s)$ (with gradient and Hessian at $s = 0$ given by $\tilde{g}_k$ and $\tilde{H}_k$, respectively), with $\tilde{\sigma}_k^m = \max\{\|\tilde{g}_k\|, -\lambda_{min}(\tilde{H}_k)\}$, which is fully quadratic (for some constants $\kappa_{ef}, \kappa_{eg}, \kappa_{eh},$ and ν_2^m, which remain the same for all iterations of Algorithm 10.3) on the ball $B(x_k; \tilde{\Delta}_k)$ for some $\tilde{\Delta}_k \in (0, \mu\tilde{\sigma}_k^m]$ given by Algorithm 10.4. In such a case set[17]

$$m_k = \tilde{m}_k \quad \text{and} \quad \Delta_k = \min\{\max\{\tilde{\Delta}_k, \beta\tilde{\sigma}_k^m\}, \Delta_k^{icb}\}.$$

Otherwise, set $m_k = m_k^{icb}$ and $\Delta_k = \Delta_k^{icb}$.

Step 2 (step calculation): Compute a step s_k that sufficiently reduces the model m_k (in the sense of (10.13)) and such that $x_k + s_k \in B(x_k; \Delta_k)$.

Step 3 (acceptance of the trial point): Compute $f(x_k + s_k)$ and define

$$\rho_k = \frac{f(x_k) - f(x_k + s_k)}{m_k(x_k) - m_k(x_k + s_k)}.$$

If $\rho_k \geq \eta_1$ or if both $\rho_k \geq \eta_0$ and the model is fully quadratic (for the positive constants $\kappa_{ef}, \kappa_{eg}, \kappa_{eh},$ and ν_2^m) on $B(x_k; \Delta_k)$, then $x_{k+1} = x_k + s_k$ and the model is updated to include the new iterate into the sample set resulting in a new model $m_{k+1}^{icb}(x_{k+1} + s)$ (with gradient and Hessian at $s = 0$ given by g_{k+1}^{icb} and H_{k+1}^{icb}, respectively), with $\sigma_{k+1}^{m,icb} = \max\{\|g_{k+1}^{icb}\|, -\lambda_{min}(H_{k+1}^{icb})\}$; otherwise, the model and the iterate remain unchanged ($m_{k+1}^{icb} = m_k$ and $x_{k+1} = x_k$).

---

[17]Note that Δ_k is selected to be the number in $[\tilde{\Delta}_k, \Delta_k^{icb}]$ closest to $\beta\|\tilde{\sigma}_k^m\|$.

Step 4 (model improvement): If $\rho_k < \eta_1$, use the model-improvement algorithm to

- attempt to certify that m_k is fully quadratic on $B(x_k; \Delta_k)$,
- if such a certificate is not obtained, we say that m_k is not certifiably fully quadratic and make one or more suitable improvement steps.

Define m_{k+1}^{icb} to be the (possibly improved) model.

Step 5 (trust-region radius update): Set

$$
\Delta_{k+1}^{icb} \in
\begin{cases}
\{\min\{\gamma_{inc}\Delta_k, \Delta_{max}\}\} & \text{if } \rho_k \geq \eta_1 \text{ and } \Delta_k < \beta\sigma_k^m, \\
[\Delta_k, \min\{\gamma_{inc}\Delta_k, \Delta_{max}\}] & \text{if } \rho_k \geq \eta_1 \text{ and } \Delta_k \geq \beta\sigma_k^m, \\
\{\gamma\Delta_k\} & \text{if } \rho_k < \eta_1 \text{ and } m_k \\
& \text{is fully quadratic,} \\
\{\Delta_k\} & \text{if } \rho_k < \eta_1 \text{ and } m_k \\
& \text{is not certifiably fully quadratic.}
\end{cases}
$$

Increment k by one and go to Step 1.

We need to recall for Algorithm 10.3 the definitions of **successful, acceptable, model improving**, and **unsuccessful** iterations which we stated for the sequence of iterations generated by Algorithm 10.1. We will use the same definitions here, adapted to the fully quadratic models. We denote the set of all successful iterations by $\mathcal{S}$ and the set of all such iterations when $\Delta_k < \beta\sigma_k^m$ by $\mathcal{S}_+$.

As in the first-order case, during a model-improvement step, Δ_k and x_k remain unchanged; hence there can be only a finite number of model-improvement steps before a fully quadratic model is obtained. The comments outlined after Theorem 10.13 about possibly changing x_k at any model-improving iteration, suitably modified, apply in the fully quadratic case as well.

The criticality step can be implemented following a procedure similar to the one described in Algorithm 10.2, essentially by replacing $\|g_k\|$ by σ_k^m and by using fully quadratic models rather than fully linear ones.

Algorithm 10.4 (Criticality step: second order). *This algorithm is applied only if* $\sigma_k^{m,icb} \leq \epsilon_c$ *and at least one the following holds: the model* m_k^{icb} *is not certifiably fully quadratic on* $B(x_k; \Delta_k^{icb})$ *or* $\Delta_k^{icb} > \mu\sigma_k^{m,icb}$. *The constant* $\omega \in (0,1)$ *is chosen at Step 0 of Algorithm* 10.3.

Initialization: Set $i = 0$. Set $m_k^{(0)} = m_k^{icb}$.

Repeat Increment i by one. Improve the previous model $m_k^{(i-1)}$ until it is fully quadratic on $B(x_k; \omega^{i-1}\Delta_k^{icb})$ (notice that this can be done in a finite, uniformly bounded number of steps, given the choice of the model-improvement algorithm in Step 0 of Algorithm 10.3). Denote the new model by $m_k^{(i)}$. Set $\tilde{\Delta}_k = \omega^{i-1}\Delta_k^{icb}$ and $\tilde{m}_k = m_k^{(i)}$.

Until $\tilde{\Delta}_k \leq \mu(\sigma_k^m)^{(i)}$.

Note that if $\sigma_k^{m,icb} \leq \epsilon_c$ in the criticality step of Algorithm 10.3 and Algorithm 10.4 is invoked, the new model m_k is fully quadratic on $B(x_k; \tilde{\Delta}_k)$ with $\tilde{\Delta}_k \leq \Delta_k$. Then, by Lemma 10.26, m_k is also fully quadratic on $B(x_k; \Delta_k)$ (as well as on $B(x_k; \mu\sigma_k^m)$).

10.6   Global convergence for second-order critical points

For global convergence to second-order critical points, we will need one more order of smoothness, namely Assumption 10.4 on the Lipschitz continuity of the Hessian of f. It will be also necessary to assume that the function f is bounded from below (Assumption 10.5). Naturally, we will also assume the existence of fully quadratic models.

We start by introducing the notation

$$\sigma^m(x) \,=\, \max\left\{ \|\nabla m(x)\|, -\lambda_{min}(\nabla^2 m(x)) \right\}$$

and

$$\sigma(x) \,=\, \max\left\{ \|\nabla f(x)\|, -\lambda_{min}(\nabla^2 f(x)) \right\}.$$

It will be important to bound the difference between the true $\sigma(x)$ and the model $\sigma^m(x)$. For that purpose, we first derive a bound on the difference between the smallest eigenvalues of a function and of a corresponding fully quadratic model.

Proposition 10.14. *Suppose that Assumption* 10.4 *holds and m is a fully quadratic model on $B(x; \Delta)$. Then we have that*

$$|\lambda_{min}(\nabla^2 f(x)) - \lambda_{min}(\nabla^2 m(x))| \,\leq\, \kappa_{eh} \Delta.$$

Proof. The proof follows directly from the bound (10.17) on the error between the Hessians of m and f and the simple observation that if v is a normalized eigenvector corresponding to the smallest eigenvalue of $\nabla^2 m(x)$, then

$$
\begin{aligned}
\lambda_{min}(\nabla^2 f(x)) - \lambda_{min}(\nabla^2 m(x)) \,&\leq\, v^\top [\nabla^2 f(x) - \nabla^2 m(x)]v \\
&\leq\, \|\nabla^2 f(x) - \nabla^2 m(x)\| \\
&\leq\, \kappa_{eh} \Delta.
\end{aligned}
$$

Analogously, letting v be a normalized eigenvector corresponding to the smallest eigenvalue of $\nabla^2 f(x)$, we would obtain

$$\lambda_{min}(\nabla^2 m(x)) - \lambda_{min}(\nabla^2 f(x)) \,\leq\, \kappa_{eh} \Delta,$$

and the result follows.   □

The following lemma shows that the difference between the true $\sigma(x)$ and the model $\sigma^m(x)$ is of the order of Δ.

Lemma 10.15. *Let Δ be bounded by Δ_{max}. Suppose that Assumption* 10.4 *holds and m is a fully quadratic model on $B(x; \Delta)$. Then we have, for some $\kappa_\sigma > 0$, that*

$$|\sigma(x) - \sigma^m(x)| \,\leq\, \kappa_\sigma \Delta. \tag{10.31}$$

Proof. It follows that

$$
\begin{aligned}
|\sigma(x) - \sigma^m(x)| &= \left| \max\left\{ \|\nabla f(x)\|, \max\{-\lambda_{min}(\nabla^2 f(x)), 0\} \right\} \right. \\
&\quad \left. - \max\left\{ \|\nabla m(x)\|, \max\{-\lambda_{min}(\nabla^2 m(x)), 0\} \right\} \right| \\
&\leq \max\left\{ | \|\nabla f(x)\| - \|\nabla m(x)\| |, \right. \\
&\quad \left. |\max\{-\lambda_{min}(\nabla^2 f(x)), 0\} - \max\{-\lambda_{min}(\nabla^2 m(x)), 0\}| \right\}.
\end{aligned}
$$

The first argument $|\,\|\nabla f(x)\| - \|\nabla m(x)\|\,|$ is bounded from above by $\kappa_{eg}\Delta_{max}\Delta$, because of the error bound (10.18) between the gradients of f and m, and from the bound $\Delta \leq \Delta_{max}$. The second argument is clearly dominated by $|\lambda_{min}(\nabla^2 f(x)) - \lambda_{min}(\nabla^2 m(x))|$, which is bounded from above by $\kappa_{eh}\Delta$ because of Proposition 10.14. Finally, we need only write $\kappa_\sigma = \max\{\kappa_{eg}\Delta_{max}, \kappa_{eh}\}$, and the result follows. □

The convergence theory will require the already mentioned assumptions (Assumptions 10.4 and 10.5), as well as the uniform upper bound on the Hessians of the quadratic models (Assumption 10.6).

As for the first-order case, we begin by noting that the criticality step can be successfully executed in a finite number of improvement steps.

Lemma 10.16. *If $\sigma(x_k) \neq 0$, Step 1 of Algorithm 10.3 will terminate in a finite number of improvement steps (by applying Algorithm 10.4).*

Proof. The proof is practically identical to the proof of Lemma 10.5, with $\|g_k^{(i)}\|$ replaced by $(\sigma_k^m)^{(i)}$ and $\nabla f(x_k)$ replaced by $\sigma(x_k)$. □

We now show that an iteration must be successful if the current model is fully quadratic and the trust-region radius is small enough with respect to σ_k^m.

Lemma 10.17. *If m_k is fully quadratic on $B(x_k; \Delta_k)$ and*

$$
\Delta_k \leq \min\left\{ \frac{\sigma_k^m}{\kappa_{bhm}}, \frac{\kappa_{fod}\sigma_k^m(1-\eta_1)}{4\kappa_{ef}\Delta_{max}}, \frac{\kappa_{fod}\sigma_k^m(1-\eta_1)}{4\kappa_{ef}} \right\},
$$

then the kth iteration is successful.

Proof. The proof is similar to the proof of Lemma 10.6 for the first-order case; however, now we need to take the second-order terms into account.

First, we recall the fractions of Cauchy and eigenstep decreases (10.14),

$$
m_k(x_k) - m_k(x_k + s_k) \geq \frac{\kappa_{fod}}{2}\max\left\{ \|g_k\| \min\left\{ \frac{\|g_k\|}{\kappa_{bhm}}, \Delta_k \right\}, -\tau_k\Delta_k^2 \right\}.
$$

From the expression for σ_k^m, one of the two cases has to hold: either $\|g_k\| = \sigma_k^m$ or $-\tau_k = -\lambda_{min}(H_k) = \sigma_k^m$.

In the first case, using the fact that $\Delta_k \leq \sigma_k^m/\kappa_{bhm}$, we conclude that

$$
m_k(x_k) - m_k(x_k + s_k) \geq \frac{\kappa_{fod}}{2}\|g_k\|\Delta_k = \frac{\kappa_{fod}}{2}\sigma_k^m\Delta_k. \tag{10.32}
$$

On the other hand, since the current model is fully quadratic on $B(x_k; \Delta_k)$, we may deduce from (10.32) and the bound (10.19) on the error between the model m_k and f that

$$
\begin{aligned}
|\rho_k - 1| &\leq \left| \frac{f(x_k + s_k) - m_k(x_k + s_k)}{m_k(x_k) - m_k(x_k + s_k)} \right| + \left| \frac{f(x_k) - m_k(x_k)}{m_k(x_k) - m_k(x_k + s_k)} \right| \\
&\leq \frac{4\kappa_{ef}\Delta_k^3}{(\kappa_{fod}\sigma_k^m)\Delta_k} \\
&\leq \frac{4\kappa_{ef}\Delta_{max}}{\kappa_{fod}\sigma_k^m}\Delta_k \\
&\leq 1 - \eta_1.
\end{aligned}
$$

In the case when $-\tau_k = \sigma_k^m$, we first write

$$
m_k(x_k) - m_k(x_k + s_k) \geq -\tfrac{\kappa_{fod}}{2}\tau_k\Delta_k^2 = \tfrac{\kappa_{fod}}{2}\sigma_k^m\Delta_k^2. \tag{10.33}
$$

But, since the current model is fully quadratic on $B(x_k; \Delta_k)$, we deduce from (10.33) and the bound (10.19) on the error between m_k and f that

$$
\begin{aligned}
|\rho_k - 1| &\leq \left| \frac{f(x_k + s_k) - m_k(x_k + s_k)}{m_k(x_k) - m_k(x_k + s_k)} \right| + \left| \frac{f(x_k) - m_k(x_k)}{m_k(x_k) - m_k(x_k + s_k)} \right| \\
&\leq \frac{4\kappa_{ef}\Delta_k^3}{(\kappa_{fod}\sigma_k^m)\Delta_k^2} \\
&\leq 1 - \eta_1.
\end{aligned}
$$

In either case, $\rho_k \geq \eta_1$ and iteration k is, thus, successful. $\square$

As in the first-order case, the following result follows readily from Lemma 10.17.

Lemma 10.18. *Suppose that there exists a constant $\kappa_1 > 0$ such that $\sigma_k^m \geq \kappa_1$ for all k. Then there exists a constant $\kappa_2 > 0$ such that*

$$
\Delta_k \geq \kappa_2
$$

for all k.

Proof. The proof is trivially derived by a combination of Lemma 10.17 and the proof of Lemma 10.7. $\square$

We are now able to show that if there are only finitely many successful iterations, then we approach a second-order stationary point.

Lemma 10.19. *If the number of successful iterations is finite, then*

$$
\lim_{k \to +\infty} \sigma(x_k) = 0.
$$

Proof. The proof of this lemma is virtually identical to that of Lemma 10.8 for the first-order case, with $\|g_k\|$ being substituted by σ_k^m and $\|\nabla f(x_k)\|$ being substituted by $\sigma(x_k)$ and by using Lemmas 10.15 and 10.17. $\square$

We now prove that the whole sequence of trust-region radii converges to zero.

Lemma 10.20.

$$\lim_{k \to +\infty} \Delta_k = 0. \tag{10.34}$$

Proof. When $\mathcal{S}$ is finite the proof is as in the proof of Lemma 10.8 (the argument is exactly the same). Let us consider the case when $\mathcal{S}$ is infinite. For any $k \in \mathcal{S}$ we have

$$f(x_k) - f(x_{k+1}) \geq \eta_1 [m(x_k) - m(x_k + s_k)]$$

$$\geq \eta_1 \frac{\kappa_{fod}}{2} \max \left\{ \|g_k\| \min \left\{ \frac{\|g_k\|}{\kappa_{bhm}}, \Delta_k \right\}, -\tau_k \Delta_k^2 \right\}.$$

Due to Step 1 of Algorithm 10.3 we have that $\sigma_k^m \geq \min\{\epsilon_c, \mu^{-1}\Delta_k\}$. If on iteration k we have $\|g_k\| \geq \max\{-\tau_k, 0\} = \{-\lambda_{min}(H_k), 0\}$, then $\sigma_k^m = \|g_k\|$ and

$$f(x_k) - f(x_{k+1}) \geq \eta_1 \frac{\kappa_{fod}}{2} \min\{\epsilon_c, \mu^{-1}\Delta_k\} \min \left\{ \frac{\min\{\epsilon_c, \mu^{-1}\Delta_k\}}{\kappa_{bhm}}, \Delta_k \right\}. \tag{10.35}$$

If, on the other hand, $\|g_k\| < -\tau_k$, then $\sigma_k^m = -\tau_k$ and

$$f(x_k) - f(x_{k+1}) \geq \eta_1 \frac{\kappa_{fod}}{2} \min\{\epsilon_c, \mu^{-1}\Delta_k\}\Delta_k^2. \tag{10.36}$$

There are two subsequences of successful iterations, possibly overlapping, $\{k_i^1\}$, for which (10.35) holds, and $\{k_i^2\}$, for which (10.36) holds. The union of these subsequences contains all successful iterations. Since $\mathcal{S}$ is infinite and f is bounded from below, then either the corresponding subsequence $\{k_i^1\}$ (resp., $\{k_i^2\}$) is finite or the right-hand side of (10.35) (resp., (10.36)) has to converge to zero. Hence, $\lim_{k \in \mathcal{S}} \Delta_k = 0$, and the proof is completed if all iterations are successful. Now recall that the trust-region radius can be increased only during a successful iteration, and it can be increased only by a ratio of at most γ_{inc}. Let $k \notin \mathcal{S}$ be the index of an iteration (after the first successful one). Then $\Delta_k \leq \gamma_{inc}\Delta_{s_k}$, where s_k is the index of the last successful iteration before k. Since $\Delta_{s_k} \to 0$, then $\Delta_k \to 0$ for $k \notin \mathcal{S}$. $\quad\square$

We obtain the following lemma as a simple corollary.

Lemma 10.21.

$$\liminf_{k \to +\infty} \sigma_k^m = 0.$$

Proof. Assume, for the purpose of deriving a contradiction, that, for all k,

$$\sigma_k^m \geq \kappa_1$$

for some $\kappa_1 > 0$. Then by Lemma 10.18 there exists a constant κ_2 such that $\Delta_k \geq \kappa_2$ for all k. We obtain contradiction with Lemma 10.20. $\quad\square$

We now verify that the criticality step (Step 1 of Algorithm 10.3) ensures that a subsequence of the iterates approaches second-order stationarity, by means of the following auxiliary result.

Lemma 10.22. *For any subsequence $\{k_i\}$ such that*

$$\lim_{i \to +\infty} \sigma^m_{k_i} = 0 \tag{10.37}$$

it also holds that

$$\lim_{i \to +\infty} \sigma(x_{k_i}) = 0. \tag{10.38}$$

Proof. From (10.37), $\sigma^m_{k_i} \leq \epsilon_c$ for i sufficiently large. The mechanism of the criticality step (Step 1) then ensures that the model m_{k_i} is fully quadratic on the ball $B(x_{k_i}; \Delta_{k_i})$ with $\Delta_{k_i} \leq \mu \sigma^m_{k_i}$ for all i sufficiently large (if $\sigma^m_{k_i} \neq 0$). Now, using (10.31),

$$\sigma(x_{k_i}) = \left(\sigma(x_{k_i}) - \sigma^m_{k_i} \right) + \sigma^m_{k_i} \leq (\kappa_\sigma \mu + 1)\sigma^m_{k_i}.$$

The limit (10.37) and this last bound then give (10.38).   □

Lemmas 10.21 and 10.22 immediately give the following global convergence result.

Theorem 10.23. *Let Assumptions* 10.4, 10.5, *and* 10.6 *hold. Then*

$$\liminf_{k \to +\infty} \sigma(x_k) = 0.$$

If the sequence of iterates is bounded, this result implies the existence of at least one limit point that is second-order critical. We are, in fact, able to prove that all limit points of the sequence of iterates are second-order critical. In this proof we make use of the additional requirement on Step 5 of Algorithm 10.3, which imposes in successful iterations an increase on the trust-region radius Δ_k if it is too small compared to σ^m_k.

Theorem 10.24. *Let Assumptions* 10.4, 10.5, *and* 10.6 *hold. Then*

$$\lim_{k \to +\infty} \sigma(x_k) = 0.$$

Proof. We have established by Lemma 10.19 that in the case when $\mathcal{S}$ is finite the theorem holds. Hence, we will assume that $\mathcal{S}$ is infinite. Suppose, for the purpose of establishing a contradiction, that there exists a subsequence $\{k_i\}$ of successful or acceptable iterations such that

$$\sigma(x_{k_i}) \geq \epsilon_0 > 0 \tag{10.39}$$

for some $\epsilon_0 > 0$ and for all i (as in the first-order case, we can ignore the other iterations, since x_k does not change during such iterations). Then, because of Lemma 10.22, we obtain that

$$\sigma^m_{k_i} \geq \epsilon > 0$$

for some $\epsilon > 0$ and for all i sufficiently large. Without loss of generality, we pick ϵ such that

$$\epsilon \leq \min\left\{\frac{\epsilon_0}{2(2+\kappa_\sigma\mu)}, \epsilon_c\right\}. \tag{10.40}$$

Lemma 10.21 then ensures the existence, for each k_i in the subsequence, of a first successful or acceptable iteration $\ell_i > k_i$ such that $\sigma^m_{\ell_i} < \epsilon$. By removing elements from $\{k_i\}$, without loss of generality and without a change of notation, we thus obtain that there exists another subsequence indexed by $\{\ell_i\}$ such that

$$\sigma^m_k \geq \epsilon \text{ for } k_i \leq k < \ell_i \text{ and } \sigma^m_{\ell_i} < \epsilon \tag{10.41}$$

for sufficiently large i.

We now restrict our attention to the set $\mathcal{K}$, which is defined as the subsequence of iterations whose indices are in the set

$$\bigcup_{i\in\mathbb{N}_0}\{k \in \mathbb{N}_0 : k_i \leq k < \ell_i\},$$

where k_i and ℓ_i belong to the two subsequences defined above in (10.41).

From Lemmas 10.17 and 10.20, just as in the proof of Theorem 10.13, it follows that for large enough $k \in \mathcal{K}$ the kth iteration is either successful, if the model is fully linear, or model improving, otherwise, i.e., that there is only a finite number of acceptable iterations in $\mathcal{K}$.

Let us now consider the situation where an index k is in $\mathcal{K} \cap \mathcal{S} \setminus \mathcal{S}_+$. In this case, $\Delta_k \geq \beta\sigma^m_k \geq \beta\epsilon$. It immediately follows from $\Delta_k \to 0$ for $k \in \mathcal{K}$ that $\mathcal{K} \cap \mathcal{S} \setminus \mathcal{S}_+$ contains only a finite number of iterations. Hence, $k \in \mathcal{K} \cap \mathcal{S}$ is also in $\mathcal{S}_+$ when k is sufficiently large.

Let us now show that for $k \in \mathcal{K} \cap \mathcal{S}_+$ sufficiently large it holds that $\Delta_{k+1} = \gamma_{inc}\Delta_k$ (when the last successful iteration in $[k_i, \ell_i - 1]$ occurs before $\ell_i - 1$). We know that since $k \in \mathcal{S}_+$, then $\Delta^{icb}_{k+1} = \gamma_{inc}\Delta_k$ after execution of Step 5. However, Δ^{icb}_{k+1} may be reduced during Step 1 of the $(k+1)$st iteration (or any subsequent iteration). By examining the assignments at the end of Step 1, we see that on any iteration $k+1 \in \mathcal{K}$ the radius Δ^{icb}_{k+1} is reduced only when $\Delta_{k+1} \geq \beta\tilde{\sigma}^m_{k+1} = \beta\sigma^m_{k+1} \geq \beta\epsilon$, but this can happen only a finite number of times, due to the fact that $\Delta_k \to 0$. Hence, for large enough $k \in \mathcal{K} \cap \mathcal{S}_+$, we obtain $\Delta_{k+1} = \gamma_{inc}\Delta_k$.

Let $\mathcal{S}^i_+ = [k_i, \ell_i - 1] \cap \mathcal{S}_+ = \{j^1_i, j^2_i, \ldots, j^*_i\}$ be the set of all indices of the successful iterations that fall in the interval $[k_i, \ell_i - 1]$. From the scheme that updates Δ_k at successful iterations, and from the fact that $x_k = x_{k+1}$ and $\Delta_{k+1} = \Delta_k$ for model improving steps, we can deduce that, for i large enough,

$$\|x_{k_i} - x_{\ell_i}\| \leq \sum_{j\in\mathcal{S}^i_+}\Delta_j \leq \sum_{j\in\mathcal{S}^i_+}\left(\frac{1}{\gamma_{inc}}\right)^{j^*_i-j}\Delta_{j^*_i} \leq \frac{\gamma_{inc}}{\gamma_{inc}-1}\Delta_{j^*_i}.$$

Thus, from the fact that $\Delta_{j^*_i} \to 0$, we conclude that $\|x_{k_i} - x_{\ell_i}\| \to 0$. We therefore obtain that

$$\lim_{i\to+\infty}\|x_{k_i} - x_{\ell_i}\| = 0.$$

Now

$$\sigma(x_{k_i}) = \left(\sigma(x_{k_i}) - \sigma(x_{\ell_i})\right) + \left(\sigma(x_{\ell_i}) - \sigma^m_{\ell_i}\right) + \sigma^m_{\ell_i}.$$

The first term of the right-hand side tends to zero because of the Lipschitz continuity of $\sigma(x)$, and it is thus bounded by ϵ for i sufficiently large. The third term is bounded by ϵ by (10.41). For the second term we use the fact that, from (10.40) and the mechanism of the criticality step (Step 1) at iteration ℓ_i, the model m_{ℓ_i} is fully quadratic on $B(x_{\ell_i}; \mu\sigma_{\ell_i}^m)$. Using (10.31) and (10.41), we also deduce that the second term is bounded by $\kappa_\sigma \mu\epsilon$ (for i sufficiently large). As a consequence, we obtain from these bounds and (10.40) that

$$\sigma(x_{k_i}) \leq (2 + \kappa_\sigma \mu)\epsilon \leq \frac{1}{2}\epsilon_0$$

for i large enough, which contradicts (10.39). Hence, our initial assumption must be false, and the theorem follows. $\square$

10.7   Model accuracy in larger concentric balls

We will show next that if a model is fully linear on $B(x; \bar{\Delta})$ with respect to some (large enough) constants κ_{ef}, κ_{eg}, and v_1^m and for some $\bar{\Delta} \in (0, \Delta_{max}]$, then it is also fully linear on $B(x; \Delta)$ for any $\Delta \in [\bar{\Delta}, \Delta_{max}]$, with the same constants. This result was needed in this chapter for the analysis of the global convergence properties of the trust-region methods. Such a property reproduces what is known for Taylor models and is a clear indication of the appropriateness of the definition of fully linear models.

Lemma 10.25. *Consider a function f satisfying Assumption 6.1 and a model m fully linear, with respect to constants κ_{ef}, κ_{eg}, and v_1^m on $B(x; \bar{\Delta})$, with $x \in L(x_0)$ and $\bar{\Delta} \leq \Delta_{max}$.*

Assume also, without loss of generality, that κ_{eg} is no less than the sum of v_1^m and the Lipschitz constant of the gradient of f, and that $\kappa_{ef} \geq (1/2)\kappa_{eg}$.

Then m is fully linear on $B(x; \Delta)$, for any $\Delta \in [\bar{\Delta}, \Delta_{max}]$, with respect to the same constants κ_{ef}, κ_{eg}, and v_1^m.

Proof. We start by considering any $\Delta \in [\bar{\Delta}, \Delta_{max}]$. Then we consider an s such that $\bar{\Delta} \leq \|s\| \leq \Delta$ and let $\theta = \bar{\Delta}/\|s\|$. Since $x + \theta s \in B(x; \bar{\Delta})$ and the model is fully linear on $B(x; \bar{\Delta})$, we obtain

$$\|\nabla f(x + \theta s) - \nabla m(x + \theta s)\| \leq \kappa_{eg}\bar{\Delta}.$$

By using the Lipschitz continuity of ∇f and ∇m and the assumption that κ_{eg} is no less than the sum of the corresponding Lipschitz constants, we derive

$$\|\nabla f(x + s) - \nabla f(x + \theta s) - \nabla m(x + \theta s) + \nabla m(x + s)\| \leq \kappa_{eg}(\|s\| - \bar{\Delta}).$$

Thus, by combining the above expressions we obtain

$$\|\nabla f(x + s) - \nabla m(x + s)\| \leq \kappa_{eg}\|s\| \leq \kappa_{eg}\Delta. \qquad (10.42)$$

In the second part of the proof, we consider the function $\phi(\alpha) = f(x + \alpha s) - m(x + \alpha s)$, $\alpha \in [0, 1]$. We want to bound $|\phi(1)|$. From the fact that m is a fully linear model on $B(x; \bar{\Delta})$, we have $|\phi(\theta)| \leq \kappa_{ef}\bar{\Delta}^2$. To bound $|\phi(1)|$, we bound $|\phi(1) - \phi(\theta)|$ first by using (10.42):

$$\left| \int_\theta^1 \phi'(\alpha) d\alpha \right| \leq \int_\theta^1 \|s\| \|\nabla f(x + \alpha s) - \nabla m(x + \alpha s)\| \, d\alpha$$

$$\leq \int_\theta^1 \alpha \kappa_{eg} \|s\|^2 d\alpha = (1/2)\kappa_{eg}(\|s\|^2 - \bar{\Delta}^2).$$

Using the assumption $\kappa_{ef} \geq (1/2)\kappa_{eg}$, we finally get

$$|f(x + s) - m(x + s)| \leq |\phi(1) - \phi(\theta)| + |\phi(\theta)| \leq \kappa_{ef} \|s\|^2 \leq \kappa_{ef} \Delta^2. \quad \square$$

Similar to the linear case, we will now show that if a model is fully quadratic on $B(x; \bar{\Delta})$ with respect to some (large enough) constants κ_{ef}, κ_{eg}, κ_{eh}, and v_2^m and for some $\bar{\Delta} \in (0, \Delta_{max}]$, then it is also fully quadratic on $B(x; \Delta)$ for any $\Delta \in [\bar{\Delta}, \Delta_{max}]$, with the same constants.

Lemma 10.26. *Consider a function f satisfying Assumption 6.2 and a model m fully quadratic, with respect to constants κ_{ef}, κ_{eg}, κ_{eh}, and v_2^m on $B(x; \bar{\Delta})$, with $x \in L(x_0)$ and $\bar{\Delta} \leq \Delta_{max}$.*

Assume also, without loss of generality, that κ_{eh} is no less than the sum of v_2^m and the Lipschitz constant of the Hessian of f, and that $\kappa_{eg} \geq (1/2)\kappa_{eh}$ and $\kappa_{ef} \geq (1/3)\kappa_{eg}$.

Then m is fully quadratic on $B(x; \Delta)$, for any $\Delta \in [\bar{\Delta}, \Delta_{max}]$, with respect to the same constants κ_{ef}, κ_{eg}, κ_{eh}, and v_2^m.

Proof. Let us consider any $\Delta \in [\bar{\Delta}, \Delta_{max}]$. Consider, also, an s such that $\bar{\Delta} \leq \|s\| \leq \Delta$, and let $\theta = (\bar{\Delta}/\|s\|)$. Since $x + \theta s \in B(x; \bar{\Delta})$, then, due to the model being fully quadratic on $B(x; \bar{\Delta})$, we know that

$$\|\nabla^2 f(x + \theta s) - \nabla^2 m(x + \theta s)\| \leq \kappa_{eh} \bar{\Delta}.$$

Since $\nabla^2 f$ and $\nabla^2 m$ are Lipschitz continuous and since κ_{eh} is no less than the sum of the corresponding Lipschitz constants, we have

$$\|\nabla^2 f(x + s) - \nabla^2 f(x + \theta s) - \nabla^2 m(x + \theta s) + \nabla^2 m(x + s)\| \leq \kappa_{eh}(\|s\| - \bar{\Delta}).$$

Thus, by combining the above expressions we obtain

$$\|\nabla^2 f(x + s) - \nabla^2 m(x + s)\| \leq \kappa_{eh} \|s\| \leq \kappa_{eh} \Delta. \tag{10.43}$$

Now let us consider the vector function $g(\alpha) = \nabla f(x + \alpha s) - \nabla m(x + \alpha s)$, $\alpha \in [0, 1]$. From the fact that m is a fully quadratic model on $B(x; \bar{\Delta})$ we have $\|g(\theta)\| \leq \kappa_{eg} \bar{\Delta}^2$. We are interested in bounding $\|g(1)\|$, which can be achieved by bounding $\|g(1) - g(\theta)\|$ first. By applying the integral mean value theorem componentwise, we obtain

$$\|g(1) - g(\theta)\| = \left\| \int_\theta^1 g'(\alpha) d\alpha \right\| \leq \int_\theta^1 \|g'(\alpha)\| d\alpha.$$

Now using (10.43) we have

$$\int_{\theta}^{1} \|g'(\alpha)\| d\alpha \leq \int_{\theta}^{1} \|s\| \|\nabla^2 f(x + \alpha s) - \nabla^2 m(x + \alpha s)\| d\alpha$$

$$\leq \int_{\theta}^{1} \alpha \kappa_{eh} \|s\|^2 d\alpha = (1/2)\kappa_{eh}(\|s\|^2 - \bar{\Delta}^2).$$

Hence, from $\kappa_{eg} \geq 1/2\kappa_{eh}$ we obtain

$$\|\nabla f(x+s) - \nabla m(x+s)\| \leq \|g(1) - g(\theta)\| + \|g(\theta)\| \leq \kappa_{eg}\|s\|^2 \leq \kappa_{eg}\Delta^2. \quad (10.44)$$

Finally, we consider the function $\phi(\alpha) = f(x + \alpha s) - m(x + \alpha s)$, $\alpha \in [0, 1]$. From the fact that m is a fully quadratic model on $B(x; \bar{\Delta})$, we have $|\phi(\theta)| \leq \kappa_{ef}\bar{\Delta}^3$. We are interested in bounding $|\phi(1)|$, which can be achieved by bounding $|\phi(1) - \phi(\theta)|$ first by using (10.44):

$$\left| \int_{\theta}^{1} \phi'(\alpha) d\alpha \right| \leq \int_{\theta}^{1} \|s\| \|\nabla f(x + \alpha s) - \nabla m(x + \alpha s)\| d\alpha$$

$$\leq \int_{\theta}^{1} \alpha^2 \kappa_{eg} \|s\|^3 d\alpha = (1/3)\kappa_{eg}(\|s\|^3 - \bar{\Delta}^3).$$

Hence, from $\kappa_{ef} \geq (1/3)\kappa_{eg}$ we obtain

$$|f(x+s) - m(x+s)| \leq |\phi(1) - \phi(\theta)| + |\phi(\theta)| \leq \kappa_{ef}\|s\|^3 \leq \kappa_{ef}\Delta^3.$$

The proof is complete. □

10.8 Trust-region subproblem

An extensive and detailed analysis as to how the trust-region subproblem (10.2) can be solved more or less exactly when the model function is quadratic is given in [57, Chapter 7] for the ℓ_∞ and ℓ_2 trust-region norms. Of course, this is at some computational cost over the approximate solutions (satisfying, for instance, Assumptions 10.1 and 10.2), but particularly in the context of modestly dimensioned domains, such as one might expect in derivative-free optimization, the additional work may well be desirable at times because of the expected faster convergence rate. Although we will not go into the same level of detail, it does seem appropriate to at least indicate how one can solve the more popular ℓ_2-norm trust-region subproblem.

The basic driver, as one might expect for such a relatively simple problem, is the optimality conditions. However, for this specially structured problem we have much more than just the first-order necessary conditions in that we are able to characterize the global solution(s) of a (possibly) nonconvex problem.

Theorem 10.27. *Any global minimizer s_* of $m(x + s) = m(x) + s^\top g + \frac{1}{2}s^\top H s$, subject to $\|s\| \leq \Delta$, satisfies the equation*

$$[H + \lambda_* I]s_* = -g, \quad (10.45)$$

where $H + \lambda_* I$ is positive semidefinite, $\lambda_* \geq 0$, and

$$\lambda_*(\|s_*\| - \Delta) = 0. \tag{10.46}$$

If $H + \lambda_* I$ is positive definite, then s_* is unique.

If Δ is large enough and H is positive definite, the complementarity conditions (10.46) are satisfied with $\lambda_* = 0$ and the unconstrained minimum lies within the trust region. In all other circumstances, a solution lies on the boundary of the trust region and $\|s_*\| = \Delta$. Suppose that H has an eigendecomposition

$$H = QEQ^\top,$$

where E is a diagonal matrix of eigenvalues $\lambda_1 \leq \lambda_2 \leq \cdots \leq \lambda_n$, and Q is an orthogonal matrix of associated eigenvectors. Then

$$H + \lambda I = Q(E + \lambda I)Q^\top.$$

Theorem 10.27 indicates that the value of λ we seek must satisfy $\lambda \geq -\lambda_1$ (as only then is $H + \lambda I$ positive semidefinite), and, if $\lambda > -\lambda_1$, the model minimizer is unique (as this ensures that $H + \lambda I$ is positive definite).

Suppose that $\lambda > -\lambda_1$. Then $H + \lambda I$ is positive definite, and thus (10.45) has a unique solution,

$$s(\lambda) = -[H + \lambda I]^{-1} g = -Q(E + \lambda I)^{-1} Q^\top g.$$

However, the solution we are looking for depends upon the nonlinear inequality

$$\|s(\lambda)\| \leq \Delta.$$

Now

$$\|s(\lambda)\|^2 = \|Q(E + \lambda I)^{-1} Q^\top g\|^2 = \|(E + \lambda I)^{-1} Q^\top g\|^2 = \sum_{i=1}^n \frac{\gamma_i^2}{(\lambda_i + \lambda)^2},$$

where γ_i is $(Q^\top g)_i$, the ith component of $Q^\top g$. It is now apparent that if $\lambda > -\lambda_1$, then $\|s(\lambda)\|$ is a continuous, nonincreasing function of λ on $(-\lambda_1, +\infty)$ that tends to zero as λ tends to $+\infty$. Moreover, provided $\gamma_j \neq 0$, then $\lim_{\lambda \to -\lambda_j} \|s(\lambda)\| = +\infty$. Thus, provided $\gamma_1 \neq 0$, $\|s(\lambda)\| = \Delta$ for a unique value of $\lambda \subset (-\lambda_1, +\infty)$.

When H is positive definite and $\|H^{-1}g\| \leq \Delta$ the solution corresponds to $\lambda = 0$, as we have already mentioned. Otherwise, when H is positive definite and $\|H^{-1}g\| > \Delta$ there is a unique solution to (10.45) in $(0, +\infty)$.

When H is not positive definite and $\gamma_1 \neq 0$ we need to find a solution to (10.45) with $\lambda > -\lambda_1$. Because of the high nonlinearities in the neighborhood of $-\lambda_1$ it turns out to be preferable to solve the so-called secular equation

$$\frac{1}{\|s(\lambda)\|} = \frac{1}{\Delta},$$

which is close to linear in the neighborhood of the optimal λ. Because of the near linearity in the region of interest, it is reasonable to expect fast convergence of Newton's method.

But, as is well known, an unsafeguarded Newton method may fail to converge, so care and ingenuity must be taken to safeguard the method.

When H is not positive definite and $\gamma_1 = 0$ we have the so-called hard case, since there is no solution to (10.45) in $(-\lambda_1, +\infty)$ when $\Delta > \|s(-\lambda_1)\|$. However, there is a solution at $\lambda = -\lambda_1$, but it includes an eigenvector of H corresponding to the eigenvalue λ_1, which thus has to be estimated.

The reader is referred to [57] for all the (considerable) details. A more accessible but less complete reference is [178, Chapter 4].

10.9 Other notes and references

Trust-region methods have been designed since the beginning of their development to deal with the absence of second-order partial derivatives and to incorporate quasi-Newton techniques. The idea of minimizing a quadratic interpolation model within a *region of validity* goes back to Winfield [227, 228] in 1969. Glad and Goldstein [106] have also suggested minimizing regression quadratic models obtained by sampling over sets defined by positive integer combinations of $D_\oplus$, as in directional direct-search methods. However, the design and analysis of trust-region methods for derivative-free optimization, when both first- and second-order partial derivatives are unavailable and hard to approximate directly, is a relatively recent topic. The first attempts in these directions have been presented by Powell in the 5th Stockholm Optimization Days in 1994 and in the 5th SIAM Conference on Optimization in 1996, using quadratic interpolation. Conn and Toint [58] in 1996 have reported encouraging numerical results for quadratic interpolation models. Around the same time, Elster and Neumaier [88] developed and analyzed an algorithm based on the minimization of quadratic regression models within trust regions built by sampling over box-type grids, also reporting good numerical results.

Conn, Scheinberg, and Toint [59] (see also [57]) introduced the criticality step and designed and analyzed the first interpolation-based derivative-free trust-region method globally convergent to first-order critical points. Most of the other issues addressed in this chapter, including the appropriate incorporation of fully linear and fully quadratic models, global convergence when acceptance of iterates is based on simple decrease of the objective function, and global convergence for second-order critical points, were addressed by Conn, Scheinberg, and Vicente [62]. This paper provided the first comprehensive analysis of global convergence of trust-region derivative-free methods to second-order stationary points. It was mentioned in [57] that such analysis could be simply derived from the classical analysis for the derivative-based case. However, as we remarked during this chapter, the algorithms in [57, 59] are not as close to a practical one as the one described in this chapter, and, moreover, the details of adjusting a "classical" derivative-based convergence analysis to the derivative-free case are not as trivial as one might expect, even without the additional "practical" changes to the algorithm. As we have seen in Sections 10.5 and 10.6, it is not necessary to increase the trust-region radius on every successful iteration, as is done in classical derivative-based methods to ensure lim-type global convergence to second-order critical points (even when iterates are accepted based on simple decrease of the objective function). In fact, as described in these sections, in the case of the second-order analysis, the trust region needs to be increased only when it is much smaller than the measure of stationarity, to allow large steps when the current iterate is far from a stationary point and the trust-region radius is small.

Other authors have addressed related issues. Colson [55] and Colson and Toint [56] showed how to take advantage of partial separability of functions in the development and implementation of interpolation-based trust-region methods. The wedge algorithm of Marazzi and Nocedal [163] and the least Frobenius norm updating algorithm of Powell [191] will be covered in detail in Chapter 11.

Finally, we would like to point out that derivative-based trust-region methods have been analyzed under the influence of inexactness of gradient values [51, 57] and inexactness of function values [57]. The influence of inexact function values, in particular, is relevant also in the derivative-free case since the objective function can be subject to noise or inaccuracy.

10.10   Exercises

1. Prove that when H_k has a negative eigenvalue, $s_k^{\rm E}$ is the minimizer of the quadratic model $m_k(x_k + s)$ along that direction and inside the trust region $B(0; \Delta_k)$.

2. Prove that if s_k satisfies a fraction of optimal decrease condition,

$$m_k(x_k) - m_k(x_k + s_k) \geq \kappa_{fod}[m_k(x_k) - m_k(x_k + s_k^*)],$$

where $\kappa_{fod} \in (0, 1]$ and s_k^* is an optimal solution of (10.2), then it also satisfies (10.14).

3. Show that when the sequence of iterates is bounded, Theorem 10.12 (resp., Theorem 10.23) implies the existence of one limit point of the sequence of iterates $\{x_k\}$ that is a first-order (resp., second-order) stationary point.

4. Prove Lemma 10.16.

Chapter 11

Trust-region interpolation-based methods

11.1 Common features and considerations

In this chapter we present several practical algorithms for derivative-free optimization based on the trust-region framework described in Chapter 10. Although the original algorithms differ from what is presented here, we try to preserve what we see as the main distinguishing ideas, whilst casting the algorithms in a form which is as close as possible to the convergent framework of Chapter 10. What these algorithms have in common is the use of the trust region and of the quadratic models based on polynomial interpolation. The differences between the algorithms lie mainly in the handling of the sample (interpolation) set and in the building of the corresponding model.

The following is a list of questions one needs to answer when designing a trust-region interpolation-based derivative-free method. We will see how these questions are answered by the different methods discussed in this chapter.

1. How many points should be included in the sample set? We know that a quadratic model is desirable, but a completely determined quadratic model based on polynomial interpolation requires $(n+1)(n+2)/2$ function evaluations. It might be too expensive to require such accuracy at each step; hence it may be desirable to build models based on fewer interpolation points, as discussed in Chapter 5. Naturally, another question arises: should the number of points in the sample set be static throughout the algorithm or should it be dynamic? For instance, the algorithm described in Section 11.2 allows a dynamic number of points, whereas the algorithm described in Section 11.3 requires the sample set to have exactly $p_1 \in [n+2, (n+1)(n+2)/2]$ points at each iteration, with p_1 fixed across all iterations.

2. Should the sample set be Λ-poised in $B(x_k; \Delta_k)$ at each iteration and, if not, how should this requirement be relaxed? To enforce Λ-poisedness, even occasionally, one needs to develop a criterion for accepting points into a sample set. This criterion is typically based on some threshold value. What value for the threshold should be chosen and when and how should it be improved if it is chosen badly? If the threshold is chosen too strictly, it might not be possible to find a suitable sample set, and if it is

chosen too loosely, the resulting set could be badly poised. Here we will rely on the material from Chapter 6.

3. Should the sample set always be contained in $B(x_k; \Delta_k)$? The framework in Chapter 10 assumes this,[18] but it might require recomputing too many sample points each time the iterate changes or the trust-region radius is reduced. Hence, the requirement of the sample sets lying in the current trust region may be relaxed in a practical algorithm. In this case, it could be restricted to $B(x_k; r\Delta_k)$, with some fixed $r \geq 1$, or it could be not restricted at all, but updated in such a way that points which are far from x_k are replaced by points in $B(x_k; \Delta_k)$ whenever appropriate.

4. The framework of Chapter 10 allows us to accept new iterates, if any amount of decrease is achieved and the model is sufficiently good, by setting $\eta_0 = 0$ in the notation of Algorithms 10.1 and 10.3. In all the algorithms discussed in this chapter we will allow accepting new iterates based on a simple decrease condition, since it is a universally desirable feature for a derivative-free algorithm. However, each of these algorithms can be trivially modified to accept only new iterates based on a sufficient decrease condition.

11.2   The "DFO" approach

The algorithm described in this section makes use of the machinery developed in the book for the purposes of proving global convergence. It will rely on the material of Chapters 6 and 10. The following are the main distinguishing features of the algorithm, developed by Conn, Scheinberg, and Toint [59, 61] and referred to as "DFO" (derivative-free optimization).

1. It makes use of as many sample points (up to $(n+1)(n+2)/2$) as are available that pass the criterion for Λ-poisedness (described below). In other words the number of points in the sample set Y_k is dynamic.

2. The model is a minimum Frobenius norm quadratic interpolation model as described in (the beginning of) Section 5.3.

3. It always maintains a Λ-poised set of sample points in $B(x_k; r\Delta_k)$ (in the linear or minimum Frobenius norm interpolation sense or in the quadratic interpolation sense), with r a small scaling factor greater than or equal to 1. Making $r \geq 2$ makes the algorithm more practical since it allows the sample points to remain in the current sample sets for a few subsequent iterations even if the current iteration is moved or the trust-region radius is reduced.

4. The original algorithm described in [59] relies on NFPs to measure and maintain Λ-poisedness of sample sets. But the model-improvement algorithm used is very similar to the pivoting algorithms described in Chapter 6. In this section, we discuss a globally convergent version that relies on pivoting algorithms to maintain its sample sets.

---

[18]The presentation of Chapter 10 could have been adapted to handle the case where the sample set is contained in $B(x_k; r\Delta_k)$, with $r > 0$ a constant fixed across all iterations.

5. At each iteration, the "DFO" algorithm updates the sample set Y_k via the following steps relying on the pivotal algorithms (Algorithms 6.4, 6.5, and 6.6) described in Section 6.3:

(i) At the beginning of the algorithm, two different threshold values are selected: $0 < \xi_{acc} < 1/4$ for the acceptance of a point into an interpolation set and $\xi_{imp} > 1$ for the improvement of the current interpolation set. The value of ξ_{acc} is typically smaller than 0.1. The region within which the pivot polynomials are optimized is a ball of radius Δ_k centered at the current iterate. If we initiate Algorithms 6.4, 6.5, and 6.6 with our monomial basis, then the value of the first pivot is always one; hence it is of no matter which interpolation point is associated with the first pivot. Given any set (possibly, not poised) of interpolation points that contains the current iterate x_k, the algorithms can assign x_k to the first pivot and then select the remaining points so that the resulting set is Λ-poised. This guarantees that the current iterate (and the center of the trust region) can always be included in the well-poised sample set.

(ii) Consider a sample point $x_k + s_k$ generated during the trust-region minimization step. If this point provides a decrease of the objective function (see Step 3 of Algorithm 11.1 below), then it becomes a new iterate. The center of the trust region is moved, and Algorithm 6.5 is applied (with threshold ξ_{acc}) to all available sample points that lie in the ball of radius $r\Delta_k$ centered at $x_k + s_k$. This procedure will either select a well-poised set of up to $(n+1)(n+2)/2$ points or will stop short due to the lack of suitable sample points.

(iii) If the new point $x_k + s_k$ does not provide a decrease of the objective function (see Step 3 of Algorithm 11.1 below), then it may still be desirable to add it to the interpolation set, since it brings information which was clearly not provided by the old model. Moreover, the possibly costly function evaluation has already been made. Consequently, we simply add $x_k + s_k$ to Y_k and apply Algorithm 6.5. If in the end some points of the new Y_k were unused, then those points are discarded. Notice that it may happen that $x_k + s_k$ is discarded and the resulting Y_k remains the same as before. This is acceptable from the point of view of the global convergence of the algorithm. It is also acceptable from a practical point of view because it happens very rarely. To avoid such a situation some heuristic approaches can be applied within Algorithm 6.5. For instance, one may put some preference on the points that are closer to the current iterate in the case where a pivot polynomial has similar absolute values at several points.

(iv) When a model-improvement step is desired after a trust-region step has been computed, then a new point is computed solely with the poisedness of the sample set in mind. If the current interpolation set Y_k is not complete (for linear or minimum Frobenius norm interpolation or quadratic interpolation, depending on the desired accuracy), then one step of Algorithm 6.4 is applied to maximize the absolute value of the appropriate pivot polynomial and increase the size of the interpolation set by one point. Otherwise, we apply one step of Algorithm 6.6 to replace one point of Y_k. The replacement is accepted only if the value of the last pivot polynomial is increased by at least a factor of ξ_{imp}. Otherwise, it is considered that the model-improvement step failed.

6. The pivotal algorithms assume that the set of interpolation points is always shifted and scaled to lie in a ball of radius 1 around the origin. Hence, every time the center or the radius of the trust region changes, the new shifting and scaling have to be introduced and the pivot polynomials have to be computed for the new shifted and scaled set. However, in practice, it is sufficient to change the shifting and scaling only every several iterations. Then one can simply update the set of pivot polynomials whenever no shifting and scaling occurs, which saves considerable computation for larger interpolation set sizes.

Now we present the modified "DFO" algorithm for both the situations where one would like to attain global convergence to first-order stationary points or to second-order stationary points. From the theory of Chapters 3–6, we know that a model is fully linear (FL), resp., fully quadratic (FQ), in $B(x_k; r\Delta_k)$ if it is based on a Λ-poised set with at least $n + 1$, resp., $(n + 1)(n + 2)/2$, points in $B(x_k; r\Delta_k)$—Definitions 6.1 and 6.2 would have to take into account the trust-region radius factor r. When we can guarantee that a model is FL, resp., FQ, we say that the model is certifiably fully linear (CFL), resp., certifiably fully quadratic (CFQ).

Algorithm 11.1 (Modified "DFO" algorithm).

Step 0 (initialization): Choose an initial point x_0, a maximum radius $\Delta_{max} > 0$, and an initial trust-region radius $\Delta_0 \in (0, \Delta_{max}]$. Choose a set Y_0 and compute the initial model $m_0(x)$.

The constants $\eta_1, \gamma, \gamma_{inc}, \epsilon_c, \beta$, and μ are also chosen and satisfy the conditions $\eta_1 \in (0, 1), 0 < \gamma < 1 < \gamma_{inc}, \epsilon_c > 0$, and $\mu > \beta > 0$. Choose positive pivot thresholds ξ_{acc} and ξ_{imp} and the trust-region radius factor $r \geq 1$. Set $k = 0$.

Step 1 (criticality step): This step is as in Algorithm 10.2 or 10.4, depending on the desired convergence result. Note that m_k, Y_k, and Δ_k might change in the course of this step.

Step 2 (step calculation): This step is as in Algorithm 10.1 or 10.3.

Step 3 (acceptance of the trial point): Check if $m_k(x)$ is CFL/CFQ, which, again, is guaranteed if Y_k contains at least $n + 1$, resp., $(n + 1)(n + 2)/2$, points in $B(x_k; r\Delta_k)$.

Compute $f(x_k + s_k)$ and define

$$\rho_k = \frac{f(x_k) - f(x_k + s_k)}{m_k(x_k) - m_k(x_k + s_k)}.$$

If $\rho_k \geq \eta_1$ or if $\rho_k > 0$ and the model is CFL/CFQ, then $x_{k+1} = x_k + s_k$. Otherwise, the model and the iterate remain unchanged ($x_{k+1} = x_k$). Apply the procedure described in items 5(ii) and 5(iii) above to include $x_k + s_k$ and update the sample set. Let the new model and sample set be m_{k+1} and Y_{k+1}.

Step 4 (model improvement): If $\rho_k < \eta_1$, then attempt a model improvement by suitable improvement steps (described in 5(iv) above). Define m_{k+1} and Y_{k+1} to be the (possibly improved) model and sample set.

Step 5 (trust-region radius update): This step is as in Algorithm 10.1 or 10.3. Increment k by one and go to Step 1.

Notes on the algorithm

Due to the use of Algorithms 6.4, 6.5, and 6.6 and the fact that Y_k contains only points which provide acceptable pivot values in those algorithms, we know that as long as Y_k lies in $B(x_k; r\Delta_k)$ it is always Λ-poised, for some unknown, but fixed, Λ. It is not difficult to observe that the model-improvement procedure generates, in a finite and uniformly bounded number of steps, an interpolation set which is Λ-poised for linear or minimum Frobenius norm interpolation or quadratic interpolation, depending on the choice of desired accuracy. From the material in Chapters 3–6, we know that the models that are used are FL or FQ, and the rest of the algorithm fits closely to the framework discussed in Chapter 10. Global convergence to first- or second-order stationary points follows as a simple conclusion.

At each iteration of the algorithm only one or two new sample points are generated; hence at most two function evaluations are required.

The algorithm is very flexible in the number of sample points one can use per iteration. In fact it is easy to modify this algorithm to allow for the use of least-squares regression models.

Aside from the function evaluations and the trust-region step the main computational effort is in computing the pivot polynomials. Generally, if $\mathcal{O}(n^2)$ points are used, then the pivotal algorithm may require $\mathcal{O}(n^6)$ operations. By keeping and reusing the pivot polynomials whenever possible we can reduce the empirical complexity of many iterations; however, some iterations may still require a large numerical effort. For small values of n this effort is often negligible compared to the expense of one function evaluation. For larger values of n, it makes sense to restrict the number of points for each model to $\mathcal{O}(n)$. The models based on the minimum Frobenius norm of the change of the model Hessian work well for a fixed number of points of $\mathcal{O}(n)$. It is natural to use minimum Frobenius norm Lagrange polynomials in the context of such models, which is what is done by the algorithm described in the next section.

11.3 Powell's methods

We now describe an algorithm which is a modification of the algorithms developed by Powell in [188, 189, 190, 191, 192]. We will combine here the features of Powell's most successful algorithm in practice with the trust-region framework of Chapter 10. The trust-region management that Powell is using is different from the one followed in this book. In particular the trust-region radius and the radius of the interpolation set are not the same or related by a constant multiple. While the radius of the interpolation set eventually converges to zero (in theory) the trust-region radius is allowed to remain bounded away from zero. In some respects, the two-region approach is closer to classical trust-region methods (see [57]) for derivative-based optimization. The reason for the trust-region radius to remain bounded away from zero is to allow relatively large steps close to optimality in order to achieve a superlinear local rate of convergence. The theoretical implication of such a choice for our framework is that to obtain a lim-type global convergence result as in Theorem 10.13 one needs to impose a sufficient decrease condition on the trust-region step acceptance. It is possible to state and analyze the framework of Chapter 10 for the case when the trust-region radius is not converging to zero, but we leave that to the interested reader.

In fact the trust-region management in Powell's methods is more complicated than simply keeping two different radii. The smaller of the two radii is also used to force the interpolation points to be sufficiently far apart to avoid the influence of noise in the function values. Also, the trust-region updating step is more complicated than that of the framework of Chapter 10. These additional schemes are developed to improve the practical performance of the algorithms. They do not significantly affect the main global convergence results but make the analysis quite complex. We present here Powell's method of handling interpolation sets via minimum Frobenius norm Lagrange polynomials embedded into the framework of Chapter 10. The model-improvement algorithm is augmented by a test and a possible extra step which helps ensure that FL models can be constructed in a finite uniformly bounded number of steps. For full details on the original Powell methods we refer the reader to [188, 189, 190, 191, 192], in particular his UOBYQA [189] and NEWUOA [192] methods. The following are the distinguishing features of the algorithm:

1. All sample sets on all iterations contain $p_1 \in [n+2, (n+1)(n+2)/2]$ points. The value $p_1 = 2n+1$ is a natural choice and is the default value in the NEWUOA software package [192].

2. The models are quadratic interpolation models with the remaining degrees of freedom taken by minimizing the Frobenius norm of the *change* in the model Hessian, with respect to the model used in the previous iteration, as described in (the last part of) Section 5.3.

3. Sample set maintenance is based on Algorithm 6.3 modified to handle the minimum Frobenius norm Lagrange polynomials as discussed in Chapter 6.

4. There is no need to involve scaling, as Lagrange polynomials scale automatically. Shifts also do not have any effect. However, shifting the points with respect to the current center of the trust region has an important effect on numerical accuracy and hence is performed regularly.

5. Updates of the sample set in the original Powell algorithms are performed via the following steps:

    (i) The set of minimum Frobenius norm Lagrange polynomials $\ell_i(x), i = 0, \ldots, p$, is maintained at every iteration.

    (ii) If the trust-region minimization of the kth iteration produces a step s_k which is not too short compared to the maximum distance between the sample points and the current iterate, then the function f is evaluated at $x_k + s_k$ and the new point becomes the next iterate x_{k+1} if the reduction in f is sufficient, or just positive and the model is guaranteed to be FL. The quality of the model m_k is established at the end of the preceding iteration. If the new point $x_k + s_k$ is accepted as the new iterate, it is included into Y_k, by removing the point y^i such that the distance $\|x_k - y^i\|$ and the value $|\ell_i(x_k + s_k)|$ are both, in the sense that follows, as large as possible. The trade off between these two objectives is achieved by maximizing the weighted absolute value $w_i |\ell_i(x_k + s_k)|$, where w_i reflects the distance $\|x_k - y^i\|$. In fact, in [191], instead of maximizing $|\ell_i(x_k + s_k)|$, Powell proposes optimizing the coefficient of the rank-two update of the system defining the Lagrange polynomials, to explicitly improve the conditioning of that system. However, operating with the values of the Lagrange polynomial serves the same purpose and makes it easier for us to use the theory developed in Chapters 5 and 6.

(iii) When the step s_k is rejected, the new point $x_k + s_k$ can still be accepted into Y_k, by removing the point y^i such that the value $w_i|\ell_i(x_k + s_k)|$ is maximized, where w_i reflects the distance $\|x_k - y^i\|$, as long as either $|\ell_i(x_k + s_k)| > 1$ or $\|y^i - x_k\| > r\Delta_k$.

(iv) If the improvement in the objective function is not sufficient, and it is believed that the model needs to be improved, then the algorithm chooses a point in Y_k which is the furthest from x_k and attempts to replace it with a point which maximizes the absolute value of the corresponding Lagrange polynomial in the trust region (or in the smaller interpolation set region, as is done in [188, 189, 190, 191, 192]).

For global convergence we need the criticality step (see Chapter 10), where one may need to construct an FL model. An analogue of this step can be found in Powell's work, and is related to improving geometry when the step s_k is much smaller than Δ_k, which occurs when the gradient of the model is small relative to the Hessian. Here we use the same step that was used in the globally convergent framework of Chapter 10; that is, we use the size of the gradient as the criticality test. Scaling with respect to the size of the Hessian is also possible, as long as arbitrarily small or large scaling factors are not allowed. The modified algorithm is presented below.

Algorithm 11.2 (Minimum Frobenius norm Lagrange polynomial-based algorithm).

Step 0 (initialization): Select $p_1 \in [n+2, (n+1)(n+2)/2]$. Choose an initial point x_0, a maximum radius $\Delta_{max} > 0$, and an initial trust-region radius $\Delta_0 \in (0, \Delta_{max}]$. Choose the trust-region radius factor $r \geq 1$. Choose a well-poised set Y_0 with cardinality p_1. Compute the minimum Frobenius norm Lagrange polynomials $\ell_i(x)$, $i = 0, \ldots, p$, associated with Y_0 and the corresponding quadratic model m_0.

Select a positive threshold value for the improvement step, $\Lambda > \max\{|\ell_i(x)| : i = 0, \ldots, p, x \in B(x_0; \Delta_0)\}$.

The constants η_1, γ, γ_{inc}, ϵ_c, τ, β, and μ are also given and satisfy the conditions $\eta_1 \in (0, 1), 0 < \gamma < 1 < \gamma_{inc}, \epsilon_c > 0, 0 < \tau < 1$, and $\mu > \beta > 0$. Set $k = 0$.

Step 1 (criticality step): This step is as in Algorithm 10.1 (the model-improvement algorithm relies on minimum Frobenius norm Lagrange polynomials). Note that m_k, Y_k, and Δ_k might change in the course of this step.

Step 2 (step calculation): This step is as in Algorithm 10.1.

Step 3 (acceptance of the trial point): If $\|s_k\| \geq \tau \max\{\|y^j - x_k\| : y^j \in Y_k\}$, then compute y^i, where $i \in \text{argmax}_j\{w_j|\ell_j(x_k + s_k)| : y^j \in Y_k\}$ and $w_j > 0$ are weights chosen to give preference to points that are further from x_k.

Compute $f(x_k + s_k)$ and define

$$\rho_k = \frac{f(x_k) - f(x_k + s_k)}{m_k(x_k) - m_k(x_k + s_k)}.$$

Set $x_{k+1} = x_k + s_k$ if $\rho_k \geq \eta_1$ or if $\rho_k > 0$ and it is known that m_k is CFL in $B(x_k; r\Delta_k)$. Include $x_k + s_k$ into the sample set Y_{k+1} by replacing y^i. Otherwise, $x_{k+1} = x_k$, and if either $\|y^i - x_k\| > r\Delta_k$ or $|\ell_i(x_k + s_k)| > 1$, then accept $x_k + s_k$ into the sample set Y_{k+1} by replacing y^i.

Step 4 (model improvement): (This step is executed only if $x_{k+1} = x_k$ after Step 3.) Choose $y^i \in \text{argmax}_j \{\|y^j - x_k\| : y^j \in Y_k\}$ and find a new $y_*^i \in \text{argmax}\{|\ell_i(x)| : x \in B(x_k; \Delta_k)\}$. If $\|y^i - x_k\| > r\Delta_k$ or if $|\ell_i(y_*^i)| > \Lambda$, then replace y^i by y_*^i in Y_{k+1}. Otherwise, consider the next furthest point from x_k and repeat the process. If eventually a point y^i is found in Y_{k+1} such that $\max\{|\ell_i(x)| : x \in B(x_k; \Delta_k)\} > \Lambda$, then this point is replaced. If no such point is found, then there is no need for improvement because Y_k is Λ-poised in $B(x_k; r\Delta_k)$. Hence, we conclude that m_{k+1} based on Y_{k+1} is CFL in $B(x_k; r\Delta_k)$.

Step 5 (trust-region radius update): This step is as in Algorithm 10.1. Update the model m_k to obtain m_{k+1} and recompute the minimum Frobenius norm Lagrange polynomials. Increment k by one and go to Step 1.

Notes on the algorithm

The steps for updating the interpolation set described above are sufficient to guarantee that FL models can be constructed in a finite, uniformly bounded number of steps, as required by the convergence analysis in Chapter 10. It is easy to see that the model-improvement step first attempts to replace any interpolation point that is too far from the current iterate (outside the $r\Delta_k$ radius). Clearly, this can be accomplished in at most p steps. Once all the interpolation points are close enough, then the model-improvement step checks the Λ-poisedness of the interpolation set, by maximizing the absolute values of the minimum Frobenius norm Lagrange polynomials one by one. If the set is already Λ-poised, then there is nothing to improve. If the set is not Λ-poised, then one point in the set is replaced. We know from Chapter 6 and from Theorem 6.3 that after a finite and uniformly bounded number of steps, a Λ-poised set is obtained in $B(x_k; r\Delta_k)$ and, hence, an FL model is obtained. Notice that Step 3 may change the interpolation set, but, in the case when the trust-region step is successful, it is not important (from a theoretical perspective) what this change does to the model. On the other hand, when the current iterate does not change, the change of Y_k allowed by Step 3 may only either improve the poisedness of Y_k or replace a far away point of Y_k by $x_k + s_k$. Hence, if the model m_k is FL at the beginning of an unsuccessful Step 3, then m_{k+1} is so at the end of it; that is, an unsuccessful Step 3 may only improve the model, while Step 4 is guaranteed to produce an FL model in a finite number or iterations. Hence, we conclude that the global convergence theory of Chapter 10 (for first-order stationary points) applies to Algorithm 11.2.

In the original methods proposed by Powell [189, 192] there is no explicit check in the model-improvement step that $|\ell_i(y_*^i)| > \Lambda > 1$; instead the model-improvement step is far more complex, but it is aimed at replacing the old interpolation points by the points within the trust region which give a large value for the corresponding Lagrange polynomials. One of the features of this procedure is to avoid the exact global optimization of the Lagrange polynomials. Consider, now, replacing an interpolation point by another point for which the absolute value of the appropriate Lagrange polynomial exceeds $\Lambda > 1$, rather than seeking the exact global maximizer of such absolute value. From the discussion after the proof of Theorem 6.3 we know that such a step provides an improvement of the poisedness of the interpolation set. On the other hand, we do not know if such replacements steps result in a model-improvement algorithm that terminates in a uniformly bounded finite number of steps. It is possible to modify the model-improvement algorithm to allow several such "cheap" improvement steps and switch to global optimization only when the maximum number of consecutive cheap steps is exceeded.

In the original Powell methods a simple decrease step is always accepted as a new iterate. We allow this in Algorithm 11.2 also but only when we can guarantee that the sample set Y_k on which the model m_k was constructed is Λ-poised in $B(x_k; r\Delta_k)$. Hence we may need to have either an extra check or a flag which can be set by Step 4 to indicate that a Λ-poised set is at hand. There is a trade off between how often such a check is performed and how often a simple decrease can be accepted. The exact implementation is a matter of practical choice.

Although not guaranteed, it is expected that the interpolation set remains well poised throughout the algorithm as long as it is in $B(x_k; r\Delta_k)$. In practice one can start with a simple well-poised interpolation set, such as is used by Powell:

$$Y_0 = \{x_0, x_0 + \Delta e_1, \ldots, x_0 + \Delta e_n, x_0 - \Delta e_1, \ldots, x_0 - \Delta e_n\},$$

where e_i is the ith vector of the identity (note that this is nothing else than (complete) polling using the positive basis $D_\oplus$). The advantage of this set is that due to its special structure it takes $\mathcal{O}(n^2)$ operations to construct the initial set of minimum Frobenius norm Lagrange polynomials [192].

11.4   Wedge methods

The two methods we have considered so far in this chapter are closely related in that they generate two kinds of sample points, those that are aimed at reducing the function value and those that are aimed at improving the poisedness of the sample set. A natural question is: is it possible to achieve both of these objectives at once, by generating one kind of points? These two objectives may be conflicting, but is it possible to find a point which improves the value of the model while maintaining an acceptable level of poisedness? An algorithm based on this idea was proposed by Marazzi and Nocedal in [163]. This algorithm follows the approach of attempting to generate points which *simultaneously* provide sufficient decrease for the model (and, thus, hopefully, provide decrease for the objective function) and satisfy the Λ-poisedness condition. At every iteration, the trust-region subproblem minimization is augmented by an additional constraint which does not allow the new point to lie near a certain manifold. This manifold is defined by a subset of the sample set that is fixed for the given iteration and allows for all possible nonpoised sets that contain the fixed subset of sample points. In the original method the constraint which defined the proximity to such a manifold (in the linear case) has the shape of a wedge, hence the name of the method. To provide theoretical guarantees for the method we have to replace the "wedge"-shaped constraint by a "strip"-shaped one. We will use the terminology "wedge method" to refer to its origin.

Let Y_k be the interpolation set at iteration k, and let $m_k(x_k + s)$ be the model based on Y_k,

$$m_k(x_k + s) = m_k(x_k) + g_k^\top s + \frac{1}{2}s^\top H_k s.$$

Before a trust-region step is taken we need to identify which point of Y_k is going to be replaced with the new point. Notice that the other two methods of this chapter first make a trust-region step and then decide which point of Y_k to replace with the new point, if appropriate. To choose a point to leave Y_k we use the following algorithm, which is a combination of Algorithms 6.5 and 6.6.

Algorithm 11.3 (Selection of the outgoing point in Y_k).

Initialization: Shift and scale Y_k and consider the new set $\hat{Y}_k = \{(y^i - x_k)/\Delta_k, i = 0, \ldots, p\}$. Set $u_i(x) = \bar{\phi}_i(x), i = 0, \ldots, p$. Choose $\xi > 0$ and $r \geq 1$. Assume that $\hat{Y}_k$ contains p_1 poised points. Set $i_k = p$.

For $i = 0, \ldots, p - 1$

   1. **Point selection:** Find $j_i = \text{argmax}_{j \in \{i, \ldots, p\}: \|\hat{y}^j\| \leq r} |u_i(\hat{y}^j)|$.

   2. **Point acceptance:** If $|u_i(\hat{y}^{j_i})| < \xi$, then $i_k = i$ and stop. Otherwise, swap points $\hat{y}^i$ and $\hat{y}^{j_i}$ in the set $\hat{Y}_k$.

   3. **Gaussian elimination:** For $j = i + 1, \ldots, p$

   $$u_j(x) \ \leftarrow \ u_j(x) - \frac{u_j(\hat{y}^i)}{u_i(\hat{y}^i)} u_i(x).$$

This algorithm selects the points in $\hat{Y}_k$ that lie in $B(0; r)$ and which give pivot values of at least ξ. It also computes the first i_k pivot polynomials. The index i_k is the index of the first point that needs to be replaced. By the logic of the algorithm such a point either does not lie in $B(0; r)$ or does not give a large enough pivot value. In case such points do not exist, then $i_k = p$ and the result of Algorithm 11.3 is similar to the result of one round of Algorithm 6.6; in that case $\hat{y}^p$ is simply a good point to be replaced.

The purpose of Algorithm 11.3 is to identify points which should be replaced either because they are far away from the current iterate or because replacing them may make the poisedness of $\hat{Y}_k$ better. Once such a point $\hat{y}^{i_k}$ is identified and the corresponding pivot polynomial is computed, the following problem is solved to produce the trust-region step:

$$\min_{s \in \mathbb{R}^n} \quad m_k(x_k) + g_k^\top s + \frac{1}{2}s^\top H_k s$$
$$\text{s.t.} \quad \|s\| \leq \Delta_k, \tag{11.1}$$
$$|u_{i_k}(s/\Delta_k)| \geq \xi.$$

The above problem is feasible for small enough ξ; hence an optimal solution exists. However, what we are interested in is an approximate solution, which is feasible and satisfies the fraction of Cauchy decrease condition (10.9). We will show that such a solution also exists, for small enough ξ, in the section of the analysis of the wedge algorithm. The actual approach to (approximately) solve such a problem can be found in [163]. We will not repeat the details here and will simply assume that if such a solution exists, then it can be found. This might require a large computational effort, but it does not require any function evaluations for f, which is normally the main concern of a derivative-free algorithm.

Recall the example of an "ideal" interpolation set illustrated in Figure 3.4. We shift and scale this set so that it lies in $B(0; 1)$. Let us apply Algorithm 6.5 to construct the pivot polynomials and then consider the constraint $|u_p(s/\Delta_k)| \geq \xi$ for the last quadratic pivot polynomial. In Figure 11.1 we can see the area of the trust region which is being discarded by this constraints for the cases when $\xi = 0.01$ and $\xi = 0.1$.

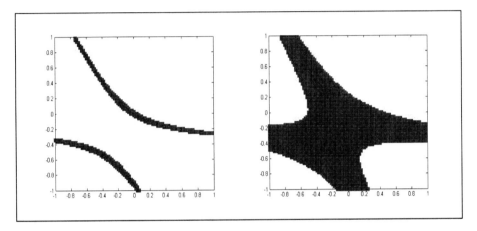

Figure 11.1. *Areas of the (squared) trust region which are forbidden by the wedge constraint $|u_p(s/\Delta_k)| \geq \xi$ for the cases $\xi = 0.01$ (left) and $\xi = 0.1$ (right). The example refers to the sample set of Figure 3.4.*

The following is a summary of the main distinguishing features of the algorithm.

1. The algorithm, as originally proposed in [163], uses exactly $n + 1$ points in the linear case and $(n + 1)(n + 2)/2$ points in the quadratic case. It is possible to extend it to use minimum Frobenius norm models, but we will not do so here.

2. Unlike the two methods discussed in the previous sections, in this case the point which will leave Y_k is selected *before* each trust-region minimization step.

3. The algorithm that we present here aims at maintaining a Λ-poised set of sample points as long as the points are in $B(x_k; r\Delta_k)$, with $r \geq 1$. Due to the use of Algorithm 11.3, and assuming that the trust-region subproblem (11.1) can be solved successfully, we can show that such a step can be obtained after a finite number of iterations (see point 5 below).

The original wedge algorithm does not employ Algorithm 11.3; it simply selects the point which is the furthest from the current iterate as the point that should be replaced. For the global convergence theory to apply, we need to make sure that the interpolation set can become Λ-poised in $B(x_k; r\Delta_k)$ after a finite number of improvement steps before the trust-region radius can be reduced (in case the trust-region step is unsuccessful). If we insist that all interpolation points lie in $B(x_k; \Delta_k)$ in order to consider the model FL or FQ, then we would have to replace a lot of points each time the iterate changes or the trust region is reduced. Making r large, say $r \geq 2$, makes the algorithm more practical.

4. There are no direct attempts to improve geometry, but there is a heuristic way of updating the balance between the required poisedness of the sample set and the required fraction of the sufficient reduction in the model (see [163]).

5. If i_k produced by Algorithm 11.3 is smaller than p, then, after the trust-region step (with the wedge constraint), the new Y_k may not be Λ-poised in $B(x_k; \Delta_k)$. However, if the wedge trust-region subproblem is solved successfully (namely, a solution for (11.1) is found for the current value of ξ_k and it satisfies a fraction of Cauchy decrease), then, if x_k and Δ_k remain unchanged, then on the next iteration i_k increases by 1. Hence, eventually either a successful trust-region step is achieved or i_k equals p. In this case we know that if the wedge trust-region subproblem (11.1) is solved successfully, then the resulting new model is CFL or CFQ (depending on p) in $B(x_k; \Delta_k)$. Notice that it may not be so in $B(x_{k+1}; \Delta_{k+1})$, but, unless sufficient decrease of the objective function is obtained by the trust-region step, $B(x_{k+1}; \Delta_{k+1}) = B(x_k; \Delta_k)$ and a CFL/CFQ model is obtained for certain.

6. The criticality step is crucial for the global convergence properties, just as in the cases of the other two algorithms described in this chapter. A criticality step needs to provide an FL/FQ model in a sufficiently small ball around the current iterate. It is possible to construct such a set by repeatedly solving the problem (11.1). However, since many interpolation points may need to be replaced, the intermediate models are likely to be inaccurate. A more practical approach is to generate a Λ-poised set directly, without consideration for the model decreases, by applying Algorithm 6.5, for instance.

We present the modified wedge algorithm for the quadratic interpolation case, but the analysis that follows mainly concerns global convergence to first-order stationary points.

Algorithm 11.4 (Modified wedge algorithm).

Step 0 (initialization): Choose an initial point x_0, a maximum radius $\Delta_{max} > 0$, and an initial trust-region radius $\Delta_0 \in (0, \Delta_{max}]$. Choose a set Y_0 and compute the initial model $m_0(x)$.

The constants η_1, γ, γ_{inc}, ϵ_c, β, and μ are also given and satisfy the conditions $\eta_1 \in (0, 1)$, $0 < \gamma < 1 < \gamma_{inc}$, $\epsilon_c > 0$, and $\mu > \beta > 0$. Choose a positive pivot threshold ξ_0, a Cauchy decrease coefficient $0 < \kappa_{fcd} < 1$, and a trust-region radius factor $r \geq 1$. Set $k = 0$.

Step 1 (criticality step): This step is as in Algorithm 10.1 (note that m_k, Y_k, and Δ_k might change in the course of this step).

Step 2 (step calculation): Apply Algorithm 11.3 to Y_k to select the pivot polynomial $u_{i_k}(x)$. Solve problem (11.1) to obtain s_k.

Step 3 (wedge management): Check if the new solution satisfies the fraction of Cauchy decrease condition:

$$m_k(x_k) - m_k(x_k + s_k) \geq \frac{\kappa_{fcd}}{2} \|g_k\| \min\left\{ \frac{\|g_k\|}{\|H_k\|}, 1 \right\}.$$

If so, let $y^{i_k} = x_k + s_k$ and proceed to the next step. Otherwise, the wedge constraint is relaxed by reducing $\xi_k \leftarrow \xi_k/2$ and return to Step 2.

Step 4 (acceptance of the trial point): Compute $f(x_k + s_k)$ and define

$$\rho_k = \frac{f(x_k) - f(x_k + s_k)}{m_k(x_k) - m_k(x_k + s_k)}.$$

If $\rho_k \geq \eta_1$ or if $\rho_k > 0$ and $Y_k \subset B(x_k; r\Delta_k)$, then $x_{k+1} = x_k + s_k$. Otherwise, $x_{k+1} = x_k$.

Step 5 (trust-region radius update): This step is as in Algorithm 10.1. Update the model m_k based on the new sample set Y_k. Increment k by one and go to Step 1.

Notes on the algorithm

The computational effort per iteration of Algorithm 11.4 is $\mathcal{O}(p^3)$ for each Step 2, to which we need to add the computational effort of solving a wedge subproblem. Solving the wedge subproblem can be expensive. It would be justifiable if the overall number of function evaluations is decreased. Marazzi and Nocedal [163] have reported on computational results which showed that the method is comparable with other derivative-free methods, but there was no particularly strong advantage, and other derivative-free methods, in particular that of Powell [192], have recently improved significantly. However, we note that Marazzi and Nocedal's implementation was relying on building complete quadratic models, rather than minimum Frobenius norm models. The latter are used in DFO and NEWUOA and often give an advantage to a trust-region interpolation-based algorithm. Hence another implementation of the wedge method might perform better.

As with the other algorithms we discussed above, there are numerous ways in which the wedge algorithm can be improved. For instance, one can consider replacing several different points in Y_k and solving several instances of the wedge trust-region subproblem (11.1) to obtain the best trade off between geometry and model decrease. This will increase the cost per iteration, but may decrease the overall number of iterations, and hence the overall number of function evaluations.

Validity of the wedge algorithm for the first-order case

We have described a version of the wedge method which uses quadratic interpolation models. However, Steps 1 and 3 indicate that the method tries only to achieve global convergence to a first-order stationary point. From the results of Chapter 10, we can conclude that all we need to show in order to establish such a convergence is that an FL model can be achieved after a finite, uniformly bounded number of unsuccessful iterations. Below we will show that if only a fraction of Cauchy decrease is required from the approximate solution to (11.1), then this is indeed the case. Moreover, we can show that an FQ model can also be achieved in a finite, uniformly bounded number of unsuccessful iterations while keeping the fraction of Cauchy decrease requirement.

However, to establish global convergence to a second-order stationary point one can no longer be satisfied with only the fraction of Cauchy decrease, and the fraction of optimal decrease (10.13) has to be considered. Thus, second-order criteria would have to be applied in Steps 1 and 3. This complicates the proofs and, as we will discuss at the end of the chapter, requires an additional safeguard step to be introduced into the algorithm.

Consider a step of Algorithm 11.3 (which is a modification of Algorithm 6.5 described in Chapter 6). Recall the vector v, the vector of the coefficients (with respect to the natural basis $\bar{\phi}$) of the polynomial $u_i(x)$ which is computed during the pivot computation. We know that $\|v\|_\infty \geq 1$. In general we know that for any $0 < \xi < 1/4$ there exists $\hat{s}_k$ such that $\|\hat{s}_k\| \leq 1$ and that $|v^\top \bar{\phi}(\hat{s}_k)| \geq \xi$ (see Lemmas 3.10 and 3.12). The purpose of the analysis below is to show that there exists a constant $\xi > 0$, such that for all iterations it is possible to find a step s_k simultaneously satisfying the following:

- $\|s_k\| \leq \Delta_k$,

- s_k provides at least a fixed fraction of Cauchy decrease of the model,

- $|v^\top \bar{\phi}(s_k/\Delta_k)| \geq \xi$.

From this it will follow that the model can be made FL or FQ by the wedge method after a finite, uniformly bounded number of unsuccessful iterations.

Let us consider the unscaled trust-region subproblem, without the wedge constraint. We index the elements of the unscaled problem by "u" and omit the index k from now on:

$$\min_{s_u \in \mathbb{R}^n} \quad c + g_u^\top s_u + \tfrac{1}{2} s_u^\top H_u s_u \;=\; m_u(s_u)$$
$$\text{s.t.} \quad \|s_u\| \leq \Delta.$$

If we scale the trust region so it has radius one, we have

$$\min_{s \in \mathbb{R}^n} \quad c + g^\top s + \tfrac{1}{2} s^\top H s \;=\; m(s)$$
$$\text{s.t.} \quad \|s\| \leq 1,$$

where $s = s_u/\Delta$, $g = \Delta g_u$, and $H = \Delta^2 H_u$.

Note that if s satisfies a fraction of Cauchy-decrease-type condition of the form

$$m(0) - m(s) \geq \frac{\kappa_{fcd}}{2} \|g\| \min\left\{\frac{\|g\|}{\|H\|}, 1\right\}, \tag{11.2}$$

then $s_u = \Delta s$ also satisfies a fraction of Cauchy-decrease-type condition of the form

$$m_u(0) - m_u(s_u) \geq \frac{\kappa_{fcd}}{2} \|g_u\| \min\left\{\frac{\|g_u\|}{\|H_u\|}, \Delta\right\}.$$

The Cauchy step itself (defined as the minimizer of $m(s)$ along $-g$ and inside the trust region) satisfies inequality (11.2) with $\kappa_{fcd} = 1$.

We want to show that there exists an approximate solution for the problem

$$\min_{s\in\mathbb{R}^n} \quad m(s)$$
$$\text{s.t.} \quad \|s\| \leq 1, \tag{11.3}$$
$$|\phi(s)^\top v| \geq \xi > 0,$$

yielding a decrease on $m(s)$ as good as a fixed fraction of the decrease obtained by the Cauchy step. The value of ξ is not yet determined, but it must be bounded away from zero for all problem data considered (meaning v, g, and H).

For a moment let us instead consider the following problem for a fixed $\kappa_{fcd} \in (0,1)$:

$$\max_{s\in\mathbb{R}^n} \quad |\phi(s)^\top v|$$
$$\text{s.t.} \quad \|s\| \leq 1, \tag{11.4}$$
$$m(0) - m(s) \geq \tfrac{\kappa_{fcd}}{2}\|g\|\min\left\{\tfrac{\|g\|}{\|H\|},1\right\}.$$

If we can show that the optimal value for this problem is always above a positive threshold $\xi_* > 0$, for all vectors v such that $\|v\|_\infty \geq 1$ and for all vectors g and symmetric matrices H, then we have shown that there exists an approximate solution for problem (11.3) satisfying a fraction of Cauchy decrease for small enough $\xi > 0$. To do so we need to show that the polynomial $\phi(s)^\top v$ does not nearly vanish on the feasible set of (11.4). The difficulty is that when $\|g\|/\|H\| \to 0$ the feasible set of (11.4) may converge to a singular point, on which a suitably chosen polynomial may vanish. What we will show is that for the problem (11.3) arising during the wedge algorithm it always holds that $0 < g_{min} \leq \|g\|/\|H\|$, where $g_{min} \in (0,1)$ is a fixed constant.

Recall that $g = g_u\Delta$ and $H = H_u\Delta^2$, and this implies that

$$\frac{\|g\|}{\|H\|} = \frac{\|g_u\|}{\Delta\|H_u\|}.$$

From the standard assumption on the boundedness of the model Hessian we know that $\|H_u\| \leq \kappa_{bhm}$. Recall also that $\Delta \leq \Delta_{max}$. The criticality step of the modified wedge algorithm ensures that any time the wedge trust-region subproblem is solved at Step 2 we have $\|g_u\| \geq \min\{\epsilon_c, \mu^{-1}\Delta\}$, and

$$\frac{\|g\|}{\|H\|} = \frac{\|g_u\|}{\Delta\|H_u\|} \geq g_{min}, \tag{11.5}$$

where

$$g_{min} = \min\left\{\frac{\epsilon_c}{\Delta_{max}\kappa_{bhm}}, \frac{1}{\mu\kappa_{bhm}}\right\}.$$

Now we prove that is possible to achieve good uniform geometry and, simultaneously, a fraction of Cauchy decrease for all g and H such that $g_{min}\|H\| \leq \|g\|$, with $g_{min} \in (0,1)$.

Theorem 11.1. *Let $g_{min} \in (0,1)$ be a given constant. There exists a positive constant ξ_* depending on both g_{min} and κ_{fcd} such that the optimal value of (11.4) satisfies*

$$\xi(v;g,H) \geq \xi_* > 0$$

for all vectors $v \in \mathbb{R}^p$ such that $\|v\|_\infty \geq 1$ and for all vectors $g \in \mathbb{R}^n$ and symmetric matrices $H \in \mathbb{R}^{n \times n}$ such that

$$g_{min} \leq \frac{\|g\|}{\|H\|}.$$

Proof. We start by defining

$$a(v; g, H) = \max \left\{ |\phi(s)^\top v| : \|s\| \leq 1, \quad m(0) - m(s) \geq \frac{\kappa_{fcd}}{2} \|g\| \right\}$$

when $\|g\| \geq \|H\|$ and

$$b(v; g, H) = \max \left\{ |\phi(s)^\top v| : \|s\| \leq 1, \quad m(0) - m(s) \geq \frac{\kappa_{fcd}}{2} \frac{\|g\|^2}{\|H\|} \right\}$$

when $g_{min}\|H\| \leq \|g\| \leq \|H\|$. The optimal value of (11.4) is given by

$$\xi(v; g, H) = \begin{cases} a(v; g, H) & \text{when } \|g\| \geq \|H\|, \\ b(v; g, H) & \text{when } g_{min}\|H\| \leq \|g\| \leq \|H\|. \end{cases}$$

We will prove that the following minimum is attained:

$$\min_{(v; g, H): \|v\|_\infty \geq 1, \ \|g\| \geq \|H\|} a(v; g, H) > 0.$$

We can perform the change of variables $g' = g/\|g\|$ and $H' = H/\|g\|$. Thus, minimizing $a(v; g, H)$ in g and H, with $\|g\| \geq \|H\|$, is equivalent to minimizing $a'(v; g', H')$ in g' and H', with $\|g'\| = 1$ and $\|H'\| \leq 1$, where

$$a'(v; g', H') = \max \left\{ |\phi(s)^\top v| : \|s\| \leq 1, \quad -g'^\top s - \frac{1}{2} s^\top H' s \geq \frac{\kappa_{fcd}}{2} \right\}.$$

Since $-s^\top H' s \geq -\|s\|^2$ when $\|H'\| \leq 1$ and $\kappa_{fcd} < 1$, it is always possible, for any g' and H' such that $\|g'\| = 1$ and $\|H'\| \leq 1$, to find s (by setting $s = -tg'$ with t slightly less than 1) such that both constraints that define $a'(v; g', H')$ are satisfied strictly. Thus, the feasible set always has an interior; hence a ball can be inscribed in this feasible region, whose positive radius depends continuously on g and H. Since g and H range over a compact set, so does the radius of the inscribed ball. Hence, there exists a smallest positive radius r_a^* such that the feasible set that defines $a'(v; g', H')$ always contains a ball of that radius, say $B(x; r_a^*)$, for some $x \in B(0; 1)$ dependent on g' and H'.

Similarly to the proof of Lemma 3.10 we can show that there exists $\xi_*^a > 0$ such that

$$\min_{x \in B(0;1)} \max_{s \in B(x; r_a^*)} |v^\top \phi(s)| \geq \xi_*^a,$$

by claiming that the above quantity is a norm on v. Hence, $a'(v; g', H') \geq \xi_*^a$ for any v, g', and H' such that $\|v\|_\infty \geq 1$, $\|g'\| = 1$, and $\|H'\| \leq 1$.

We also need to prove that the following minimum is attained:

$$\min_{(v; g, H): \|v\|_\infty \geq 1, \ g_{min}\|H\| \leq \|g\| \leq \|H\|} b(v; g, H) > 0.$$

Once again, the structure of the problem appears to be favorable since $m(0) - m(s) \geq (\kappa_{fcd}/2) \|g\|^2/\|H\|$ is equivalent to

$$-\left(\frac{g}{\|H\|}\right)^\top s - \frac{1}{2}s^\top\left(\frac{H}{\|H\|}\right)s \geq \frac{\kappa_{fcd}}{2}\frac{\|g\|^2}{\|H\|^2}.$$

The change of variables in this case is $g' = g/\|H\|$ and $H' = H/\|H\|$. Thus, minimizing $b(v; g, H)$ in v, g, and H, with $g_{min}\|H\| \leq \|g\| \leq \|H\|$, is equivalent to minimizing $b'(v; g', H')$ in v, g', and H', this time with $g_{min} \leq \|g'\| \leq 1$ and $\|H'\| = 1$, where

$$b'(v; g', H') = \max\left\{|\phi(s)^\top v| : \|s\| \leq 1, \quad -g'^\top s - \frac{1}{2}s^\top H's \geq \frac{\kappa_{fcd}}{2}\|g'\|^2\right\}.$$

Since $-s^\top H's \geq -\|s\|^2$ when $\|H'\| = 1$ and $\kappa_{fcd} < 1$, it is always possible, for any g' and H' such that $g_{min} \leq \|g'\| \leq 1$ and $\|H'\| = 1$, to find s (by setting $s = -tg'$ with t slightly less than 1) such that both constraints that define $b'(v; g', H')$ are satisfied strictly. Thus, the feasible set of the problem that defines $b'(v; g', H')$ always has an interior. Hence there is a smallest radius r_b^* such that the feasible region always contains a ball of at least such a radius. Repeating the arguments for $a'(v; g', H')$ we conclude that there exists $\xi_*^b > 0$ such that $b'(v; g', H') \geq \xi_*^b$ for all v, g', and H', such that $\|v\|_\infty \geq 1$, $g_{min} \leq \|g'\| \leq 1$, and $\|H'\| = 1$.

The value of ξ_* that we are looking for is therefore

$$\xi_* = \min\left\{\xi_*^a, \xi_*^b\right\} > 0. \quad \square$$

Validity of the wedge algorithm for the second-order case

To be able to prove global convergence to second-order critical points for the wedge method one first of all needs to strengthen the fraction of Cauchy decrease requirement by the fraction of the eigenstep decrease requirement. Recalling the scaled and unscaled trust-region subproblems, we can see that if s satisfies a fraction of eigenstep-decrease-type condition of the form

$$m(0) - m(s) \geq \frac{\kappa_{fed}}{2}\max\{-\lambda_{\min}(H), 0\}, \tag{11.6}$$

then $s_u = \Delta s$ also satisfies a fraction of eigenstep-decrease-type condition of the form

$$m_u(0) - m_u(s_u) \geq \frac{\kappa_{fed}}{2}\max\{-\lambda_{\min}(H_u), 0\}\Delta^2.$$

When $\lambda_{min}(H) \geq 0$, any step s such that $m(0) \geq m(s)$, and in particular the Cauchy step, satisfies inequality (11.6) for $\kappa_{fed} > 0$. When $\lambda_{min}(H) < 0$, the eigenstep (defined as an eigenvector u of H associated with the minimal eigenvalue of H, appropriately scaled such that $\|u\| = 1$ and $u^\top g \leq 0$) satisfies inequality (11.6) with $\kappa_{fed} = 1$.

The analogue of problem (11.4) for the second-order case is

$$\begin{aligned}
\max_{s\in\mathbb{R}^n} \quad & |\phi(s)^\top v| \\
\text{s.t.} \quad & \|s\| \leq 1, \\
& m(0) - m(s) \geq \frac{\kappa_{fcd}}{2}\|g\| \min\left\{\frac{\|g\|}{\|H\|}, 1\right\}, \\
& m(0) - m(s) \geq \frac{\kappa_{fed}}{2}\max\{-\lambda_{min}(H), 0\},
\end{aligned} \tag{11.7}$$

with $\kappa_{fcd}, \kappa_{fed} \in (0, 1)$.

We can show an extended result of Theorem 11.1 for the second-order case.

Theorem 11.2. *Let $\sigma_{min} \in (0, 1)$ be a given constant. There exists a positive constant ξ_* depending on σ_{min}, κ_{fcd}, and κ_{fed} such that the optimal value of (11.7) satisfies*

$$\xi(v; g, H) \geq \xi_* > 0$$

for all vectors $v \in \mathbb{R}^p$ such that $\|v\|_\infty \geq 1$ and for all vectors $g \in \mathbb{R}^n$ and symmetric matrices $H \in \mathbb{R}^{n \times n}$ such that

$$\sigma_{min} \leq \max \left\{ \frac{\|g\|}{\|H\|}, \frac{-\lambda_{min}(H)}{\|H\|} \right\}. \tag{11.8}$$

However, this result is not enough anymore because, unlike the first-order case, there is no guarantee that the bound σ_{min} in Theorem 11.2 exists. We know that the scaled quantity $\|g\|/\|H\|$ is bounded away from zero in the first-order case due to the criticality step. In the second-order case, what we obtain from the criticality step is

$$\max \left\{ \frac{\|g\|}{\|H\|}, \frac{1}{\Delta} \frac{-\lambda_{min}(H)}{\|H\|} \right\} = \frac{\sigma_u^m}{\Delta \|H_u\|} \geq \min \left\{ \frac{\epsilon_c}{\Delta_{max}\kappa_{bhm}}, \frac{1}{\mu\kappa_{bhm}} \right\}. \tag{11.9}$$

Unfortunately, the boundedness away from zero of the quantity involving Δ, given by $\max\{\|g\|/\|H\|, -\lambda_{min}(H)/(\Delta\|H\|)\}$, does not imply the boundedness away from zero of $\max\{\|g\|/\|H\|, -\lambda_{min}(H)/\|H\|\}$, which is what we need in Theorem 11.2. When Δ is small, it is possible that both $\|g\|/\|H\|$ and the most negative eigenvalue of $H/\|H\|$ approach zero. In this case, the part of the trust region where a fraction of eigenstep decrease can be obtained may shrink to a region of empty interior. This means that a polynomial may nearly vanish on such a region and, hence, there is no threshold on the pivot value that can be guaranteed to be achievable. Therefore, there might be no step computed in Step 2 of Algorithm 11.4 for which a fraction of eigenstep decrease is attained.

To show this in more detail let us assume that after the criticality step one has

$$g_u = \begin{bmatrix} 0 \\ \Delta^{\frac{3}{2}} \end{bmatrix} \quad \text{and} \quad H_u = \begin{bmatrix} -\Delta^{\frac{1}{2}} & 0 \\ 0 & 1 \end{bmatrix}.$$

We can see that $\Delta \leq \mu\sigma_u^m$ holds for sufficiently small Δ, since

$$\Delta \leq \mu \max\{\|g_u\|, -\lambda_{min}(H_u)\} = \mu\Delta^{\frac{1}{2}}.$$

Since $\|H_u\| \leq \max\{\Delta_{max}^{\frac{1}{2}}, 1\}$, we see that the model Hessian is bounded. Now, when we consider the scaled quantities $\|g\|$ and $\|H\|$, we obtain, for $\Delta \leq 1$,

$$\frac{\|g\|}{\|H\|} = \frac{\|g_u\|}{\Delta \|H_u\|} = \Delta^{\frac{1}{2}}$$

and

$$\frac{-\lambda_{min}(H)}{\|H\|} = \Delta^{\frac{1}{2}},$$

and therefore the right-hand side of (11.8) converges to zero when Δ approaches zero. Hence, the current criticality step does not guarantee the existence of a positive lower bound on $\max\{\|g\|/\|H\|, -\lambda_{min}(H)/\|H\|\}$. Notice that if the Hessian of f is nonsingular at optimality, then so are all the model Hessians close to optimality and this situation does not occur. For the situation of a singular Hessian, a different criticality step may need to be devised that can provide necessary bounds, while maintaining all other properties required for global convergence.

11.5   Other notes and references

The DFO algorithm described in [59, 61] was implemented by Scheinberg as an open source package. UOBYQA [189] and NEWUOA [192] were implemented by Powell and are distributed by the author. WEDGE is Marazzi's MATLAB® [1] implementation of the Wedge method [163] and is available for free download. Berghen [33] and Berghen and Bersini [34] have implemented a version of the UOBYQA algorithm of Powell [189] in a parallel environment. Their code is called CONDOR. The web addresses of these packages are given in the appendix. The paper [34] also contains numerical comparisons among UOBYQA, CONDOR, and DFO. Uğur et al. [221] report numerical results of CONDOR and DFO on a practical application.

Fasano, Morales, and Nocedal [89] studied the performance of a trust-region interpolation-based algorithm that dispenses with any control of the geometrical positions of the sample points. They reported good numerical results compared to NEWUOA when using, in both algorithms, quadratic models built by $(n+1)(n+2)/2$ points. It seems that, although the condition number of the matrices of the interpolation systems grows considerably during the course of the iterations, it tends to stabilize at tolerable levels and the algorithm is able to make progress towards the solution. However, further research is needed to better understand such a behavior, and more testing is necessary, in particular, for the more practical scenario where the number of points for model building is linear in n.

The area is still quite active, and new papers are likely to appear soon (see, e.g., Custódio, Rocha, and Vicente [69] and Wild [225]).

Part III

Review of other topics

Chapter 12

Review of surrogate model management

Chapters 12 and 13 briefly cover issues related to derivative-free optimization not yet addressed in this book. The main intention is to organize a number of relevant topics in a systematic form and to provide a range of pointers to places in the literature where the material is described with the appropriate depth. We will necessarily oversimplify some of the issues for the sake of brevity. Occasionally, a little more detail is given, especially when the technicalities allow us to easily illustrate the points we wish to highlight. The current chapter addresses surrogate modeling and rigorous optimization frameworks to handle surrogate models.

12.1 Surrogate modeling

In engineering modeling it is frequently the case that the function to be optimized is expensive to evaluate. The problem to be (approximately) solved may require extensive simulation of systems of differential equations, possibly associated with different disciplines, or may involve other time-consuming numerical computations. In the notation of this book, the true function is denoted by $f(x)$, where $f : \mathbb{R}^n \to \mathbb{R}$.

Engineers frequently consider models of the true function, and indeed they often consider the problem to already be a model and use the term "true model" for what we call the true function and then consider the model of the model to be a surrogate model. In order not to confuse the various models, we will usually continue to use the term true function for the underlying problem functions rather than referring to them as true models. A surrogate model can serve several purposes; in particular it can be used only for modeling and analysis, in order to gain insight about problem features and behavior without many expensive evaluations. More interestingly for our context, a surrogate model can take the place of the true function for purposes of optimization. Some of the methods for derivative-free optimization are based upon choosing suitable, in some ways essentially simple, models as surrogate models for the true function. A surrogate model is typically less accurate or has less quality than the true function, but it is cheaper to evaluate or consumes fewer computing resources. Several evaluations of the surrogate model can still be less expensive than one evaluation of the true function (true model). We will use the notation $sm(x)$ for surrogate models, where $sm : \mathbb{R}^n \to \mathbb{R}$.

Following suggestions of other authors (in particular as adopted in the PhD theses by Serafini [206] and Søndergaard [209]), we will the adopt the (necessarily simplified) point of view of classifying surrogate models as either *functional* or *physical*. We will summarize next the various approaches for building physical and functional surrogate models. More comprehensive reviews are made in Keane and Nair [139] and Søndergaard [209].

Physical surrogate models

By physical surrogate models we mean surrogate models built from a physical or numerical simplification of the true problem functions. One can think, for instance, of a coarser mesh discretization in numerical PDEs or of a linearization of term sources or equations as ways of obtaining physical surrogate models. Another popular approach is reduced basis methods (see, e.g., [139]). Physical surrogate models are in many circumstances based on some knowledge of the physical system or phenomena being modeled, and thus any particular such model is difficult to exploit across different problems. Some authors [43, 175] called them *mechanistic models*.

There are other procedures not directly based on simplified physics to build physical surrogate models from true functions with physical meaning. These procedures typically involve some form of correction, scaling or alignment of the available surrogate model using information (function values or gradients) of the true functions. One example is the β-correlation method [118]. Another, more rigorous, example is the space-mapping method which will be briefly described in Section 12.2. The multipoint method [219] is a process of generating physical surrogate models by partitioning the system into a number of individual subsystems.

If derivatives or approximations to derivatives of the true function are not used in physical surrogate models these might not exhibit the trends of the true function. In such cases, one may have to resort to optimizing the possibly expensive true function without derivatives which may be computationally problematic due to the curse of dimensionality.

Functional surrogate models

Functional surrogate models, on the other hand, are algebraic representations of the true problem functions. Interpolant and regression polynomial models (see Chapters 3–5) can be classified as functional models. One can say that functional models are typically based on the following components: a class of basis functions, a procedure for sampling the true functions, a regression or fitting criterion, and some deterministic or stochastic mathematical technique to combine them all. Functional surrogate models have a mathematical nature different from the true, original functions. The knowledge of "truth" is not directly embodied in their mathematical structure, but it is revealed implicitly in the values of their coefficients. Functional surrogate models are thus generic and empirical and not (at least entirely) specific to a class of problems. They are strongly dependent on samples of the true function and are applicable to a wide variety of problems.

Taylor-based models can be seen as either physical or functional, or a mixture of both. In fact, if the gradients are approximated, for instance, by finite differences, Taylor models are clearly functional. But if the gradients require some intrinsic procedure like the simulation of the adjoint equations, then their nature may be physical. Note also that most of the physical models are not absolutely pure in the sense that they may contain some empirical elements, like parameters adjusted to experimental data.

Among the most common classes of basis functions used to build functional surrogate models are low-order polynomials (see Chapters 3–5), radial basis functions (see [49, 50] and the review below), adaptive regression splines (see [121]), and wavelets (see [72]). The most widely used criterion for fitting the data is based on least-squares regression. In the statistical community, regression models are commonly used in response surface methodology. Artificial neural networks in turn are nothing other than nonlinear regression models [121].

There are also single-point techniques specifically developed to extract as much information as possible from what is known at a given point (such as the value of the true function and its gradient). Taylor models are the most obvious of these approaches, but others exist such as reciprocal, modified reciprocal, conservative, and posynomial approximation models (see [11] and the references therein).

Radial basis functions

So far in this book, we have treated in detail only polynomial (linear and quadratic) interpolation and regression models. Now we give some details concerning the class of models given by radial basis functions, which have frequently been used in derivative-free optimization [179, 181, 226]. The earliest reference to radial basis functions appears to be Duchon [86], but they are currently an active research field in approximation theory (see, for example, [49, 50, 185]).

One of the attractive features of radial basis functions is their ability to model well the curvature of the underlying function. Another key feature is that the coefficient matrix defined by the interpolation conditions is nonsingular under relatively weak conditions (see the discussion below). In fact, one can guarantee useful bounds on the norm estimates and condition numbers of the interpolation matrices (see, e.g., [27] and [50, Section 5.3]). The interpolation linear systems that arise when modeling with radial basis functions are typically dense, and their solution by iterative algorithms has been investigated. However, this is relevant only when the number of interpolation points is large, which is not really an issue in our context.

In order to interpolate a function f whose values on a set $Y = \{y^0, \dots, y^p\} \subset \mathbb{R}^n$ are known, one can use a radial basis functional surrogate model of the form

$$sm(x) = \sum_{i=0}^{p} \lambda_i \phi(\|x - y^i\|), \qquad (12.1)$$

where $\phi : \mathbb{R}_+ \to \mathbb{R}$ and $\lambda_0, \dots, \lambda_p \in \mathbb{R}$. The term *radial basis* stems from the fact that $\phi(\|x\|)$ is constant on any sphere centered at the origin in $\mathbb{R}^n$. For $sm(x)$ to be twice continuously differentiable, the radial basis function $\phi(x)$ must be both twice continuously differentiable and have a derivative that vanishes at the origin (see the exercises).

Some of the most popular (twice continuously differentiable) radial basis functions are the following:

- cubic $\phi(r) = r^3$,

- Gaussian $\phi(r) = e^{-\frac{r^2}{\rho^2}}$,

- multiquadric of the form $\phi(r) = (r^2 + \rho^2)^{\frac{3}{2}}$,

- inverse multiquadric of the form $\phi(r) = (r^2 + \rho^2)^{-\frac{1}{2}}$,

where ρ^2 is any positive constant.

Since in many applications it is desirable that the linear space spanned by the basis functions include constant functions (and the standard radial basis choices do not give this property when n is finite), it turns out to be useful to augment the radial basis function model in (12.1) by adding a constant term. Similarly, if one wants to include linear functions (and/or other suitable low-order polynomial functions), one adds a low-order *polynomial tail* of degree $d-1$ that one can express as $\sum_{j=0}^{q} \gamma_j p_j(x)$, where $p_j \in \mathcal{P}_n^{d-1}$, $j = 0, \ldots, q$, are the basis functions for the polynomial and $\gamma_0, \ldots, \gamma_q \in \mathbb{R}$. The new surrogate model is now of the form

$$sm(x) = \sum_{i=0}^{p} \lambda_i \phi(\|x - y^i\|) + \sum_{j=0}^{q} \gamma_j p_j(x).$$

Furthermore, the coefficients λ are required to satisfy

$$\sum_{i=0}^{p} \lambda_i p_j(y^i) = 0, \quad j = 0, \ldots, q.$$

These, plus the interpolation conditions $sm(y^i) = f(y^i), i = 0, \ldots, p$, give the linear system

$$\begin{bmatrix} \Phi & P \\ P^\top & 0 \end{bmatrix} \begin{bmatrix} \lambda \\ \gamma \end{bmatrix} = \begin{bmatrix} f(Y) \\ 0 \end{bmatrix}, \tag{12.2}$$

where $\Phi_{ij} = \phi(\|y^i - y^j\|)$ and $P_{ij} = p_j(y^i)$ for $i, j \in \{0, \ldots, p\}$, and $f(Y)$ is the vector formed by the values $f(y^0), \ldots, f(y^p)$.

The symmetric system (12.2) has a unique solution for the examples of ϕ given above, provided P has full rank and $d \geq 2$. Such a property is a consequence of the fact that, for the examples above, ϕ is *conditionally positive definite* of order d with $d = 2$ (see the exercises). One says that ϕ is *conditionally positive definite* of order d when $\sum_{i,j=0}^{p} \phi(\|y^i - y^j\|)\lambda_i \lambda_j$ is positive for all distinct points $y^0, \ldots, y^p$ and $\lambda \neq 0$ satisfying $\sum_{i=0}^{p} \lambda_i p_j(y^i) = 0, j = 0, \ldots, q$, where the p's represent a basis for $\mathcal{P}_n^{d-1}$. If ϕ is conditionally positive definite of order d, then it is so for any larger order. For instance, in the case of Gaussian and inverse multiquadric radial basis functions, the matrix Φ is already a positive definite one (and so trivially conditionally positive definite of order 2).

The approaches by Oeuvray [179] and Oeuvray and Bierlaire [181], and Wild, Regis, and Shoemaker [226] for derivative-free optimization use cubic radial basis functions and linear polynomial tails:

$$sm(x) = \sum_{i=0}^{p} \lambda_i \|x - y^i\|^3 + c + g^\top x. \tag{12.3}$$

In this case poisedness is equivalent to the existence of $n+1$ affinely independent (and thus distinct) points in the interpolation set. If the number of interpolation points is $p+1$, then the model has $p+n+2$ parameters, $p+1$ for the radial basis terms and $n+1$ for the linear polynomial terms. However, when the number of points is $n+1$ (or less), the solution of the interpolation system gives rise to a linear polynomial, since all the parameters λ_i, $i = 0,\ldots,p$, are zero (see the second block equation in (12.2)). Consequently, the simplest nonlinear model $sm(x)$ of the form (12.3) is based on $n+2$ interpolation points and has $2n+3$ parameters.

It is easy to show that the model (12.3), built on a poised sample set contained in a ball of radius Δ, provides an error in function values of order Δ^2 and in gradient of order Δ, as happens for linear polynomial interpolation or regression, and where the constant in the upper bounds depends on the size of λ. Wild, Regis, and Shoemaker [226] studied how to rigorously manage the conditioning of the matrix arising from the system (12.2) to derive uniform upper bounds for λ.

Kriging models

Functional models may incorporate a stochastic component, which may make them better suited for global optimization purposes. A popular example is Kriging. Roughly speaking, Kriging models are decomposed into two components. The first component is typically a simple model intended to capture the trend in the data. The other component measures the deviation between the simple model and the true function.

In the following example, we consider the simple model to be a constant for ease of illustration. Assume then that the true function is of the form

$$f(x) = \beta + z(x),$$

where $z(x)$ follows a stationary Gaussian process of mean 0 and variance σ^2. Let the covariance between two sample points y and w be modeled as follows:

$$R(y,w) = \mathbf{Cov}(z(y),z(w)),$$

where $R(y,w)$ is such that $R(Y,Y)$ is positive definite for any set Y of distinct sample points. A popular choice is to model the covariance by radial basis functions $R(y,w) = \phi(\|y-w\|)$.

Given a set Y of distinct sample points, the Kriging model $ksm(x)$ is defined as the expected value of the true function given the observed values $f(Y)$:

$$ksm(x) = E(f(x)|Y).$$

One can prove that $ksm(x)$ is of the form (see the exercises at the end of the chapter)

$$ksm(x) = \hat{\beta} + R(Y,x)^{\top} R(Y,Y)^{-1}(f(Y) - \hat{\beta}e), \tag{12.4}$$

where e is a vector of ones and

$$\hat{\beta} = \frac{e^{\top} R(Y,Y)^{-1} f(Y)}{e^{\top} R(Y,Y)^{-1} e}.$$

It is useful to think of $\hat{\beta}$ as an approximation to β and of $R(Y,x)^{\top} R(Y,Y)^{-1}(f(Y) - \hat{\beta}e)$ as some average approximation to $z(x)$.

Kriging was originally developed in the geostatistics community (see Matheron [166]) and used first for computer experiments by Currin et al. [67]. A good survey article referenced by many researchers is [203].

Design of experiments and response surface methodologies

A number of sampling procedures have been derived for the construction of functional surrogate models, depending on the origin of the true function, the cost of its evaluation, and the smoothness of the model, among other issues.

The techniques developed in Part I of this book, for instance, provide a rigorous and efficient deterministic sampling procedure for interpolant and regression polynomial models. In fact, we have seen in Chapters 3–5 how the poisedness or conditioning of the sample set affects the quality of the polynomial models. We have introduced a measure of well poisedness (called Λ-poisedness) and showed in Chapter 6 how to ensure Λ-poisedness (with $\Lambda > 1$ not very large) when building and updating sample sets for these polynomial models.

In the statistical literature the process of determining the location of the points in the sampling space is called *design of experiments*, where one of the main goals focuses on reducing the noise of the experiments. The sample sets are derived by techniques like Latin or factorial designs, supposed to spread well the points in the sampling space by typically placing them at hypercube-type vertices. In the language of this book, such sampling sets are likely to be well poised for linear interpolation or regression. There are more elaborated criteria to choose the sampling locations, such as D-optimality, where the sample set Y is chosen so that quantities related to $|\det((Y^\top Y)^{-1})|$ are minimized (see Driessen et al. [85] for a trust-region approach based on such a type of sampling).

General references for design of experiments are [144, 203, 204, 208]. In particular, *design and analysis of computer experiments* (DACE) [203] is a widely used statistical framework for using Kriging models in the context of simulation-based experiments, which, as we have already mentioned, often uses radial basis functions for correlation between sampling points. DACE also stands for a MATLAB® [1] toolbox [159], which provides tools for sampling (by Latin hypercubes) and building Kriging surrogate models. Latin hypercube sampling was developed by McKay, Conover, and Beckman [168] (see also [211]). Another technique with interesting spread-out-type sampling properties is called orthogonal arrays, motivated from the series of papers by Rao in the 1940s (see, for instance, the papers [182, 213], the book [124], and the references therein).

Response surface methodology (RSM) is a framework to minimize a function (resulting from the response of a system) by sequentially building and minimizing functional surrogate models. The first RSM proposed is reported in the paper by Box and Wilson [44]. An RSM typically starts by selecting the most important and relevant variables or parameters by applying some analysis of variance. Variables contributing little to the variation in the response may be left out. Then some heuristic optimization iterative procedure is started where at each iteration a functional surrogate model $sm_k(x)$ is built (for instance, by Kriging) and used to produce a direction of potential descent for the true function f. The current iterate can then be improved by minimizing the true function along this direction. Such an iterative procedure is terminated when the iterates approach a point $\bar{x}$. An RSM can then be concluded by minimizing a potentially more accurate functional surrogate model (like a second-order polynomial model) locally around $\bar{x}$, without concern about noise. For a comprehensive coverage of RSMs see the books by Box and Draper [43] and Myers and Montgomery [175] and the survey paper of Simpson et al. [208].

One can regard other RSMs as one-shot optimization approaches, where a functional surrogate model is first built and then minimized to determine the optimum (see [203]). Such approaches, however, are likely to require dense sampling to produce meaningful results. However, the statistical nature of many applications allows modeling techniques like

Kriging to assign confidence intervals to the models generated and to identify outliers and lack of fit. A review paper that considers existing approaches for using response surfaces in the context of global optimization is given in [137].

12.2 Rigorous optimization frameworks to handle surrogate models

For the purpose of optimization, surrogate models are often incorporated in some form of model management framework. RSM, mentioned above, is a popular surrogate model management framework. We will review next rigorous optimization frameworks to incorporate surrogate models which retain the global convergence properties of the underlying algorithms.

Handling surrogate models in direct search

Booker et al. [40] introduced a rigorous framework to deal with surrogate models in the context of directional direct-search algorithms. The idea is simple and based on the flexibility of the search step of these algorithms, which is free of the stricter poll requirements.

Suppose that one has a surrogate model $sm(x)$ of the true function $f(x)$. We have in mind situations where $sm(x)$ is less accurate and cheaper to evaluate than $f(x)$. Suppose also that, besides having ways of building such an initial surrogate model $sm(x) = sm_0(x)$, one also has ways of possibly improving or *recalibrating* the surrogate model along the course of the optimization process. One can think of a process where a sequence of surrogate models $\{sm_k(x)\}$ is built by successive recalibrations. Every time f is evaluated at new points, those values can be used to recalibrate and hence improve the quality of the surrogate models. In Chapter 11 we have seen a number of ways of updating interpolation and regression polynomial models within the framework of trust-region interpolation-based methods. These polynomial models and updating techniques could also be applied in the context of a search step in directional direct search.

We now describe one possible framework to handle surrogate models in directional direct search along the spirit of [40] (see Algorithm 12.1 below). Two plausible alternatives are given for using a surrogate model in the search step, but several others are possible, and they may be more appropriate given the specifics of the problem at hand. Note that the poll step itself might benefit from surrogate modeling by, for instance, ordering the evaluation of the true function in $P_k = \{x_k + \alpha_k d : d \in D_k\}$ according to increasing values of $sm_k(\cdot)$.

Algorithm 12.1 (Handling surrogate models in a directional direct-search framework).

Initialization: Initialize as in Algorithm 7.2 *and, in addition, form an initial surrogate model* $sm_0(\cdot)$.

For $k = 0, 1, 2, \ldots$

    1. **Search step:** Try to compute a point x with $f(x) < f(x_k)$ by evaluating the function f a finite number of times, by means, for instance, of one of the following two alternatives:

(a) Evaluate $sm_k(\cdot)$ on a set of points $Y_k = \{y_k^1, \ldots, y_k^{p_k}\}$. Order Y_k by increasing values: $sm(y_k^1) \leq \cdots \leq sm(y_k^{p_k})$. Start evaluating f in Y_k along this order.

(b) Apply some finite optimization process to minimize $sm_k(\cdot)$ possibly in some feasible set. Let y_k be the approximated minimizer. Evaluate $f(y_k)$.

If such a point x is found, then set $x_{k+1} = x$, declare the iteration and the search step successful, and skip the poll step.

2. **Poll step:** Poll as in Algorithm 7.2. *(The surrogate model $sm_k(\cdot)$ could be used to order the poll set just before polling.)*

3. **Model calibration:** *Form $sm_{k+1}(\cdot)$, possibly by recalibrating $sm_k(\cdot)$.*

4. **Mesh parameter update:** Update the mesh size parameter as in Algorithm 7.2.

A natural question that now arises is how to guarantee that such a surrogate model management framework yields a globally convergent algorithm, in other words, an algorithm which converges to a first-order stationary point regardless of the starting point. Fortunately, we are already equipped to properly answer this question. In fact, we have seen in Chapter 7 two ways of guaranteeing global convergence for directional direct-search methods.

One such way was to ask the iterates to lie in integer lattices (see Section 7.5). Such a globalization scheme allows the algorithm to take steps based on a simple decrease of the objective function. The set of positive bases (or of positive spanning sets) used by the algorithm must then satisfy the requirements of Assumption 7.8, and the step size parameter must follow the rules of Assumption 7.10. But what is relevant in the context of surrogate model management is the requirement stated in Assumption 7.9, which restricts the search step of this class of algorithms to evaluate points only in the mesh M_k (defined, for instance, by (7.8) or (7.15)). In the context of Algorithm 12.1 one can accomplish such a condition in different ways. If, for instance, one selects the first alternative in the search step, then Y_k must be chosen as a subset of M_k, which does not seem to be a problem at all even if we want to generate Y_k with some level of randomness. In the second possibility, we could project y_k onto the mesh M_k. Such a projection could be computationally expensive for some choices of positive bases, but it is a trivial task for others. For instance, it is trivial to project onto M_k if we choose $\mathcal{D} = \{D_\oplus\}$, where $D_\oplus$ is the set of unit coordinate vectors given in (7.1). Note that there is no guarantee that the projected point, say x, provides a decrease of the form $f(x) < f(x_k)$, even when y_k does. If $f(x) \geq f(x_k)$ and $f(y_k) < f(x_k)$, then either the iteration is considered unsuccessful or y_k is taken and the geometrical considerations needed for convergence are ignored.

The other way to enforce global convergence for directional direct-search methods is to impose a sufficient decrease condition on the acceptance of new points, both in the search step and in the poll step (see Section 7.7). In this case, the set of positive bases (or of positive spanning sets) used by the algorithm in the poll step is required only to satisfy Assumption 7.3. To impose sufficient decrease, Algorithm 12.1 would have to be modified appropriately. In particular, the search step would accept a point x only when $f(x) < f(x_k) - \rho(\alpha_k)$, where $\rho(\cdot)$ is a forcing function (for example, consider $\rho(t) = t^{1+a}$ with $a > 0$ or see the rigorous definition given in Section 7.7).

What we describe in Algorithm 12.1 is a simplified, though potentially effective, way of handling surrogate models in directional direct search. The reader interested in this topic is pointed to Booker et al. [40], as well as to Abramson, Audet, and Dennis [7], Marsden et al. [165], Serafini [206], and Torczon and Trosset [218], where other algorithmic devices were developed and tested. One extension of this framework for constrained optimization is proposed in Audet et al. [14].

Despite the fact that we have been focusing mainly on directional direct search (Chapter 7), it is also possible to consider ways, perhaps not as effective, of enhancing simplicial direct search (Chapter 8) by exploring the availability of surrogate modeling. We could, for instance, test the "quality" of the expanded point by first evaluating the current surrogate model. If the surrogate model predicts a bad performance of the true function f for such a point, it could be immediately disregarded. One could also, for example, minimize the current surrogate model and take that point as a center of a (possibly needed) simplex restart.

Handling nonlinear models in trust-region methods

In Chapter 10 we have considered trust-region methods based on derivative-free trust-region models. We assumed there that the trust-region models were quadratic (or linear) functions of the form (see (10.1))

$$m_k(x_k + s) = m_k(x_k) + s^\top g_k + \frac{1}{2} s^\top H_k s, \tag{12.5}$$

so that we could easily compute a Cauchy step and guarantee that it satisfies (10.10):

$$m_k(x_k) - m_k(x_k + s_k) \geq \frac{\kappa_{fcd}}{2} \|g_k\| \min \left\{ \frac{\|g_k\|}{\|H_k\|}, \Delta_k \right\}, \tag{12.6}$$

with $\kappa_{fcd} = 1$ (see Theorem 10.1). Thus, any step that achieves a fraction $\kappa_{fcd} \in (0, 1)$ of Cauchy decrease (see (10.9) in Assumption 10.1) would necessarily verify (12.6).

A natural question that arises is what happens when the trust-region model is not quadratic, i.e., when the model is not of the form (12.5). It is well known that (12.6) is a very mild condition, but it could still happen that an approximated solution of a trust-region subproblem (10.2) defined by a nonlinear model does not satisfy (12.6), when $\kappa_{fcd} \in (0, 1)$ is fixed across all iterations.

If $m_k(x_k + s)$ is not a quadratic function in s, then Theorem 10.1 is not directly applicable. On the other hand, the basis of the Cauchy step is that it determines the minimum of the model along the model steepest descent direction, so it is natural to be interested in determining the equivalent of Theorem 10.1 for more general models. One way to achieve this is to use a backtracking algorithm along the model steepest descent direction, where the backtracking is from the boundary of the trust region (as suggested in Conn, Gould, and Toint [57, Section 6.3.3]). In this approach one considers choosing the smallest (nonnegative integer) j such that

$$x_{k+1} = x_k + \beta^j s, \quad \text{where} \quad s = -\frac{\Delta_k}{\|g_k\|} g_k \quad \text{and} \quad \beta \in (0, 1), \tag{12.7}$$

so that one attains a sufficient decrease of the form

$$m_k(x_{k+1}) \leq m_k(x_k) + \kappa_c \beta^j s^\top g_k, \tag{12.8}$$

where $\kappa_c \in (0, 1)$. The proof is relatively straightforward and is based upon a standard line-search condition. If one considers the right-hand side of (12.8) with $\kappa_c = 1$, then this is just a first-order approximation to the model. Thus, more generally, the right-hand side in (12.8) is a linear approximation with a slope between the gradient model and the horizontal (slope zero). From (12.7) we see that (12.8) becomes

$$m_k(x_{k+1}) - m_k(x_k) \leq -\kappa_c \beta^j \Delta_k \|g_k\|. \tag{12.9}$$

Using the mean value theorem on the left-hand side, this is equivalent to

$$-\beta^j \Delta_k \|g_k\| + \frac{1}{2} \beta^{2j} \Delta_k^2 g_k^\top \nabla^2 m_k(y_{k,j}) g_k / \|g_k\|^2 \leq -\kappa_c \beta^j \Delta_k \|g_k\|$$

for some $y_{k,j} \in [x_k, x_k + \beta^j s]$. Thus, assuming $\|\nabla^2 m_k(y_{k,j})\| \leq \kappa_{bhm}$, we conclude that (12.9) is satisfied, provided

$$\frac{\beta^j \Delta_k}{\|g_k\|} \leq \frac{2(1 - \kappa_c)}{\kappa_{bhm}}.$$

Thus, we indeed find a j_k satisfying (12.9) such that $\beta^{j_k} > 2(1 - \kappa_c)\beta \|g_k\|/(\kappa_{bhm} \Delta_k)$. By defining $s_k^{\mathrm{AC}} = \beta^{j_k} s$ as the approximate Cauchy step, we obtain

$$m_k(x_k) - m_k(x_k + s_k^{\mathrm{AC}}) \geq \kappa_c \beta^{j_k} \Delta_k \|g_k\|.$$

On the other hand, if the approximate Cauchy step takes us to the boundary, we immediately deduce from (12.8) that the decrease in the model exceeds or is equal to $\kappa_c \Delta_k \|g_k\|$, and so we can conclude that

$$m_k(x_k) - m_k(x_k + s_k^{\mathrm{AC}}) \geq \bar{\kappa}_c \|g_k\| \min\left\{ \frac{\|g_k\|}{\kappa_{bhm}}, \Delta_k \right\}$$

for a suitably defined $\bar{\kappa}_c > 0$.

Finally, we point out that something similar can be done in the computation of a step satisfying a fraction of the eigenstep decrease for nonquadratic trust-region models (as suggested in [57, Section 6.6.2]). In this case, one computes a damped eigenstep of the form $s_k^{\mathrm{AE}} = \beta^j s_k^{\mathrm{AE}}$, $\beta \in (0, 1)$, so that $m_k(x_k + s_k^{\mathrm{AE}}) \leq m_k(x_k) + \kappa_e \tau_k \beta^{2j} \|s_k^{\mathrm{E}}\|^2$, where $\kappa_e \in (0, 1/2)$ and τ_k is the smallest (assumed negative) eigenvalue of the Hessian of the model at x_k. It is easy to show that such a condition can be satisfied in a finite number of backtrack steps (reductions by β) and that the selected j_k yields a lower bound of the form $\beta^{j_k} \geq (2\kappa_e - 1)\beta \tau_k/(\nu \Delta_k)$, where ν is a Lipschitz constant for the Hessian of the models. From this one can readily show that the approximated eigenstep s_k^{AE} satisfies a condition of the form (10.12) for a given constant $\bar{\kappa}_e > 0$.

Another way of dealing with nonquadratic trust-region models was suggested by Alexandrov et al. [11]. Their idea is simple and related to the previously described one. In this approach, one applies a trust-region method to an auxiliary constrained problem of minimizing $h(s) = m_k(x_k + s)$ in the variables s subject to $\|s\| \leq \Delta_k$. At each iteration of such an algorithm, we would have a trust-region subproblem, consisting of minimizing a quadratic model of $h(s)$ subject to two ℓ_2-norm constraints (one is the auxiliary problem constraint $\|s\| \leq \Delta_k$ and the other is the internal trust-region restriction of the algorithm being applied to the auxiliary problem). Note that one can still guarantee a fraction of Cauchy decrease for such a trust-region subproblem (see Heinkenschloss [125]). We are also guaranteed under the standard assumptions that, at the very first iteration of a run of such an algorithm (starting from the initial iterate $s^0 = 0$ and a trust-region radius $\delta^0 = \Delta_k$), a successful step s^j is obtained. One can then prove that such a step yields an appropriate fraction of Cauchy decrease for the original problem.

Handling surrogate models by space mapping

It is often the case in many engineering applications that the function $f(x)$ considered for optimization is of the form $H(F(x))$, where $F : \mathbb{R}^n \to \mathbb{R}^m$ describes a response of a system and $H : \mathbb{R}^m \to \mathbb{R}$ is some merit function, for instance, a norm. Similarly, suppose we have a surrogate model $S(x)$ for $F(x)$ and thus a surrogate model $sm(x) = H(S(x))$ for $f(x)$. More generally, one can have a family of surrogate models $S(x; p)$ for $F(x)$, parametrized by some $p \in \mathbb{R}^p$, and a corresponding family of surrogate models $sm(x; p) = H(S(x; p))$ for $f(x)$.

One can then think of an iterative optimization process for the approximated minimization of $f(x)$ consisting of extracting better parameters p by aligning $S(x; p)$ to $F(x)$, followed by the minimization of $sm(x; p)$ for the extracted parameters. The parameters extracted are called space-mapping parameters. Such a process is summarized now.

Algorithm 12.2 (A space-mapping approach).

Initialization: Choose $x^{(0)}$.

For $k = 0, 1, 2, \ldots$

    1. **Parameter extraction:** Compute $p^{(k)}$ as a solution for

$$\min_p \| S(x^{(k)}; p) - F(x^{(k)}) \|. \tag{12.10}$$

    2. **Minimizing the surrogate:** Compute $x^{(k+1)}$ as a solution for

$$\min_x sm(x; p^{(k)}) = H(S(x; p^{(k)})). \tag{12.11}$$

In the above formulations we assumed for simplicity that the domains of x (in both S and F) and of p are unrestricted. One could consider F restricted to $\Omega_F \subset \mathbb{R}^n$, S restricted to $\Omega_S \subset \mathbb{R}^n$, and p restricted to $\Omega_p \subset \mathbb{R}^p$, in which case the minimizations in (12.10) and (12.11) are restricted to Ω_p and $\Omega_F \cap \Omega_S$, respectively. We can also consider Ω_S dependent on the parameters $p^{(k)}$. Another natural generalization is to extend the fitting in (12.10) to all previous iterates:

$$\min_p \sum_{i=0}^{k} \omega_i^k \| S(x^{(i)}; p) - F(x^{(i)}) \|,$$

where $\omega_k^k = 1$ and ω_i^k, $i = 0, \ldots, k-1$, are nonnegative factors weighting the contribution of the previous surrogate models in the current space-mapping parameter extraction.

If the Lipschitz constants of the optimal mappings in (12.10) and (12.11) are sufficiently small (with respect to the variables x and the parameters p, respectively), one can show that the process given in Algorithm 12.2 generates a convergent sequence $\{(x^{(k)}, p^{(k)})\}$

(see Koziel, Bandler, and Madsen [148]). If, in addition, one imposes other conditions, including the fact that the space-mapping parameter extraction is exact at the limit point (x^*, p^*), meaning that $S(x^*; p^*) = F(x^*)$, then x^* is an optimal solution for the original problem of minimizing the true function $f(x) = H(F(x))$ (see also [148]).

The approach given above corresponds to what has been called *implicit* space mapping because of the implicit dependence of S on the space-mapping parameters p. The name *space-mapping* is associated with the mapping $P : \Omega_F \to \Omega_S$, defined by

$$P(x_f) \in \operatorname*{argmin}_{x \in \Omega_S} \|S(x) - F(x_f)\|. \tag{12.12}$$

The original space-mapping approach consisted of the minimization of the space-mapping surrogate model $smsm(x) = sm(P(x)) = H(S(P(x)))$. In practice, one needs to regularize the definition of the space mapping to guarantee existence and uniqueness of solution in (12.12). Secant updates have been derived to approximate the Jacobian of $P(x)$ based only on responses of the system (i.e., evaluations of F). It is thus possible to form a surrogate model of the form

$$S(x; B^{(k)}) = S[P(x^{(k)}) + B^{(k)}(x - x^{(k)})],$$

where $B^{(k)}$ is a secant update for the Jacobian of P (e.g., a Broyden update; see [29] and also the modification suggested in [128]). One can fit the original space-mapping approach into the framework of Algorithm 12.2 by considering $p^{(k)} \equiv B^{(k)}$ and by looking at the secant update as a form of approximately solving (12.10). Problem (12.11) is not solved exactly but locally, by forming a quadratic model of $H(S(x; B^{(k)}))$ centered around $x^{(k)}$ and minimizing it within a trust region. The result is a space-mapping-based trust-region framework for the minimization of $smsm(x)$ (suggested first in [25], and whose global convergence properties have been analyzed in [162, 223]).

The space-mapping technique was introduced first by Bandler et al. [28] in 1994, and has been developed along different directions and generalized to a number of contexts, some of which we have already discussed (some others are the *input* and the *output* space-mapping approaches). Surveys on the topic can be found, for instance, in the thesis by Søndergaard [209] and in the review papers of Bakr et al. [26] and Bandler, Koziel, and Madsen [30]. The interested reader is further referred to the special issue on surrogate modeling and space mapping that has been edited by Bandler and Madsen [31].

12.3   Exercises

1. Show that if $\phi(r)$ is twice continuously differentiable and $\phi'(0) = 0$, then $h(x) = \phi(\|x\|)$ is also twice continuously differentiable. In particular, one has $\nabla h(0) = 0$ and $\nabla^2 h(0) = \phi''(0)I$.

2. Prove that when the radial basis function ϕ is conditionally positive definite of order d and P has full rank, then the matrix of the system (12.2) is nonsingular.

3. Show that the model (12.3), built on a poised sample set contained in a ball of radius Δ, yields an error in function values of order Δ^2 and in the gradient of order Δ (the constant in the upper bounds should depend on the size of λ).

4. To obtain the expression (12.4) consider $ksm(x) = \beta + R(Y, x)^\top \gamma$ and solve $ksm(Y) = f(Y)$ and $e^\top \gamma = 0$ for β and γ.

Chapter 13

Review of constrained and other extensions to derivative-free optimization

In this chapter we briefly consider the case of constrained nonlinear optimization problems written in the form

$$
\begin{aligned}
\min_{x \in \mathbb{R}^n} \quad & f(x) \\
\text{s.t.} \quad & x \in \Omega, \\
& h_i(x) \leq 0, \quad i = 1, \ldots, m_h,
\end{aligned}
\tag{13.1}
$$

where $f, h_i : \Omega \subseteq \mathbb{R}^n \to \mathbb{R} \cup \{+\infty\}$ are functions defined on a set or domain Ω. We will briefly review the existing approaches for solving problems of the form (13.1) without computing or directly estimating the derivatives of the functions f and h_i, $i = 1, \ldots, m_h$, but assuming that the derivatives of the functions that algebraically define Ω are available.

We emphasize the different nature of the constraints that define Ω and the constraints defined by the functions h_i, $i = 1, \ldots, m_h$. The first type of constraints are typically simple bounds of the form $l \leq x \leq u$ or linear constraints of the form $Ax \leq b$. For instance, the mass of a segment of a helicopter rotor blade has to be nonnegative, or a wire used on a circuit must have a width that is bounded from below and above by manufacturability considerations. In addition, there might be linear constraints, such as a bound on the total mass of the helicopter blade.

The objective function f (and in some cases the constraint functions h_i) is often not defined outside Ω; hence (a possible subset of) the constraints defining Ω have to be satisfied at all iterations in an algorithmic framework for which the objective function (and/or some h_i's) is (are) evaluated. Such constraints are not *relaxable*. In contrast, *relaxable* constraints need only be satisfied approximately or asymptotically.[19]

If Ω is defined by linear constraints or simple bounds, then it is often easy to treat any constraints that define Ω as unrelaxable constraints. In theory Ω may also include general nonlinear constraints, whose derivatives are available. We call all the constraints that define Ω "constraints with available derivatives." Whether to treat these constraints as relaxable or not often depends on the application and the algorithmic approach.

---

[19]Other authors refer to relaxable and unrelaxable constraints as soft and hard constraints, or as open and closed constraints, respectively.

In many applications there are constraints of the form $h_i(x) \le 0$, $i = 1, \ldots, m_h$, where the functions h_i are computed in the same manner as f, i.e., by a black box which does not provide derivatives. In circuit tuning, for instance, a bound on delay or the power of the circuit may be such a constraint. We will call such constraints "derivative free." Clearly, the form of these constraints is general enough to model any type of constraints, including equality constraints which can be converted in two inequalities with opposite signs, although this may not be a good idea in practice (see below).

It is often the case that a derivative-free optimization problem has several functions that the user is trying to optimize, all of whose values, at a given point x, are computed by the same call to the black-box simulator. One of these functions is typically used as an objective function in a given formulation, while others are treated as constraints with a bound on their value. In this case the constraints are usually relaxable. There are possible situations where the derivative-free constraints need to be treated as unrelaxable. These situations require the availability of a feasible starting point, are quite difficult to address in general, and are, fortunately, rare in practice.

An extreme case of derivative-free unrelaxable constraints that does occur in practice are the so-called "hidden" constraints. Hidden constraints are not part of the problem specification/formulation, and their manifestation comes in the form of some indication that the objective function could not be evaluated. For example, the objective function $f(x)$ may be computed by a simulation package which may not converge for certain (unknown a priori) values of input parameters, failing to produce the objective function value. So far these constraints are treated in practical implementations by a heuristic approach or by using the extreme barrier function approach.

Most of the theoretical work on the constrained cases has been done in the framework of directional direct search (see Chapter 7, for the unconstrained case), and the area is still a subject of intense research.

13.1 Directional direct-search methods

A significant number of derivative-free methods for constrained problems are feasible methods, in the sense that the iterates produced are always kept feasible. Feasible approaches might be preferred for several reasons. On the one hand, the constraints of the problem might not be relaxable and the objective function value cannot be evaluated outside the feasible region. On the other hand, generating a sequence of feasible points allows the iterative process to be terminated prematurely, a procedure commonly applied when the objective function is very expensive to evaluate. In such cases there would be the guarantee of feasibility for the best point tested so far.

There are several methodologies to solve constrained nonlinear optimization problems. A popular approach in the 1960s and 1970s to deal with constraints consisted of using penalty and barrier functions, originally called exterior and interior penalty functions, respectively. The (exterior) penalty function typically consists of adding to the objective function a measure of infeasibility multiplied by a penalty parameter. The resulting penalty method allows infeasible iterates. Barrier functions have a different nature: they ensure feasibility by adding, to the objective function, a function (multiplied by a barrier parameter) that approaches $+\infty$ when (strictly) feasible points approach the frontier of the feasible region. Barrier methods with suitable step sizes, thus, preserve feasible iterates.

Since it may be desirable that a derivative-free algorithm generates feasible iterates, the barrier approaches are particularly appealing. The feasibility may be enforced with respect only to Ω or with respect to the whole feasible region of problem (13.1):

$$X = \{x \in \Omega : h_i(x) \leq 0, \; i = 1, \ldots, m_h\}.$$

Derivative-free algorithms (and in particular directional direct-search methods) can be applied not to f directly but to the *extreme barrier* function f_Ω or f_X, where f_S, for any S, is defined by

$$f_S(x) = \begin{cases} f(x) & \text{if } x \in S, \\ +\infty & \text{otherwise.} \end{cases} \qquad (13.2)$$

It is not necessary (in many of the existing approaches) to evaluate f at infeasible points. Rather, the value of the extreme barrier function is set to $+\infty$ at such points— and here we should recall that direct-search methods compare function values rather than building models. Clearly, such an approach could be inappropriate for methods based on interpolation or regression.

When all constraint derivatives are available

We now consider the case when $X = \Omega$; that is, there are no derivative-free constraints:

$$\begin{aligned} \min_{x \in \mathbb{R}^n} \quad & f(x) \\ \text{s.t.} \quad & x \in \Omega. \end{aligned} \qquad (13.3)$$

Directional direct-search methods for unconstrained optimization, described in Chapter 7, are directly applicable to the minimization of f_Ω. However, the extreme barrier technique cannot be applied using an arbitrary positive spanning set. In fact, a descent direction for the objective function (for instance, a direction that makes an acute angle with the negative gradient of a continuously differentiable function) may not be feasible. To guarantee global convergence in the constrained setting the directions chosen must therefore reflect properly the geometry of the feasible region near the current iterate.

When the constraints amount to simple bounds on the values of the variables,

$$\Omega = \{x \in \mathbb{R}^n : l \leq x \leq u\}, \qquad (13.4)$$

where $l \in (\{-\infty\} \cup \mathbb{R})^n$ and $u \in (\mathbb{R} \cup \{+\infty\})^n$, then (a subset of) the positive spanning set $D_\oplus$ (given in (7.1)) reflects adequately the feasible region near any feasible point. As a consequence of this simple fact, a directional direct-search method applied to (13.2) that includes $D_\oplus$ among the vectors used for polling is globally convergent for first-order stationary points when f is continuously differentiable (see, e.g., Lewis and Torczon [152]). A first-order stationary point for (13.3) when Ω is of the form given in (13.4) is a point $x \in \Omega$ such that $(\nabla f(x))_i = 0$ if $l_i < x_i < u_i$, $(\nabla f(x))_i \geq 0$ if $x_i = l_i$, and $(\nabla f(x))_i \leq 0$ if $x_i = u_i$.

To illustrate the simplicity of the resulting coordinate-search method we describe it formally in Algorithm 13.1. A search step could have been included, but it is omitted to keep the presentation simple.

Algorithm 13.1 (Coordinate-search method with bounds).

Initialization: Let $\Omega = \{x \in \mathbb{R}^n : l \leq x \leq u\}$ and $D_\oplus = [I - I] = [e_1 \cdots e_n - e_1 \cdots - e_n]$. Choose $x_0 \in \Omega$ and $\alpha_0 > 0$.

For $k = 0, 1, 2, \ldots$

1. **Poll step:** Order the poll set $P_k = \{x_k + \alpha_k d : d \in D_\oplus\}$. Start evaluating f_Ω at the poll points following the order determined. If a poll point $x_k + \alpha_k d_k$ is found such that $f_\Omega(x_k + \alpha_k d_k) < f_\Omega(x_k) = f(x_k)$, then stop polling, set $x_{k+1} = x_k + \alpha_k d_k$, and declare the iteration and the poll step successful. Otherwise, declare the iteration (and the poll step) unsuccessful and set $x_{k+1} = x_k$.

2. **Parameter update:** If the iteration was successful, set $\alpha_{k+1} = \alpha_k$ (or $\alpha_{k+1} = 2\alpha_k$). Otherwise, set $\alpha_{k+1} = \alpha_k/2$.

Note that in the simple-bounded case it is very easy to check if a point is outside Ω. In those cases one sets f_Ω to $+\infty$ right away, saving one evaluation of f.

Under the presence of more general constraints for which the derivatives are known, it becomes necessary to identify the set of active constraints, or, more precisely, the set of quasi-active constraints, in order to construct an appropriate set of poll positive generators. Suppose, for the purpose of the current discussion, that

$$\Omega = \left\{x \in \mathbb{R}^n : c_i(x) \leq 0, \ i = 1, \ldots, m_c\right\}. \tag{13.5}$$

(Equality constraints, when included in the original problem formulation, could be converted into two inequalities. However, such a procedure can introduce degeneracy and complicate the calculation of a feasible point. One alternative is to get rid of the equalities by eliminating some of the problem variables. Others are available for specific cases, like in the linearly constrained case [151].) Given a point $x \in \Omega$ and a parameter $\epsilon > 0$, the index set of the ϵ-active constraints is defined by $I(x;\epsilon) = \{i \in \{1, \ldots, m_c\} : c_i(x) \geq -\epsilon\}$. Note that when a constraint is linear ($c_i(x) = a_i^\top x + b_i$) it can be properly scaled so that it is ϵ-active at a point x if and only if the distance from x to the hyperplane $\{x \in \mathbb{R}^n : a_i^\top x + b_i = 0\}$ does not exceed ϵ (see [9]).

Given $x \in \Omega$, we call $N(x;\epsilon)$ the cone positively generated by the vectors $\nabla c_i(x)$ for $i \in I(x;\epsilon)$:

$$N(x;\epsilon) = \left\{\sum_{i \in I(x;\epsilon)} \lambda_i \nabla c_i(x) : \lambda_i \geq 0, i \in I(x;\epsilon)\right\}.$$

The polar cone $T(x;\epsilon) = N(x;\epsilon)^\circ$ is then defined by

$$T(x;\epsilon) = \left\{v \in \mathbb{R}^n : w^\top v \leq 0 \ \forall w \in N(x;\epsilon)\right\}.$$

For proper choices of ϵ and under a constraint qualification, $x + T(x;\epsilon)$ approximates well the local geometry of the feasible region near x (see, e.g., [178]). This property allows an algorithm to make feasible displacements from x along any direction chosen in $T(x;\epsilon)$. Note that in nonlinear programming, $T(x;0)$ and $N(x;0)$ are, respectively, the tangent and normal cones for Ω at x (again, under the presence of a constraint qualification such as, for instance, linearity of the active constraints or linear independence of the gradients of the functions defining the active constraints).

If at a given iteration k, $N(x_k; \epsilon) = \{0\}$, then $T(x_k; \epsilon) = \mathbb{R}^n$, which allows the problem to be locally seen as unconstrained. In the context of a directional direct-search method, such an occurrence suggests the use in the poll step of a positive spanning set for $\mathbb{R}^n$. In the case $N(x_k; \epsilon) \neq \{0\}$, the set of poll directions must reflect well the local feasible region near the iterate x_k and therefore must contain the positive generators of $T(x_k; \epsilon)$. Let us consider for simplicity only the nondegenerate case where $N(x_k; \epsilon)$ has a set of linear independent generators (denoted by the columns of the matrix N_k). Let us also assume that a full QR decomposition of the matrix N_k has been performed:

$$N_k = \left[\begin{array}{cc} Y_k & Z_k \end{array}\right] \left[\begin{array}{c} R_k \\ 0 \end{array}\right],$$

where Y_k forms an orthonormal basis for the range of N_k and Z_k forms an orthonormal basis for the null space of $N_k^\top$. The matrix R_k is upper triangular and nonsingular. Then the following is a set of positive generators for $T(x_k; \epsilon)$ (May [167]):

$$\left[\begin{array}{cccc} Z_k & -Z_k & Y_k R_k^{-\top} & -Y_k R_k^{-\top} \end{array}\right].$$

Note that $[Z_k \ -Z_k]$ positively spans the null space of $N_k^\top$. In fact, other positive bases or positive spanning sets of the null space of $N_k^\top$ could be used.

In the linearly constrained case, Lewis and Torczon [154] have shown that a construction of this type provides the positive generators for all the cones of the form $T(x_k; \varepsilon)$ for all $\varepsilon \in [0, \epsilon]$. In addition, they pointed out that if the linear algebra is performed via Gaussian elimination and N_k has rational entries, then the positive generators have rational entries too (which allows the consideration of integer lattices). The extension to equality and inequality linear constraints is considered in Lewis, Shepherd, and Torczon [151]. The case of linear degenerate constraints has been studied by Abramson et al. [9], Lewis, Shepherd, and Torczon [151], and Price and Coope [193]. The degenerate case poses additional computational difficulties when the number of nonredundant constraints is high.

In the linearly constrained case, a number of directional direct-search approaches have been investigated, in particular the following three:

- The set D_k contains positive generators for all the cones $T(x_k; \varepsilon)$, for all $\varepsilon \in [0, \epsilon_*]$, where $\epsilon_* > 0$ is independent of the iteration counter k, and infeasible poll points are dealt with by the extreme barrier function (see Lewis and Torczon [154]). In this case, global convergence can be attained by simple decrease with integer lattices.

- The set D_k is a positive generator set for the cone $T(x_k; \epsilon_k)$, the parameter ϵ_k is reduced at unsuccessful iterations, polling at feasible points is enforced by projection onto the feasible set, and a sufficient decrease condition is imposed to accept new iterates (see Algorithm 1 of Lucidi, Sciandrone, and Tseng [161]).

- The set D_k is a positive generator set for the cone $T(x_k; \epsilon_k)$, the parameter ϵ_k is set to be of $\mathcal{O}(\alpha_k)$ where α_k is the step size or mesh parameter, a sufficient decrease condition is imposed to accept new iterates, and infeasible poll points are dealt with by the extreme barrier function (see Algorithm 5.1 of Kolda, Lewis, and Torczon [147]).

The initial point must lie in Ω, which is relatively easy to enforce in the linearly constrained case. For all these algorithmic frameworks, it can be shown that the sequence of iterates generated has a limit point satisfying the first-order necessary conditions (the first-order Karush–Kuhn–Tucker conditions; see, for instance, [178]) for problem (13.3) when the constraints in (13.5) are linear ($c_i(x) = a_i^\top x + b_i$, $i = 1, \ldots, m_c$). These conditions are satisfied at a point $x \in \Omega$ if there exists a nonnegative vector $\lambda \in \mathbb{R}^{m_c}$ of Lagrange multipliers such that

$$\nabla f(x) + \sum_{i=1}^{m_c} \lambda_i \nabla c_i(x) = 0, \quad \lambda_i c_i(x) = 0, \; i = 1, \ldots, m_c.$$

Other approaches go back to the work by May [167], who extended Mifflin's algorithm [171] (see the notes in Chapter 9) to linearly constrained problems, proving global convergence to first- and second-order stationary points in the nondegenerate case. May explored simplex gradients and Hessians of the type considered in [171] along the positive generators of the tangent cones, proving convergence to first- and second-order stationary points.

The general nonlinear constrained case, for which the derivatives of the functions defining the constraints are known, has been studied by Lucidi, Sciandrone, and Tseng [161]. They suggested two globally convergent algorithms (one already mentioned above) with the following common features: (i) use of sufficient decrease for accepting new iterates; (ii) projection of poll points onto the feasible set (which could be expensive, especially when Ω is nonconvex); (iii) projection of the search direction onto a tangent cone of a nearby point; (iv) requirement of feasibility for the initial point; (v) permission to apply magical steps [57], i.e., any steps that are able to compute points for which the value of the objective function is less than or equal to the value of the accepted point. In the first algorithm, the set D_k is a positive generator set for the cone $T(x_k; \epsilon_k)$ and the parameter ϵ_k is reduced at unsuccessful iterations. In the second algorithm, D_k contains positive generators for all the cones $T(x_k; \varepsilon)$, for all $\varepsilon \in [0, \epsilon_*]$, where $\epsilon_* > 0$ is independent of the iteration counter k. The convergence theory requires some regularity of the constraints, which is satisfied, for instance, in the linearly constrained case or when the Mangasarian–Fromovitz constraint qualification (see, e.g., [178]) holds at every feasible point. For the second algorithm (apparently the more efficient of the two), the authors were able to prove that all limit points of the sequence of iterates satisfy the first-order necessary conditions.

Recently, Dreisigmeyer [81, 82, 84] has studied direct-search methods of directional and simplicial types for equality-constrained problems by treating the (twice continuously differentiable) constraints as implicitly defining a Riemannian manifold. In [83] he studied the case where no derivatives for the constraints are available by treating them implicitly as Lipschitz manifolds.

When derivative-free constraints are present

Now we turn our attention to the situation where there are derivative-free constraints of the type $h_i(x) \leq 0$, $i = 1, \ldots, m_h$. The directional direct-search approaches reported here treat derivative-free constraints as relaxable (unless stated otherwise).

Lewis and Torczon [155] suggested an approach based on an augmented Lagrangian method. The augmented Lagrangian function in their approach incorporates only the functions corresponding to nonlinear constraints (see [146, 155]). The augmented Lagrangian method then considers the solution of a sequence of subproblems where the augmented Lagrangian function is minimized subject to the remaining constraints (bounds on the variables or more general linear constraints). Original nonlinear inequality constraints must be converted into equalities by means of slack variables. Each problem can then be approximately solved using an appropriate directional direct-search method. This application of augmented Lagrangian methods yields global convergence results to first-order stationary points of the same type as those obtained under the presence of derivatives (see [146, 155]).

Liuzzi and Lucidi [157] developed and analyzed an algorithm for inequality constrained problems, based on a nonsmooth exact penalty function and the imposition of sufficient decrease. Linear constraints are handled separately by the use of positive generators. They have proved that a subset of the set of limit points of the sequence of iterates satisfies the first-order necessary conditions of the original problem.

We point out that is not clear how the augmented Lagrangian or exact penalty approaches can handle general unrelaxable constraints, other than the linear ones. However, these methods allow one to start infeasible with respect to the relaxable constraints.

Audet and Dennis [16] suggested using the filter technique within the framework of directional direct search to solve nonsmooth instances of (13.1). Filter methods were recently developed by Fletcher and Leyffer [96] (see the survey by Fletcher, Leyffer, and Toint [97]) as an efficient technique to globalize constrained nonlinear optimization algorithms. Roughly speaking, the filter methodology considers the original problem as a bi-objective optimization problem, which attempts to simultaneously minimize the objective function and a measure of infeasibility, but where the latter has some form of priority. The algorithm by Audet and Dennis [16] handles linear inequalities separately and applies the extreme barrier approach to all unrelaxable constraints. Using the nonsmooth analysis of Clarke [54], the authors have shown that their filter directional direct-search method generates a subsequence of iterates converging to a first-order stationary point of a modified problem, possibly different from the original nonsmooth problem when the derivatives of the constraints cannot be used. This discrepancy resulted from using a finite number of positive spanning sets. In the special case where the functions defining the constraints are continuously differentiable, Dennis, Price, and Coope [75] considered the use of an infinite number of positive spanning sets for polling. Their filter approach guarantees global convergence to a first-order stationary point of the original problem under strict differentiability of the objective function at the limit point. Here the concept of an envelope around the filter is used as means of measuring sufficient decrease.

In the context of directional direct search, one of the most general approaches for the case where the derivatives of the function defining the constraints are unavailable is the mesh adaptive direct search (MADS) method of Audet and Dennis [19], already described in Section 7.6 for unconstrained optimization, which uses sets of poll directions whose union is asymptotically dense in $\mathbb{R}^n$. This algorithm is applicable to problems with general derivative-free constraints, including the situation where constraints are hidden. MADS considers all constraints as unrelaxable and requires a feasible starting point. Infeasibility is ruled out by means of the extreme barrier for the whole feasible set. The theory guarantees a limit point satisfying the first-order necessary conditions for the original (possibly nonsmooth) problem. A more recent approach by these authors is called mesh adaptive direct search with a progressive barrier (see [20]), exhibiting global convergence properties similar to MADS. It allows the handling of both types of constraints, by combining MADS techniques and the extreme barrier for unrelaxable constraints with nondominance filter-

248 Chapter 13. Constrained and other extensions to derivative-free optimization

type concepts for the relaxable constraints. An interesting feature is that a constraint can be considered relaxable until it becomes feasible, whereupon it is transferred to the set of unrelaxable constraints.

We also point out that asynchronous parallel directional direct-search methods based on the generating set search framework [145] have been proposed by Griffin, Kolda, and Lewis [114] for linearly constrained optimization and by Griffin and Kolda [113] for more general nonlinear derivative-free optimization problems.

13.2 Trust-region interpolation-based methods

Most of the analysis of derivative-free methods for constrained optimization has been done within the framework of directional direct search. Several practical approaches were developed to handle constraints by trust-region interpolation-based methods (outlined in Chapters 10 and 11 for unconstrained optimization). For simplicity we assume in this section that all derivative-free constraints are relaxable and all other constraints are not relaxable, which is often the case in practice.

Conceptually speaking, the extension of these methods to handle constraints is relatively straightforward for the unrelaxable constraints with available derivatives ($x \in \Omega$). The simplest approach is to intersect Ω with the trust region $B(x_k; \Delta_k)$ and generate only feasible sample points. Depending on the shape of Ω, the corresponding trust-region subproblem may become difficult. It may become especially challenging to perform global optimization of Lagrange polynomials, as discussed in Chapters 6 and 11. In general, maintaining a Λ-poised sample set needs additional care in the presence of constraints. However, when Ω is defined by linear or box constraints, the theory should be reasonably easy to adapt.

In practice the approach of simply using a local solution to the constrained trust-region subproblem is often sufficient [61]. A careful practical approach to the constrained trust-region subproblem was developed in [33, 34]. No specific convergent theory exists so far for the case of unrelaxable constraints with available derivatives.

The case of derivative-free relaxable constraints is treated differently by the existing model-based methods. First, we notice that relaxable constraints have no effect on the issue of poisedness, since a poised set can be computed without any consideration of relaxable constraints. Second, the simple idea of combining the objective function and the relaxable constraints into a merit function (such as a penalty function) allows the convergence theory for the unconstrained case to apply directly. Moreover, this simple minded approach is often very useful when the derivative-free constraint functions h_i are in fact additional objective functions and the user has good intuition about how to combine f and h_i into a sensible penalty function. However, such an approach may not generate a truly feasible solution, even asymptotically.

A perhaps more theoretically sound approach is to use, for instance, a sequential quadratic programming (SQP) framework. In SQP one has to build quadratic models for the Lagrangian function (and thus one needs quadratic models for the functions defining the constraints). Such models are then minimized within trust regions and subject to some form of models of the constraints. The models of the constraints can be linear, which makes the trust-region subproblem easier to solve, or they can be quadratic, if a quadratic model is used for f. In this case the trust-region subproblem becomes significantly harder, but the constraint representation becomes more accurate. For expensive function evaluations this may be a reasonable trade off.

It is important to notice that quadratic models of the constraints often are available at little additional cost. It is typically the case in practical applications that the values of the constraint functions are computed by the same black-box call as the value of the objective function. This means that if we have a model of f based on a Λ-poised set Y and we have the set of Lagrange polynomials for Y, then we also have the values of all of the constraint functions at all points of Y and we can compute the quadratic models of the constraints by using the fundamental property of Lagrange polynomials, described in Lemma 3.5.

The significant additional cost of using quadratic models of the constraints may come from the trust-region subproblem. Depending on the underlying algorithm, the shape of the constraints, and the cost of the function evaluations, this may or may not be acceptable. Also, since an iterate might be infeasible with respect to the models of the derivative-free constraints, the constrained trust-region subproblem may be infeasible. When an iterate is infeasible with respect to the models of the derivative-free constraints, a model merit function can be optimized or a model filter methodology employed. Whatever is used has to be consistent with the approach used for the original problem.

The trust-region interpolation-based code of Scheinberg (see [59, 61]), called DFO, is based on the type of approach described above. DFO allows trust-region subproblems to include quadratic models of the derivative-free constraints. Globalization is handled by using f as the merit function. Powell [186] suggested a method that models the objective and constraint functions by linear interpolation (on which his COBYLA software package was based). The CONDOR code of Berghen [33] (see also [34]) is based on the UOBYQA algorithm of Powell [189] and also handles constraints. In his PhD thesis, Colson [55] applied the filter technique to avoid merit functions and management of penalty parameters. His trust-region interpolation-based method follows the trust-region SQP filter algorithm suggested and analyzed in [95], where each SQP step is decomposed into normal and tangential components. Another approach we are aware of is by Brekelmans et al. [46]. Their algorithm framework consists of solving a sequence of nonlinear (not necessarily quadratic) trust-region subproblems, building linear interpolation models for all functions whose derivatives are unavailable. The globalization strategy is based on the filter mechanism. Currently, there is no convergence theory developed for trust-region interpolation-based methods.

13.3   Derivative-free approaches for global optimization, mixed-integer programming, and other problems

Directional direct-search methods have been adapted for derivative-free global optimization [13, 123, 222]. In fact, it is an attractive idea to use the search step of the search-poll framework of Algorithm 7.2 to incorporate a dissemination method or heuristic for global optimization purposes, since such schemes could provide a wider exploration of the variable domain or feasible region but would not disturb the convergence theory. Vaz and Vicente [222] selected a particle swarm method for this purpose. The resulting algorithm has been shown to be competitive for bound-constrained problems (in terms of both efficiency and robustness) when compared to other global optimization solvers. In this case, the poll step acts at the best particle and contributes to the efficiency of the overall method by allowing the particles to be dropped out as the search focuses around a point. Audet, Béchard, and Le Digabel [13] followed a similar idea, incorporating a variable neighborhood search in the search step of MADS [19] to better explore the feasible region in terms of global optimality. Pattern search methods have also been combined with evolutionary techniques (see the work by Hart [119, 120]), and the resulting evolutionary pattern search method compared favorably against evolutionary algorithms.

A new approach to global optimization of derivative-free unconstrained problems in the presence of noise has been developed by Ferris and Deng in [92]. They select promising areas from which to start local derivative-free search by applying ideas from classification used in machine learning.

DIviding RECTangles (DIRECT) is an optimization algorithm designed to search for global minima of a function over a bound-constrained domain, motivated by a modification to Lipschitzian optimization and proposed by Jones, Perttunen, and Stuckman [136]. The algorithm does not use derivatives of the objective function, and it relies on the iteration history to determine future sample locations. In each iteration, DIRECT first tries to identify hyperrectangles that have the potential to contain a global minimizer and then divides such hyperrectangles into smaller ones. The objective function is evaluated at the centers of the new hyperrectangles. DIRECT tries to balance the global and local searches. See also [91, 93, 101, 102].

Multilevel coordinate search (MCS) is another approach developed for the global optimization of a function in a bound-constrained domain without derivatives. MCS was developed by Huyer and Neumaier [134] and is also inspired in the DIRECT method [136]. The algorithm contains local enhancements based on the minimization of quadratic interpolation models.

Derivative-free global optimization methods based on radial basis functions have also been proposed by several authors (see [36, 117, 138] and the sequence of papers by Regis and Shoemaker [196, 197, 198, 199] which includes extensions to the constrained and parallel cases).

Finally, we would like to point out that directional direct-search methods have also been applied to mixed integer variable problems, with or without constraints, by Audet et al. in [4, 5, 17], to linearly constrained finite minimax problems, by Liuzzi, Lucidi, and Sciandrone [158], and to multiobjective optimization, by Audet, Savard, and Zghal [23].

Appendix

Software for derivative-free optimization

We list below a number of software packages developed for derivative-free optimization.

Chapter 1: Introduction

Benchmarking Derivative-Free Optimization Algorithms
`http://www.mcs.anl.gov/~more/dfo`

Chapter 7: Directional direct-search methods

APPSPACK
Asynchronous parallel pattern search
`http://software.sandia.gov/appspack`

Iterative Methods for Optimization: MATLAB® Codes
Hooke–Jeeves and multidirectional search methods
`http://www4.ncsu.edu/~ctk/matlab_darts.html`

The Matrix Computation Toolbox
Multidirectional search and alternating directions methods
`http://www.maths.manchester.ac.uk/~higham/mctoolbox`

NOMAD
Generalized pattern search and mesh adaptive direct search
`http://www.gerad.ca/NOMAD`
`http://www.afit.edu/en/ENC/Faculty/MAbramson/NOMADm.html`

SID-PSM
Generalized pattern search guided by simplex derivatives
`http://www.mat.uc.pt/sid-psm`

Chapter 8: Simplicial direct-search methods

`fminsearch`
MATLAB implementation of the Nelder–Mead method
`http://www.mathworks.com/access/helpdesk/help/techdoc/ref/`
`fminsearch.html`

Iterative Methods for Optimization: MATLAB Codes
Nelder–Mead method
`http://www4.ncsu.edu/~ctk/matlab_darts.html`

The Matrix Computation Toolbox
Nelder–Mead method
`http://www.maths.manchester.ac.uk/~higham/mctoolbox`

Chapter 9: Line-search methods based on simplex derivatives

Implicit Filtering
Implicit filtering method
`http://www4.ncsu.edu/~ctk/iffco.html`

Chapter 11: Trust-region interpolation-based methods

BOOSTERS
Trust-region interpolation-based method (based on radial basis functions)
`http://roso.epfl.ch/rodrigue/boosters.htm`

CONDOR
Trust-region interpolation-based method (version of UOBYQA in parallel)
`http://www.applied-mathematics.net/optimization/`
`CONDORdownload.html`

DFO
Trust-region interpolation-based method (see Section 11.2)
`http://www.coin-or.org/projects.html`

ORBIT
Trust-region interpolation-based method (based on radial basis functions)
`http://www.mcs.anl.gov/~wild/orbit`

UOBYQA, NEWUOA
Trust-region interpolation-based methods (see Section 11.3)
`mjdp@cam.ac.uk`

WEDGE
Trust-region interpolation-based method (see Section 11.4)
`http://www.ece.northwestern.edu/~nocedal/wedge.html`

Chapter 12: Review of surrogate model management

DACE
Design and analysis of computer experiments
`http://www2.imm.dtu.dk/~hbn/dace`

Chapter 13: Review of constrained and other extensions to derivative-free optimization

Section 13.1: Directional direct-search methods; and Section 13.2: Trust-region interpolation-based methods

The software packages APPSPACK, CONDOR, DFO, and NOMAD also deal with constrained derivative-free optimization. SID-PSM solves constrained problems too but requires derivatives for the constraints.

Section 13.3: Derivative-free approaches for global optimization, mixed-integer programming, and other problems

DIRECT
DIRECT – A Global Optimization Algorithm
`http://www4.ncsu.edu/~ctk/Finkel_Direct`

MATLAB Toolbox 2.4, The MathWorks™
Genetic Algorithm and Direct Search Toolbox 2.4
`http://www.mathworks.com/products/gads`

MCS
Global optimization by multilevel coordinate search
`http://www.mat.univie.ac.at/~neum/software/mcs/`

PSwarm
Coordinate search and particle swarm for global optimization (including a parallel version)
`http://www.norg.uminho.pt/aivaz/pswarm`

Bibliography

[1] *MATLAB®, The MathWorks^{TM}*.

[2] *Powerspice simulates circuits faster and more accurately*, Electronics, (1985), pp. 50–51.

[3] M. A. ABRAMSON AND C. AUDET, *Convergence of mesh adaptive direct search to second-order stationary points*, SIAM J. Optim., 17 (2006), pp. 606–619.

[4] M. A. ABRAMSON, C. AUDET, J. W. CHRISSIS, AND J. G. WALSTON, *Mesh adaptive direct search algorithms for mixed variable optimization*, Optim. Lett., (2008, to appear).

[5] M. A. ABRAMSON, C. AUDET, AND J. E. DENNIS, JR., *Filter pattern search algorithms for mixed variable constrained optimization problems*, Pac. J. Optim., 3 (2004), pp. 477–500.

[6] ———, *Generalized pattern searches with derivative information*, Math. Program., 100 (2004), pp. 3–25.

[7] ———, *Blackbox Optimization*, to appear.

[8] M. A. ABRAMSON, C. AUDET, J. E. DENNIS, JR., AND S. LE DIGABEL, *OrthoMADS: A deterministic MADS instance with orthogonal directions*, Tech. Report G-2008-15, GERAD, École Polytechnique de Montréal, Canada, 2008.

[9] M. A. ABRAMSON, O. A. BREZHNEVA, J. E. DENNIS, JR., AND R. L. PINGEL, *Pattern search in the presence of degenerate linear constraints*, Optim. Methods Softw., 23 (2008), pp. 297–319.

[10] P. ALBERTO, F. NOGUEIRA, H. ROCHA, AND L. N. VICENTE, *Pattern search methods for user-provided points: Application to molecular geometry problems*, SIAM J. Optim., 14 (2004), pp. 1216–1236.

[11] N. M. ALEXANDROV, J. E. DENNIS, R. M. LEWIS, AND V. TORCZON, *A trust region framework for managing the use of approximation models in optimization*, Structural Optimization, 15 (1998), pp. 16–23.

[12] C. AUDET, *Convergence results for pattern search algorithms are tight*, Optim. Eng., 5 (2003), pp. 101–122.

255

[13] C. AUDET, V. BÉCHARD, AND S. LE DIGABEL, *Nonsmooth optimization through mesh adaptive direct search and variable neighborhood search*, J. Global Optim., 41 (2008), pp. 299–318.

[14] C. AUDET, A. J. BOOKER, J. E. DENNIS, JR., P. D. FRANK, AND D. W. MOORE, *A surrogate-model-based method for constrained optimization*, in AIAA Paper 2000-4891, AIAA/USAF/NASA/ISSMO Symposium on Multidisciplinary Analysis and Optimization, September 2000.

[15] C. AUDET, A. L. CUSTÓDIO, AND J. E. DENNIS, JR., *Erratum: Mesh adaptive direct search algorithms for constrained optimization*, SIAM J. Optim., 18 (2008), pp. 1501–1503.

[16] C. AUDET AND J. E. DENNIS, JR., *A pattern search filter method for nonlinear programming without derivatives*, SIAM J. Optim., 14 (2004), pp. 980–1010.

[17] ——, *Pattern search algorithms for mixed variable programming*, SIAM J. Optim., 11 (2000), pp. 573–594.

[18] ——, *Analysis of generalized pattern searches*, SIAM J. Optim., 13 (2003), pp. 889–903.

[19] ——, *Mesh adaptive direct search algorithms for constrained optimization*, SIAM J. Optim., 17 (2006), pp. 188–217.

[20] ——, *A MADS algorithm with a progressive barrier for derivative-free nonlinear programming*, Tech. Report G-2007-37, GERAD, École Polytechnique de Montréal, Canada, 2007.

[21] C. AUDET, J. E. DENNIS, JR., AND S. LE DIGABEL, *Parallel space decomposition of the mesh adaptive direct search algorithm*, Tech. Report G-2007-81, GERAD, École Polytechnique de Montréal, Canada, 2007.

[22] C. AUDET AND D. ORBAN, *Finding optimal algorithmic parameters using derivative-free optimization*, SIAM J. Optim., 17 (2006), pp. 642–664.

[23] C. AUDET, G. SAVARD, AND W. ZGHAL, *Multiobjective optimization through a series of single-objective formulations*, SIAM J. Optim., 19 (2008), pp. 188–210.

[24] A. M. BAGIROV, B. KARASÖZEN, AND M. SEZER, *Discrete gradient method: Derivative-free method for nonsmooth optimization*, J. Optim. Theory Appl., 137 (2008), pp. 317–334.

[25] M. H. BAKR, J. W. BANDLER, R. M. BIERNACKI, S. H. CHEN, AND K. MADSEN, *A trust region aggressive space mapping algorithm for EM optimization*, IEEE Trans. Microwave Theory Tech., 46 (1998), pp. 2412–2425.

[26] M. H. BAKR, J. W. BANDLER, K. MADSEN, AND J. SØNDERGAARD, *Review of the space mapping approach to engineering optimization and modeling*, Optim. Eng., 1 (2000), pp. 241–276.

[27] K. BALL, N. SIVAKUMAR, AND J. D. WARD, *On the sensitivity of radial basis interpolation to minimal data separation distance*, Constr. Approx., 8 (1992), pp. 401–426.

[28] J. W. BANDLER, R. M. BIERNACKI, S. H. CHEN, P. A. GROBELNY, AND R. H. HEMMERS, *Space mapping technique for electromagnetic optimization*, IEEE Trans. Microwave Theory Tech., 42 (1994), **pp.** 2536–2544.

[29] J. W. BANDLER, R. M. BIERNACKI, S. H. CHEN, R. H. HEMMERS, AND K. MADSEN, *Electromagnetic optimization exploiting aggressive space mapping*, IEEE Trans. Microwave Theory Tech., 43 (1995), pp. 2874–2882.

[30] J. W. BANDLER, S. KOZIEL, AND K. MADSEN, *Space mapping for engineering optimization*, SIAM SIAG/OPT Views-and-News, 17 (1) (2006), pp. 19–26.

[31] J. W. BANDLER AND K. MADSEN, *Editorial—surrogate modelling and space mapping for engineering optimization*, Optim. Eng., 2 (2002), pp. 367–368.

[32] A. BECK AND M. TEBOULLE, *A convex optimization approach for minimizing the ratio of indefinite quadratic functions over an ellipsoid*, Math. Program., (2008, to appear).

[33] F. V. BERGHEN, *CONDOR: A Constrained, Non-Linear, Derivative-Free Parallel Optimizer for Continuous, High Computing Load, Noisy Objective Functions*, PhD thesis, Université Libre de Bruxelles, Belgium, 2004.

[34] F. V. BERGHEN AND H. BERSINI, *CONDOR, a new parallel, constrained extension of Powell's UOBYQA algorithm: Experimental results and comparison with the DFO algorithm*, J. Comput. Appl. Math., 181 (2005), pp. 157–175.

[35] G. BERMAN, *Lattice approximations to the minima of functions of several variables*, J. ACM, 16 (1969), pp. 286–294.

[36] M. BJÖRKMAN AND K. HOLMSTRÖM, *Global optimization of costly nonconvex functions using radial basis functions*, Optim. Eng., 1 (2000), pp. 373–397.

[37] C. BOGANI, M. G. GASPARO, AND A. PAPINI, *Generalized pattern search methods for a class of nonsmooth problems with structure*, Preprint, 2006.

[38] A. J. BOOKER, J. E. DENNIS, JR., P. D. FRANK, D. W. MOORE, AND D. B. SERAFINI, *Managing surrogate objectives to optimize a helicopter rotor design—further experiments*, in AIAA Paper 1998–4717, 8th AIAA/ISSMO Symposium on Multidisciplinary Analysis and Optimization, St. Louis, MO, 1998.

[39] A. J. BOOKER, J. E. DENNIS, JR., P. D. FRANK, D. B. SERAFINI, AND V. TORCZON, *Optimization using surrogate objectives on a helicopter test example*, in Computational Methods for Optimal Design and Control, J. T. Borggaard, J. Burns, E. Cliff, and S. Schreck, eds., Birkhäuser, Boston, 1998, pp. 49–58.

[40] A. J. BOOKER, J. E. DENNIS, JR., P. D. FRANK, D. B. SERAFINI, V. TORCZON, AND M. W. TROSSET, *A rigorous framework for optimization of expensive functions by surrogates*, Struct. Multidiscip. Optim., 17 (1998), pp. 1–13.

[41] D. M. BORTZ AND C. T. KELLEY, *The simplex gradient and noisy optimization problems*, in Computational Methods in Optimal Design and Control, J. T. Borggaard, J. Burns, E. Cliff, and S. Schreck, eds., vol. 24 of Progr. Systems Control Theory, Birkhäuser, Boston, 1998, pp. 77–90.

[42] G. E. P. BOX, *Evolutionary operation: A method for increasing industrial productivity*, Appl. Statist., 6 (1957), pp. 81–101.

[43] G. E. P. BOX AND N. R. DRAPER, *Empirical Model-Building and Response Surfaces*, John Wiley & Sons, New York, 1987.

[44] G. E. P. BOX AND K. B. WILSON, *On the experimental attainment of optimum conditions*, J. Roy. Statist. Soc. Ser. B., 13 (1951), pp. 1–45.

[45] M. J. BOX, *A new method for constrained optimization and a comparison with other methods*, Comput. J., 8 (1965), pp. 42–52.

[46] R. BREKELMANS, L. DRIESSEN, H. HAMERS, AND D. DEN HERTOG, *Constrained optimization involving expensive function evaluations: A sequential approach*, European J. Oper. Res., 160 (2005), pp. 121–138.

[47] R. P. BRENT, *Algorithms for Minimization without Derivatives*, Prentice–Hall, Englewood Cliffs, NJ, 1973. Reissued by Dover Publications, Mineola, NY, 2002.

[48] A. G. BUCKLEY AND H. MA, *A derivative-free algorithm for parallel and sequential optimization*, Tech. Report, Computer Science Department, University of Victoria, Canada, 1994.

[49] M. D. BUHMANN, *Radial basis functions: The state-of-the-art and new results*, Acta Numer., 9 (2000), pp. 1–37.

[50] ———, *Radial Basis Functions*, Cambridge University Press, Cambridge, UK, 2003.

[51] R. G. CARTER, *On the global convergence of trust region algorithms using inexact gradient information*, SIAM J. Numer. Anal., 28 (1991), pp. 251–265.

[52] T. D. CHOI AND C. T. KELLEY, *Superlinear convergence and implicit filtering*, SIAM J. Optim., 10 (2000), pp. 1149–1162.

[53] P. G. CIARLET AND P. A. RAVIART, *General Lagrange and Hermite interpolation in R^n with applications to finite element methods*, Arch. Ration. Mech. Anal., 46 (1972), pp. 177–199.

[54] F. H. CLARKE, *Optimization and Nonsmooth Analysis*, John Wiley & Sons, New York, 1983. Reissued by SIAM, Philadelphia, 1990.

[55] B. COLSON, *Trust-Region Algorithms for Derivative-Free Optimization and Nonlinear Bilevel Programming*, PhD thesis, Département de Mathématique, FUNDP, Namur, Belgium, 2003.

[56] B. COLSON AND PH. L. TOINT, *Optimizing partially separable functions without derivatives*, Optim. Methods Softw., 20 (2005), pp. 493–508.

[57] A. R. CONN, N. I. M. GOULD, AND PH. L. TOINT, *Trust-Region Methods*, MPS-SIAM Series on Optimization, SIAM, Philadelphia, 2000.

[58] A. R. CONN AND PH. L. TOINT, *An algorithm using quadratic interpolation for unconstrained derivative free optimization*, in Nonlinear Optimization and Applications, G. D. Pillo and F. Gianessi, eds., Plenum, New York, 1996, pp. 27–47.

[59] A. R. CONN, K. SCHEINBERG, AND PH. L. TOINT, *On the convergence of derivative-free methods for unconstrained optimization*, in Approximation Theory and Optimization, Tributes to M. J. D. Powell, M. D. Buhmann and A. Iserles, eds., Cambridge University Press, Cambridge, UK, 1997, pp. 83–108.

[60] ——, *Recent progress in unconstrained nonlinear optimization without derivatives*, Math. Program., 79 (1997), pp. 397–414.

[61] ——, *A derivative free optimization algorithm in practice*, in Proceedings of the 7th AIAA/USAF/NASA/ISSMO Symposium on Multidisciplinary Analysis and Optimization, St. Louis, MO, September 1998.

[62] A. R. CONN, K. SCHEINBERG, AND L. N. VICENTE, *Global convergence of general derivative-free trust-region algorithms to first and second order critical points*, SIAM J. Optim., (2009, to appear).

[63] ——, *Geometry of interpolation sets in derivative free optimization*, Math. Program., 111 (2008), pp. 141–172.

[64] ——, *Geometry of sample sets in derivative-free optimization: Polynomial regression and underdetermined interpolation*, IMA J. Numer. Anal., 28 (2008), pp. 721–748.

[65] I. D. COOPE AND C. J. PRICE, *Frame based methods for unconstrained optimization*, J. Optim. Theory Appl., 107 (2000), pp. 261–274.

[66] ——, *On the convergence of grid-based methods for unconstrained optimization*, SIAM J. Optim., 11 (2001), pp. 859–869.

[67] C. CURRIN, T. MITCHELL, M. MORRIS, AND D. YLVISAKER, *A Bayesian approach to the design and analysis of computer experiments*, Tech. Report ORNL-6498, Oak Ridge National Laboratory, Oak Ridge, TN, 1988.

[68] A. L. CUSTÓDIO, J. E. DENNIS, JR., AND L. N. VICENTE, *Using simplex gradients of nonsmooth functions in direct search methods*, IMA J. Numer. Anal., 28 (2008), pp. 770–784.

[69] A. L. CUSTÓDIO, H. ROCHA, AND L. N. VICENTE, *Incorporating minimum Frobenius norm models in direct search*, Tech. Report 08-51, Departamento de Matemática, Universidade de Coimbra, Portugal, 2008.

[70] A. L. CUSTÓDIO AND L. N. VICENTE, *Using sampling and simplex derivatives in pattern search methods*, SIAM J. Optim., 18 (2007), pp. 537–555.

[71] A. P. DAMBRAUSKAS, *The simplex optimization method with variable step*, Engrg. Cybernet., 1 (1970), pp. 28–36.

[72] I. DAUBECHIES, *Ten Lectures on Wavelets*, SIAM, Philadelphia, 1992.

[73] W. C. DAVIDON, *Variable metric method for minimization*, SIAM J. Optim., 1 (1991), pp. 1–17.

[74] C. DAVIS, *Theory of positive linear dependence*, Amer. J. Math., 76 (1954), pp. 733–746.

[75] J. E. DENNIS, JR., C. J. PRICE, AND I. D. COOPE, *Direct search methods for nonlinearly constrained optimization using filters and frames*, Optim. Eng., 5 (2004), pp. 123–144.

[76] J. E. DENNIS, JR., AND R. B. SCHNABEL, *Numerical Methods for Unconstrained Optimization and Nonlinear Equations*, Prentice–Hall, Englewood Cliffs, NJ, 1983. Reissued by SIAM, Philadelphia, 1996.

[77] J. E. DENNIS, JR., AND V. TORCZON, *Direct search methods on parallel machines*, SIAM J. Optim., 1 (1991), pp. 448–474.

[78] J. E. DENNIS, JR., AND D. J. WOODS, *Optimization on microcomputers: The Nelder–Mead simplex algorithm*, in New Computing Environments: Microcomputers in Large-Scale Computing, A. Wouk, ed., SIAM, Philadelphia, 1987, pp. 116–122.

[79] M. A. DINIZ-EHRHARDT, J. M. MARTÍNEZ, AND M. RAYDAN, *A derivative-free nonmonotone line search technique for unconstrained optimization*, J. Comput. Appl. Math., 219 (2008), pp. 383–397.

[80] E. D. DOLAN, R. M. LEWIS, AND V. TORCZON, *On the local convergence of pattern search*, SIAM J. Optim., 14 (2003), pp. 567–583.

[81] D. W. DREISIGMEYER, *Direct search algorithms over Riemannian manifolds*, Tech. Report LA-UR-06-7416, Los Alamos National Laboratory, Los Alamos, NM, 2006.

[82] ——, *Equality constraints, Riemannian manifolds and direct search methods*, Tech. Report LA-UR-06-7406, Los Alamos National Laboratory, Los Alamos, NM, 2006.

[83] ——, *Direct search algorithms over Lipschitz manifolds*, Tech. Report LA-UR-07-1073, Los Alamos National Laboratory, Los Alamos, NM, 2007.

[84] ——, *A simplicial continuation direct search method*, Tech. Report LA-UR-07-2755, Los Alamos National Laboratory, Los Alamos, NM, 2007.

[85] L. DRIESSEN, R. C. M. BREKELMANS, H. HAMERS, AND D. DEN HERTOG, *On D-optimality based trust regions for black-box optimization problems*, Struct. Multidiscip. Optim., 31 (2006), pp. 40–48.

[86] J. DUCHON, *Splines minimizing rotation-invariant semi-norms in Sobolev spaces*, in Constructive Theory of Functions of Several Variables, W. Schempp and K. Zeller, eds., Springer-Verlag, Berlin, 1977, pp. 85–100.

[87] R. DUVIGNEAU AND M. VISONNEAU, *Hydrodynamic design using a derivative-free method*, Struct. Multidiscip. Optim., 28 (2004), pp. 195–205.

[88] C. ELSTER AND A. NEUMAIER, *A grid algorithm for bound constrained optimization of noisy functions*, IMA J. Numer. Anal., 15 (1995), pp. 585–608.

[89] G. FASANO, J. L. MORALES, AND J. NOCEDAL, *On the geometry phase in model-based algorithms for derivative-free optimization*, Tech. Report, Optimization Center, Northwestern University, Evanston, IL, 2008.

[90] E. FERMI AND N. METROPOLIS, Tech. Report, Los Alamos Unclassified Report LA–1492, Los Alamos National Laboratory, Los Alamos, NM, 1952.

[91] M. C. FERRIS AND G. DENG, *Extension of the DIRECT optimization algorithm for noisy functions*, in Proceedings of the 2007 Winter Simulation Conference, B. Biller, S. Henderson, M. Hsieh, and J. Shortle, eds., Washington, D.C., 2007.

[92] ———, *Classification-based global search: An application to a simulation for breast cancer*, in Proceedings of the 2008 NSF Engineering Research and Innovation Conference, Knoxville, TN, 2008.

[93] D. E. FINKEL AND C. T. KELLEY, *Additive scaling and the DIRECT algorithm*, J. Global Optim., 36 (2006), pp. 597–608.

[94] R. FLETCHER, *Function minimization without derivatives—a review*, Comput. J., 8 (1965), pp. 33–41.

[95] R. FLETCHER, N. I. M. GOULD, S. LEYFFER, PH. L. TOINT, AND A. WÄCHTER, *Global convergence of trust-region SQP-filter algorithms for general nonlinear programming*, SIAM J. Optim., 13 (2002), pp. 635–659.

[96] R. FLETCHER AND S. LEYFFER, *Nonlinear programming without a penalty function*, Math. Program., 91 (2002), pp. 239–269.

[97] R. FLETCHER, S. LEYFFER, AND PH. L. TOINT, *A brief history of filter methods*, SIAM SIAG/OPT Views-and-News, 18 (1) (2006), pp. 2–12.

[98] C. FORTIN AND H. WOLKOWICZ, *The trust region subproblem and semidefinite programming*, Optim. Methods Softw., 19 (2004), pp. 41–67.

[99] K. R. FOWLER, J. P. REESE, C. E. KEES, J. E. DENNIS, JR., C. T. KELLEY, C. T. MILLER, C. AUDET, A. J. BOOKER, G. COUTURE, R. W. DARWIN, M. W. FARTHING, D. E. FINKEL, J. M. GABLONSKY, G. GRAY, AND T. G. KOLDA, *A comparison of derivative-free optimization methods for groundwater supply and hydraulic capture community problems*, Adv. in Water Resources, 31 (2007), pp. 743–757.

[100] L. FRIMANNSLUND AND T. STEIHAUG, *A generating set search method using curvature information*, Comput. Optim. Appl., 38 (2007), pp. 105–121.

[101] J. M. GABLONSKY, *Modifications of the DIRECT Algorithm*, PhD thesis, North Carolina State University, Raleigh, NC, 2001.

[102] J. M. GABLONSKY AND C. T. KELLEY, *A locally-biased form of the DIRECT algorithm*, J. Global Optim., 21 (2001), pp. 27–37.

[103] U. M. GARCÍA-PALOMARES AND J. F. RODRÍGUEZ, *New sequential and parallel derivative-free algorithms for unconstrained optimization*, SIAM J. Optim., 13 (2002), pp. 79–96.

[104] P. E. GILL, W. MURRAY, AND M. H. WRIGHT, *Practical Optimization*, Academic Press, San Diego, 1981.

[105] P. GILMORE AND C. T. KELLEY, *An implicit filtering algorithm for optimization of functions with many local minima*, SIAM J. Optim., 5 (1995), pp. 269–285.

[106] T. GLAD AND A. GOLDSTEIN, *Optimization of functions whose values are subject to small errors*, BIT, 17 (1977), pp. 160–169.

[107] F. GLOVER, *Tabu search—Part I*, ORSA J. Comput., 1 (1989), pp. 190–206.

[108] D. E. GOLDBERG, *Genetic Algorithms in Search, Optimization and Machine Learning*, Addison–Wesley Longman, Boston, 1989.

[109] G. H. GOLUB AND C. F. VAN LOAN, *Matrix Computations*, The John Hopkins University Press, Baltimore, London, third ed., 1996.

[110] G. A. GRAY AND T. G. KOLDA, *Algorithm 856: APPSPACK 4.0: Asynchronous parallel pattern search for derivative-free optimization*, ACM Trans. Math. Software, 32 (2006), pp. 485–507.

[111] A. GRIEWANK, *Evaluating Derivatives: Principles and Techniques of Algorithmic Differentiation*, SIAM, Philadelphia, 2000.

[112] A. GRIEWANK AND G. F. CORLISS, EDS., *Automatic Differentiation of Algorithms: Theory, Implementation, and Application*, SIAM, Philadelphia, 1991.

[113] J. D. GRIFFIN AND T. G. KOLDA, *Nonlinearly-constrained optimization using asynchronous parallel generating set search*, Tech. Report SAND2007-3257, Sandia National Laboratories, Albuqurque, NM, Livermore, CA, 2007.

[114] J. D. GRIFFIN, T. G. KOLDA, AND R. M. LEWIS, *Asynchronous parallel generating set search for linearly constrained optimization*, SIAM J. Sci. Comput., 30 (2008), pp. 1892–1924.

[115] L. GRIPPO, F. LAMPARIELLO, AND S. LUCIDI, *Global convergence and stabilization of unconstrained minimization methods without derivatives*, J. Optim. Theory Appl., 56 (1998), pp. 385–406.

[116] D. N. GUJARATI, *Basic Econometrics*, McGraw–Hill International Editions, Singapore, third ed., 1995.

[117] H.-M. GUTMANN, *A radial basis function method for global optimization*, J. Global Optim., 19 (2001), pp. 201–227.

[118] R. T. HAFTKA, *Combining global and local approximations*, AIAA J., 29 (1991), pp. 1523–1525.

[119] W. E. HART, *A generalized stationary point convergence theory for evolutionary algorithms*, in Proceedings of the 7th International Conference on Genetic Algorithms, San Francisco, 1997, pp. 127–134.

[120] W. E. HART, *Comparing evolutionary programs and evolutionary pattern search algorithms: A drug docking application*, in Proceedings of the Genetic and Evolutionary Computation Conference, Orlando, FL, 1999, pp. 855–862.

[121] T. HASTIE, R. TIBSHIRANI, AND J. FRIEDMAN, *The Elements of Statistical Learning: Data Mining, Inference, and Prediction*, Springer-Verlag, New York, 2001.

[122] S. HAYKIN, *Neural Networks: A Comprehensive Foundation*, Prentice–Hall, Upper Saddle River, NJ, 1994.

[123] A.-R. HEDAR AND M. FUKUSHIMA, *Heuristic pattern search and its hybridization with simulated annealing for nonlinear global optimization*, Optim. Methods Softw., 19 (2004), pp. 291–308.

[124] A. S. HEDAYAT, N. J. A. SLOANE, AND J. STUFKEN, *Orthogonal Arrays: Theory and Applications*, Springer-Verlag, New York, 1999.

[125] M. HEINKENSCHLOSS, *On the solution of a two ball trust region subproblem*, Math. Program., 64 (1994), pp. 249–276.

[126] N. J. HIGHAM, *Optimization by direct search in matrix computations*, SIAM J. Matrix Anal. Appl., 14 (1993), pp. 317–333.

[127] ———, *Accuracy and Stability of Numerical Algorithms*, SIAM, Philadelphia, second ed., 2002.

[128] H. HINTERMÜLLER AND L. N. VICENTE, *Space mapping for optimal control of partial differential equations*, SIAM J. Optim., 15 (2005), pp. 1002–1025.

[129] J. H. HOLLAND, *Adaptation in Natural and Artificial Systems: Introductory Analysis with Applications to Biology, Control, and Artificial Intelligence*, MIT Press, Cambridge, MA, 1992.

[130] R. HOOKE AND T. A. JEEVES, *"Direct search" solution of numerical and statistical problems*, J. ACM, 8 (1961), pp. 212–229.

[131] R. A. HORN AND C. R. JOHNSON, *Topics in Matrix Analysis*, Cambridge University Press, Cambridge, UK, 1999.

[132] P. D. HOUGH, T. G. KOLDA, AND V. J. TORCZON, *Asynchronous parallel pattern search for nonlinear optimization*, SIAM J. Sci. Comput., 23 (2001), pp. 134–156.

[133] P. D. HOUGH AND J. C. MEZA, *A class of trust-region methods for parallel optimization*, SIAM J. Optim., 13 (2002), pp. 264–282.

[134] W. HUYER AND A. NEUMAIER, *Global optimization by multilevel coordinate search*, J. Global Optim., 14 (1999), pp. 331–355.

[135] L. M. HVATTUM AND F. GLOVER, *Finding local optima of high-dimensional functions using direct search methods*, European J. Oper. Res., (2008, to appear).

[136] D. JONES, C. PERTTUNEN, AND B. STUCKMAN, *Lipschitzian optimization without the Lipschitz constant*, J. Optim. Theory Appl., 79 (1993), pp. 157–181.

[137] D. R. JONES, *A taxonomy of global optimization methods based on response surfaces*, J. Global Optim., 21 (2001), pp. 345–383.

[138] J.-E. KÄCK, *Constrained global optimization with radial basis functions*, Research Report MdH-IMa-2004, Department of Mathematics and Physics, Mälardalen University, Västerås, Sweden, 2004.

[139] A. J. KEANE AND P. NAIR, *Computational Approaches for Aerospace Design: The Pursuit of Excellence*, John Wiley & Sons, New York, 2006.

[140] C. T. KELLEY, *Detection and remediation of stagnation in the Nelder–Mead algorithm using a sufficient decrease condition*, SIAM J. Optim., 10 (1999), pp. 43–55.

[141] ——, *Iterative Methods for Optimization*, SIAM, Philadelphia, 1999.

[142] J. KENNEDY AND R. EBERHART, *Particle swarm optimization*, in Proceedings of the 1995 IEEE International Conference on Neural Networks, Perth, Australia, IEEE Service Center, Piscataway, NJ, pp. 1942–1948.

[143] S. KIRKPATRICK, C. D. GELATT, JR., AND M. P. VECCHI, *Optimization by simulated annealing*, Science, 220 (1983), pp. 671–680.

[144] J. R. KOEHLER AND A. B. OWEN, *Computer experiments*, in Handbook of Statistics, S. Ghosh and C. R. Rao, eds., vol. 13, Elsevier Science, New York, 1996, pp. 261–308.

[145] T. G. KOLDA, R. M. LEWIS, AND V. TORCZON, *Optimization by direct search: New perspectives on some classical and modern methods*, SIAM Rev., 45 (2003), pp. 385–482.

[146] ——, *A generating set direct search augmented Lagrangian algorithm for optimization with a combination of general and linear constraints*, Tech. Report SAND2006-5315, Sandia National Laboratories, Albuquerque, NM, Livermore, CA, 2006.

[147] ——, *Stationarity results for generating set search for linearly constrained optimization*, SIAM J. Optim., 17 (2006), pp. 943–968.

[148] S. KOZIEL, J. W. BANDLER, AND K. MADSEN, *Quality assessment of coarse models and surrogates for space mapping optimization*, Optim. Eng., 9 (2008), pp. 375–391.

[149] J. C. LAGARIAS, J. A. REEDS, M. H. WRIGHT, AND P. E. WRIGHT, *Convergence properties of the Nelder–Mead simplex method in low dimensions*, SIAM J. Optim., 9 (1998), pp. 112–147.

[150] T. LEVINA, Y. LEVIN, J. MCGILL, AND M. NEDIAK, *Dynamic pricing with online learning and strategic consumers: An application of the aggregation algorithm*, Oper. Res., (2008, to appear).

[151] R. M. LEWIS, A. SHEPHERD, AND V. TORCZON, *Implementing generating set search methods for linearly constrained optimization*, SIAM J. Sci. Comput., 29 (2007), pp. 2507–2530.

[152] R. M. LEWIS AND V. TORCZON, *Pattern search algorithms for bound constrained minimization*, SIAM J. Optim., 9 (1999), pp. 1082–1099.

[153] ———, *Rank ordering and positive bases in pattern search algorithms*, Tech. Report TR96-71, ICASE, NASA Langley Research Center, Hampton, VA, 1999.

[154] ———, *Pattern search methods for linearly constrained minimization*, SIAM J. Optim., 10 (2000), pp. 917–941.

[155] ———, *A globally convergent augmented Lagrangian pattern search algorithm for optimization with general constraints and simple bounds*, SIAM J. Optim., 12 (2002), pp. 1075–1089.

[156] R. M. LEWIS, V. TORCZON, AND M. TROSSET, *Direct search methods: Then and now*, J. Comput. Appl. Math., 124 (2000), pp. 191–207.

[157] G. LIUZZI AND S. LUCIDI, *A derivative-free algorithm for inequality constrained nonlinear programming*, Tech. Report 659, Istituto di Analisi dei Sistemi e Informatica "Antonio Ruberti", CNR, Rome, Italy, 2007.

[158] G. LIUZZI, S. LUCIDI, AND M. SCIANDRONE, *A derivative-free algorithm for linearly constrained finite minimax problems*, SIAM J. Optim., 16 (2006), pp. 1054–1075.

[159] S. N. LOPHAVEN, H. B. NIELSEN, AND J. SØNDERGAARD, *DACE: A MATLAB Kriging toolbox*, Tech. Report IMM-TR-2002-12, Informatics and Mathematical Modelling, Technical University of Denmark, Lyngby, Denmark, 2003.

[160] S. LUCIDI AND M. SCIANDRONE, *On the global convergence of derivative-free methods for unconstrained optimization*, SIAM J. Optim., 13 (2002), pp. 97–116.

[161] S. LUCIDI, M. SCIANDRONE, AND P. TSENG, *Objective-derivative-free methods for constrained optimization*, Math. Program., 92 (2002), pp. 37–59.

[162] K. MADSEN AND J. SØNDERGAARD, *Convergence of hybrid space mapping algorithms*, Optim. Eng., 5 (2004), pp. 145–156.

[163] M. MARAZZI AND J. NOCEDAL, *Wedge trust region methods for derivative free optimization*, Math. Program., 91 (2002), pp. 289–305.

[164] A. L. MARSDEN, *Aerodynamic Noise Control by Optimal Shape Design*, PhD thesis, Stanford University, Stanford, CA, 2004.

[165] A. L. MARSDEN, M. WANG, J. E. DENNIS, JR., AND P. MOIN, *Optimal aeroacoustic shape design using the surrogate management framework*, Optim. Eng., 5 (2004), pp. 235–262.

[166] G. MATHERON, *Principles of geostatistics*, Econ. Geol., 58 (1963), pp. 1246–1266.

[167] J. H. MAY, *Linearly Constrained Nonlinear Programming: A Solution Method That Does Not Require Analytic Derivatives*, PhD thesis, Yale University, New Haven, CT, 1974.

[168] M. D. MCKAY, W. J. CONOVER, AND R. J. BECKMAN, *A comparison of three methods for selecting values of input variables in the analysis of output from a computer code*, Technometrics, 21 (1979), pp. 239–245.

[169] K. I. M. MCKINNON, *Convergence of the Nelder–Mead simplex method to a nonstationary point*, SIAM J. Optim., 9 (1998), pp. 148–158.

[170] J. C. MEZA AND M. L. MARTINEZ, *On the use of direct search methods for the molecular conformation problem*, J. Comput. Chem., 15 (1994), pp. 627–632.

[171] R. MIFFLIN, *A superlinearly convergent algorithm for minimization without evaluating derivatives*, Math. Program., 9 (1975), pp. 100–117.

[172] J. J. MORÉ AND D. C. SORENSEN, *Computing a trust region step*, SIAM J. Sci. Statist. Comput., 4 (1983), pp. 553–572.

[173] J. J. MORÉ AND S. M. WILD, *Benchmarking derivative-free optimization algorithms*, Tech. Report ANL/MCS-P1471-1207, Mathematics and Computer Science Division, Argonne National Laboratory, Argonne, IL, 2007.

[174] P. MUGUNTHAN, C. A. SHOEMAKER, AND R. G. REGIS, *Comparison of function approximation, heuristic and derivative-based methods for automatic calibration of computationally expensive groundwater bioremediation models*, Water Resour. Res., 41 (2005), W11427.

[175] R. H. MYERS AND D. C. MONTGOMERY, *Response Surface Methodology: Process and Product in Optimization Using Designed Experiments*, John Wiley & Sons, New York, second ed., 2002.

[176] L. NAZARETH AND P. TSENG, *Gilding the lily: A variant of the Nelder–Mead algorithm based on golden-section search*, Comput. Optim. Appl., 22 (2002), pp. 133–144.

[177] J. A. NELDER AND R. MEAD, *A simplex method for function minimization*, Comput. J., 7 (1965), pp. 308–313.

[178] J. NOCEDAL AND S. J. WRIGHT, *Numerical Optimization*, Springer-Verlag, Berlin, second ed., 2006.

[179] R. OEUVRAY, *Trust-Region Methods Based on Radial Basis Functions with Application to Biomedical Imaging*, PhD thesis, Institut de Mathématiques, École Polytechnique Fédérale de Lausanne, Switzerland, 2005.

[180] R. OEUVRAY AND M. BIERLAIRE, *A new derivative-free algorithm for the medical image registration problem*, Int. J. Model. Simul., 27 (2007), pp. 115–124.

[181] ——, *BOOSTERS: A derivative-free algorithm based on radial basis functions*, International Journal of Modelling and Simulation, (2008, to appear).

[182] A. B. OWEN, *Orthogonal arrays for computer experiments, integration and visualization*, Statist. Sinica, 2 (1992), pp. 439–452.

[183] M. J. D. POWELL, *An efficient method for finding the minimum of a function of several variables without calculating derivatives*, Comput. J., 7 (1964), pp. 155–162.

[184] ——, *A new algorithm for unconstrained optimization*, in Nonlinear Programming, J. B. Rosen, O. L. Mangasarian, and K. Ritter, eds., Academic Press, New York, 1970, pp. 31–65.

[185] ——, *The theory of radial basis function approximation in 1990*, in Advances in Numerical Analysis, Vol. II: Wavelets, Subdivision Algorithms and Radial Basis Functions, W. A. Light, ed., Oxford University Press, Cambridge, UK, 1992, pp. 105–210.

[186] ——, *A direct search optimization method that models the objective and constraint functions by linear interpolation*, in Advances in Optimization and Numerical Analysis, Proceedings of the Sixth Workshop on Optimization and Numerical Analysis, Oaxaca, Mexico, S. Gomez and J.-P. Hennart, eds., vol. 275 of Math. Appl., Kluwer Academic Publishers, Dordrecht, The Netherlands, 1994, pp. 51–67.

[187] ——, *Direct search algorithms for optimization calculations*, Acta Numer., 7 (1998), pp. 287–336.

[188] ——, *On the Lagrange functions of quadratic models that are defined by interpolation*, Optim. Methods Softw., 16 (2001), pp. 289–309.

[189] ——, *UOBYQA: Unconstrained optimization by quadratic approximation*, Math. Program., 92 (2002), pp. 555–582.

[190] ——, *On trust region methods for unconstrained minimization without derivatives*, Math. Program., 97 (2003), pp. 605–623.

[191] ——, *Least Frobenius norm updating of quadratic models that satisfy interpolation conditions*, Math. Program., 100 (2004), pp. 183–215.

[192] ——, *The NEWUOA software for unconstrained optimization without derivatives*, Tech. Report DAMTP 2004/NA08, Department of Applied Mathematics and Theoretical Physics, University of Cambridge, UK, 2004.

[193] C. J. PRICE AND I. D. COOPE, *Frames and grids in unconstrained and linearly constrained optimization: A nonsmooth approach*, SIAM J. Optim., 14 (2003), pp. 415–438.

[194] C. J. PRICE, I. D. COOPE, AND D. BYATT, *A convergent variant of the Nelder-Mead algorithm*, J. Optim. Theory Appl., 113 (2002), pp. 5–19.

[195] C. J. PRICE AND PH. L. TOINT, *Exploiting problem structure in pattern-search methods for unconstrained optimization*, Optim. Methods Softw., 21 (2006), pp. 479–491.

[196] R. G. REGIS AND C. A. SHOEMAKER, *Constrained global optimization of expensive black box functions using radial basis functions*, J. Global Optim., 31 (2005), pp. 153–171.

[197] ———, *Improved strategies for radial basis function methods for global optimization*, J. Global Optim., 37 (2007), pp. 113–135.

[198] ———, *Parallel radial basis function methods for the global optimization of expensive functions*, European J. Oper. Res., 182 (2007), pp. 514–535.

[199] ———, *A stochastic radial basis function method for the global optimization of expensive functions*, INFORMS J. Comput., 19 (2007), pp. 497–509.

[200] R. T. ROCKAFELLAR, *Convex Analysis*, Princeton University Press, Princeton, NJ, 1970.

[201] H. H. ROSENBROCK, *An automatic method for finding the greatest or least value of a function*, Comput. J., 3 (1960), pp. 175–184.

[202] A. S. RYKOV, *Simplex algorithms for unconstrained optimization*, Problems Control Inform. Theory, 12 (1983), pp. 195–208.

[203] J. SACKS, W. J. WELCH, T. J. MITCHELL, AND H. P. WYNN, *Design and analysis of computer experiments*, Statist. Science, 4 (1989), pp. 409–423.

[204] T. J. SANTNER, B. J. WILLIAMS, AND W. I. NOTZ, *The Design and Analysis of Computer Experiments*, Springer-Verlag, New York, 2003.

[205] T. SAUER AND Y. XU, *On multivariate Lagrange interpolation*, Math. Comp., 64 (1995), pp. 1147–1170.

[206] D. B. SERAFINI, *A Framework for Managing Models in Nonlinear Optimization of Computationally Expensive Functions*, PhD thesis, Department of Computational and Applied Mathematics, Rice University, Houston, TX, 1998.

[207] W. H. SHAWN, *Direct search methods*, in Numerical Methods for Unconstrained Optimization, W. Murray, ed., Academic Press, London, 1972, pp. 13–28.

[208] T. W. SIMPSON, J. D. POPLINSKI, P. N. KOCH, AND J. K. ALLEN, *Metamodels for computer-based engineering design: Survey and recommendations*, Engineering with Computers, 17 (2001), pp. 129–150.

[209] J. SØNDERGAARD, *Optimization Using Surrogate Models—by the Space Mapping Technique*, PhD thesis, Informatics and Mathematical Modelling, Technical University of Denmark, Lyngby, Denmark, 2003.

[210] W. SPENDLEY, G. R. HEXT, AND F. R. HIMSWORTH, *Sequential application of simplex designs in optimisation and evolutionary operation*, Technometrics, 4 (1962), pp. 441–461.

[211] M. STEIN, *Large sample properties of simulations using Latin hypercube sampling*, Technometrics, 29 (1987), pp. 143–151.

[212] D. E. STONEKING, G. L. BILBRO, R. J. TREW, P. GILMORE, AND C. T. KELLEY, *Yield optimization using a GaAs process simulator coupled to a physical device model*, IEEE Trans. Microwave Theory Tech., 40 (1992), pp. 1353–1363.

[213] B. TANG, *Orthogonal array-based Latin hypercubes*, J. Amer. Statist. Assoc., 88 (1993), pp. 1392–1397.

[214] S. W. THOMAS, *Sequential Estimation Techniques for Quasi-Newton Algorithms*, PhD thesis, Cornell University, Ithaca, NY, 1975.

[215] PH. L. TOINT AND F. M. CALLIER, *On the accelerating property of an algorithm for function minimization without calculating derivatives*, J. Optim. Theory Appl., 23 (1978), pp. 531–547. See also the same journal 26 (1978), pp. 465–467.

[216] V. TORCZON, *On the convergence of the multidirectional search algorithm*, SIAM J. Optim., 1 (1991), pp. 123–145.

[217] ——, *On the convergence of pattern search algorithms*, SIAM J. Optim., 7 (1997), pp. 1–25.

[218] V. TORCZON AND M. W. TROSSET, *Using approximations to accelerate engineering design optimization*, in Proceedings of the 7th AIAA/USAF/NASA/ISSMO Symposium on Multidisciplinary Analysis and Optimization, St. Louis, MO, September 1998.

[219] V. V. TOROPOV, *Simulation approach to structural optimization*, Struct. Multidiscip. Optim., 1 (1989), pp. 37–46.

[220] P. TSENG, *Fortified-descent simplicial search method: A general approach*, SIAM J. Optim., 10 (1999), pp. 269–288.

[221] Ö. UĞUR, B. KARASÖZEN, M. SCHÄFER, AND K. YAPICI, *Derivative free optimization of stirrer configurations*, in Numerical Mathematics and Advanced Applications, Springer-Verlag, Berlin, 2006, pp. 1031–1039.

[222] A. I. F. VAZ AND L. N. VICENTE, *A particle swarm pattern search method for bound constrained global optimization*, J. Global Optim., 39 (2007), pp. 197–219.

[223] L. N. VICENTE, *Space mapping: Models, sensitivities, and trust-regions methods*, Optim. Eng., 4 (2003), pp. 159–175.

[224] H. WHITE, *Asymptotic Theory for Econometricians*, Academic Press, London, 1984.

[225] S. M. WILD, *MNH: A derivative-free optimization algorithm using minimal norm Hessians*, Tech. Report ORIE-1466, School of Operations Research and Information Engineering, Cornell University, Ithaca, NY, 2008.

[226] S. M. WILD, R. G. REGIS, AND C. A. SHOEMAKER, *ORBIT: Optimization by radial basis function interpolation in trust-regions*, SIAM J. Sci. Comput., 30 (2008), pp. 3197–3219.

[227] D. WINFIELD, *Function and Functional Optimization by Interpolation in Data Tables*, PhD thesis, Harvard University, Cambridge, MA, 1969.

[228] ——, *Function minimization by interpolation in a data set*, J. Inst. Math. Appl., 12 (1973), pp. 339–347.

[229] T. A. WINSLOW, R. J. TREW, P. GILMORE, AND C. T. KELLEY, *Doping profiles for optimum class B performance of GaAs MESFET amplifiers*, in Proceedings of the IEEE/Cornell Conference on Advanced Concepts in High Speed Devices and Circuits, 1991, pp. 188–197.

[230] D. J. WOODS, *An Interactive Approach for Solving Multi-objective Optimization Problems*, PhD thesis, Department of Mathematical Sciences, Rice University, Houston, TX, 1985.

[231] M. H. WRIGHT, *Direct search methods: Once scorned, now respectable*, in Numerical Analysis 1995 (Proceedings of the 1995 Dundee Biennial Conference in Numerical Analysis), D. F. Griffiths and G. A. Watson, eds., vol. 344 of Pitman Res. Notes Math. Ser., CRC Press, Boca Raton, FL, 1996, pp. 191–208.

[232] Y. YE, *A new complexity result on minimization of a quadratic function with a sphere constraint*, in Recent Advances in Global Optimization, C. A. Floudas and P. M. Pardalos, eds., Princeton University Press, Princeton, NJ, 1992, pp. 19–31.

[233] J.-H. YOON AND C. A. SHOEMAKER, *Comparison of optimization methods for ground-water bioremediation*, J. Water Resources Planning and Management, 125 (1999), pp. 54–63.

[234] W.-C. YU, *The convergent property of the simplex evolutionary techniques*, Sci. Sinica, 1 (1979), pp. 269–288.

[235] ——, *Positive basis and a class of direct search techniques*, Sci. Sinica, 1 (1979), pp. 53–67.

[236] Y.-X. YUAN, *An example of non-convergence of trust region algorithms*, in Advances in Nonlinear Programming, Y.-X. Yuan, ed., Kluwer Academic Publishers, Dordrecht, The Netherlands, 1998, pp. 205–215.

[237] W. I. ZANGWILL, *Minimizing a function without calculating derivatives*, Comput. J., 10 (1967), pp. 293–296.

Index